AF292390

Order, Disorder and Chaos in Quantum Systems

Proceedings of a conference
held at Dubna, USSR
on October 17–21, 1989

Edited by

P. Exner
H. Neidhardt

1990

Birkhäuser Verlag
Basel · Boston · Berlin

Editors' address:

Prof. P. Exner
Prof. H. Neidhardt
Laboratory of Theoretical Physics
Joint Institute for Nuclear Research
Head Post Office P.O. Box 79
Môscow
USSR

Deutsche Bibliothek Cataloguing-in-Publication Data

Order, disorder and chaos in quantum systems: proceedings of
a conference held at Dubna, USSR, on October 17–21, 1989 /
ed. by P. Exner ; H. Neidhardt. – Basel ; Boston ; Berlin :
Birkhäuser, 1990
  (Operator theory ; Vol. 46)
  ISBN-13: 978-3-0348-7308-6   e-ISBN-13: 978-3-0348-7306-2
  DOI: 10.1007/978-3-0348-7306-2
NE: Exner, Pavel [Hrsg.]; GT

Dedicated to the memory of

Mark Grigorevich Krein

1907-1989

PREFACE

The present volume collects the contributions to the conference "Order, disorder and chaos in quantum systems" which was held at Dubna last October. It is the third meeting in the series started three years ago in which we tried to put together mathematical physicists from the member and non-member countries of JINR with their colleagues from Soviet universities and institutes using this international centre as a convenient basis. As in the previous cases, new faces, subjects and ideas appeared but the spirit remained the same, relaxed and inspirative.

Among this conference contributions, a majority should be listed in the "orderly" category. Being more specific, this means mostly various aspects of the theory of Schroedinger operators that has been always a core of quantum mechanics. In spite of the fact that it is studied already for several decades, there are still many interesting problems to solve as some of the lectures collected below witness. At the same time, the theory extends to some new areas motivated by physical problems ; let us mention Schroedinger operators in complicated spatial domains appearing in some parts of solid-state physics or various models using the concept of contact interactions.

Our world is far from perfect and to keep a perfect order is difficult not only in everyday life but also in most physical systems. Theoreticians are used to take this fact into account introducing stochastic factors into their considerations. This direction is represented by several lectures of this volume though, frankly speaking, we hoped for more.

It has been known for long that even purely deterministic classical equations may yield highly irregular solutions, however, only the last two decades gave boost to study of these phenomena, now commonly called dynamical chaos, which are essential for understanding of such all-important effects as, e.g., turbulence. Naturally the question has arisen whether chaotic behaviour can be found in quantum systems too. In distinction to the classical case, unfortunately, there are some conceptual difficulties because even the definition of quantum chaos allows different approaches.

At the same time, this field offers many interesting problems. At the present stage, the investigation is concentrated mostly on analysis of simple quantum models where the chaotic behaviour could be manifested. Various relations to the topics mentioned above can be found, and we believe that in this way which does not rely entirely on numerical experiments one could achieve a deeper understanding of quantum chaotic phenomena.

We acknowledge with gratitude the access to facilities of Joint Institute for Nuclear Research and the support we got from its officials, particularly from Prof.A.N.Sissakian, in preparation of the conference. We want especially thank our colleagues Petr Šeba and Valentin Zagrebnov who helped us to organize the conference but could not from various reasons participate in editing this proceedings volume.

Dubna, February 1990

Pavel Exner
Hagen Neidhardt

TO THE MEMORY OF A GREAT MATHEMATICIAN

On the second day of the conference we learned about the death of Mark Grigorevich Krein. These sad news touched deeply each of us independently of his or her age, experience, scientific interests or nationality.

Mark Grigorevich was indisputably one of the greatest mathematicians of our century who enriched the science with many deep and beautiful concepts, methods, theorems and formulae. To name just a few of them, let us recall his contributions to functional analysis, harmonic analysis on locally compact groups or theory of self-adjoint extensions. One should mention also the eigenfunction expansions for ordinary differential operators, the indefinite-metrics spaces bearing now his name, as well as his results about stability of solutions to differential equations, inverse problems, non-selfadjoint operators and plenty of others.

In addition to his scientific merits, he was known as an outstanding teacher who managed to induce a true passion for mathematics to his students. A few older colleagues taking part in our conference were his disciples, and many younger ones were disciples of his disciples, and had also the opportunity to meet him in person. It makes their feeling of our loss even deeper.

We are convinced that only a historical perspective will allow to appreciate the great impact that the works of Professor Krein had on the development of modern mathematics in a full complexity. Reading carefully this book you can trace his ideas in a lot of the papers collected here. We deem therefore that to dedicate the present volume to Mark Grigorevich is the best way to honour his memory.

The editors

TABLE OF CONTENTS

2. <u>Point and contact interactions, self-adjoint extensions</u>

XIV

SCHROEDINGER OPERATORS : SPECTRA, SCATTERING

AND SEMICLASSICAL BEHAVIOUR

Operator Theory:
Advances and Applications, Vol. 46
© 1990 Birkhäuser Verlag Basel

NEGATIVE DISCRETE SPECTRUM OF THE SCHROEDINGER OPERATOR WITH LARGE COUPLING CONSTANT: A QUALITATIVE DISCUSSION [*)]

M.S.Birman, M.Z.Solomyak

1. For the Schrödinger operator

$$(1) \qquad -\Delta - \alpha V(x), \quad x \in \mathbb{R}^m, \quad m \geq 3, \quad \alpha > 0,$$

by $N(\alpha,a,V)$ we denote the number of eigenvalues (including their multiplicities) situated to the left from $\lambda = -a$, $a \geq 0$. In this lecture we discuss two-sided estimates and the asymptotic behavior as $\alpha \longrightarrow \infty$ of the function $N(\alpha,a,V)$. We recall that for $V \in L_{m/2}(\mathbb{R}^m)$ the upper estimate (found by Rosenblum-Lieb-Cwikel, see [1])

$$(2) \qquad N(\alpha,a,V) \leq C(m) \; \alpha^{m/2} \int V_+^{m/2} \; dx$$

and its asymptotics

____

[*)] Translated by the editors

$$(3) \qquad \lim_{\alpha \to \infty} \alpha^{-m/2} \, N(\alpha,a,V) = v_m \, (2\pi)^{-m} \int v_+^{m/2} \, dx.$$

are well-known. Here V_+ and v_m denote the positive part of the function V and the volume of the unit ball in $\mathbb{R}^m$, respectively. We note that the asymptotics (3) was derived for the first time in [3] but under very restricting conditions for the potentials.

In the last time the interest has concentrated at the case $V \notin L_{m/2}(\mathbb{R}^m)$. However, in essential only upper estimates were obtained for $N(\alpha,a,V)$. Recently, two general methods allowing to obtain these estimates were presented by the authors in [3] and [4] (see also the literature referred to there). The main goal of the present paper is to discuss the qualitative differences of the behavior of $N(\alpha,a,V)$ with respect to large α for $V \in L_{m/2}$ and $V \notin L_{m/2}$. We will call these two cases the *regular* and the *non-regular* one, respectively. In some sense the regular case could be referred to as the quasi-classical one but it seems to us that this notion is too obliging. In the non-regular case we will consider only the potentials V that ensure a powerlike behavior of the estimate, i.e. $N(\alpha,a,V) = O(\alpha^q)$, $2q > m$. For estimates in more general and exact scales the reader is referred to [4]. Throughout of this paper we use the notation

$$D_q(a,V) = \sup_{\alpha>0} \alpha^{-q} \, N(\alpha,a,V),$$

$$\Delta_q(a,V) = \limsup_{\alpha \to \infty} \alpha^{-q} \, N(\alpha,a,V),$$

$$\delta_q(a,V) = \liminf_{\alpha \to \infty} \alpha^{-q} \, N(\alpha,a,V)$$

where $a \geq 0$. If $a = 0$, then we drop "a" in the symbols N, D_q, Δ_q and δ_q .

2. Studying the function $N(\alpha,a,V)$ one uses frequently other (equivalent) definitions of it. Let us introduce the quotient of quadratic forms

$$(4) \qquad \int V|u|^2 dx \Big/ \int (|\nabla u|^2 + a|u|^2) dx.$$

The quotient (4) is defined first for $u \in C_o^\infty(\mathbb{R}^m)$, then the closure is taken in the metrics given by the quadratic form in the denominator. Thus, for $a > 0$ we have to study (4) in the Sobolev class $H^1(\mathbb{R}^m)$ while for $a = 0$ we have to consider the so-called "homogeneous" Sobolev class $\mathcal{H}^1$,

$$\mathcal{H}^1(\mathbb{R}^m) = \{u \in H^1_{loc}(\mathbb{R}^m): \int (|\nabla u|^2 + |x|^{-2}|u|^2) dx < \infty\}.$$

Let $T_a(v)$ be the self-adjoint operator generated by the quadratic form $\int V|u|^2 dx$ in the Hilbert space equipped with the metric form $\int (|\nabla u|^2 + a|u|^2) dx$ and let us denote the distribution function of its positive (discrete) spectrum by $n_+(s,a,V)$. In other words, the distribution function $n_+(s,a,V)$ represents the number of successive maxima (eigenvalues) of the quotient (4) situated to the right from $s > 0$. The relation to the original problem becomes transparent from the relation

$$N(\alpha,a,V) = n_+(\alpha^{-1},a,V), \quad a \geq 0.$$

3. Another very often used reformulation ties (4) with an integral operator. Here we restrict ourselves to the case $V \geq 0$. Setting $W = V^{1/2}$ we introduce the integral operator $S_a = S_a(W)$,

$$(5) \qquad (S_a f)(x) = (2\pi)^{-m/2} \int W(x)(|\xi|^2 + a)^{-1/2} e^{ix\xi} f(\xi)\, d\xi.$$

Denoting by $\upsilon(.,S_a)$ the distribution function of the singular numbers of the operator S_a (i.e., the eigenvalues of $(S_a^* S_a)^{1/2}$) we have

$$(6) \qquad N(\alpha,a,V) = \upsilon(\alpha^{-1/2}, S_a(V^{1/2})).$$

One method to prove the estimate (2) is based on the relation (6). To make this clear we consider the following more general integral operator

$$(7) \qquad (Gf)(x) = \int W(x)\ h(\xi)\ g(x,\xi)\ f(\xi)\ d\xi.$$

We set $|G|_p^p := \sup\limits_{s>0}\{s^p \upsilon(s,G)\}$. Cwikel has proved [5] the following assertion proposed by Simon.

PROPOSITION 1. *Set* $g(x,\xi) = e^{ix\xi}$ *in* (7). *If* $p > 2$, $W \in L_p(\mathbb{R}^m)$ *and* $h \in L_{p,w}(\mathbb{R}^m)$, *i.e.*

$$(8) \qquad |h|_p^p := \sup\limits_{t>0} t^p\ \text{mes}\{\xi \in \mathbb{R}^m:\ |h(\xi)| > t\} < \infty,$$

then the estimate

$$(9) \qquad |G|_p \le C(p)\ \|W\|_{L_p}\ |h|_p, \quad p > 2$$

is valid for the operator (7).

Applying the estimate (9) to the operator (5) for $p = m > 2$ and using the relation (6) we obtain the estimate (2). A careful analysis of the proof of Proposition 1 given in [5] leads to the following generalization.

PROPOSITION 2. *Let* $g \in L_\infty(\mathbb{R}^{2n})$ *in* (7), *and moreover, let* g *be the kernel of an integral operator acting in* $L_2(\mathbb{R}^m)$ *with the norm* $M(g)$. *If* $W \in L_p(\mathbb{R}^m)$ *and* $h \in L_{p,w}(\mathbb{R}^m)$, $p \geq 2$, *then*

$$(10) \qquad |G|_p \leq C(p) \; \|W\|_{L_p} \, |h|_p \|g\|_{L_\infty}^{\theta} \, (M(g))^{1-\theta}, \qquad \theta = 2/p.$$

REMARK. Assume that under the assumptions of Proposition 2 we have additionally $h \in L^o_{p,w}(\mathbb{R}^m)$, i.e. that together with (8) the condition

$$(11) \qquad \mathrm{mes}\{\xi \in \mathbb{R}^m : |h(\xi)| > t\} = o(t^{-p}), \quad t \to 0, \quad t \to \infty,$$

is satisfied. Then besides the estimate (10) the relation

$$(12) \qquad \upsilon(s,G) = o(s^{-p}), \quad s \to 0,$$

holds. Moreover, the relation (12) is fulfilled even if the condition (11) is valid for $t \to 0$ only.

4. For the *regular* case the following features are typical.

(a) The estimate (2) is uniform in $a \geq 0$. Its exactness is confirmed by the asymptotics (3). In (3) the *rhs* does not depend on a. Thus choosing α sufficiently large, we can have a significant part of the negative eigenvalues situated to the left from an any fixed point $\lambda < 0$.

(b) The presence of an integral in the asymptotics (3) means that all values of the potential V (more exactly, all values of V_+) contribute to the asymptotic coefficient.

(c) The asymptotic coefficient $\Delta_{m/2}(a,V)$ depends

continuously on $V \in L_{m/2}$. The class $L_{m/2}(\mathbb{R}^m)$ is separable and the set $C_o^\infty(\mathbb{R}^m)$ is dense in it.

(d) Only large values of the momentum are responsible for the coefficient $\Delta_{m/2}(a,V)$. This means exacly that the formula (3) remains true if in (5) we restrict the integral to the region $|\xi| \geq R$, $\forall R > 0$. In fact, it is sufficient to verify this for $V \in C_o^\infty(\mathbb{R}^m)$. But doing so the integral over the ball $|\xi| < R$ is a Hilbert-Schmidt operator. By the way, this consideration can be absorbed into the last assertion of the Remark of Section 3.

5. All the assertions (a)-(d) are false generally in the non-regular case, i.e. in the case $V \notin L_{m/2}$. In order to explain this we mention first of all some estimates and examples. In essential they are adopted from [3] and [4]. Let us introduce the necessary notation.

Let $\varphi \in L_{m/2,w}(\mathbb{R}^m)$, $\varphi > 0$ a.e.. We write $V \in \mathscr{L}_{q,w}(\varphi)$, $2q > m$, if the expression

$$|V|_{q,\varphi}^q := \sup_{s>0} s^q \int_{|V(x)|>s\varphi(x)} \varphi^{m/2}\, dx$$

is finite. The class $\mathscr{L}_{q,w}(\varphi)$ is complete with respect to the quasi-norm $\|.\|_{q,\varphi}$ but not separable. By $\mathscr{L}_{q,w}^o(\varphi)$ we denote its separable subspace fixed by the condition

$$\int_{V(x)>s\varphi(x)} \varphi^{m/2}(x)\, dx = o(s^{-q}), \quad s \to 0, \quad s \to \infty.$$

An example of a function V belonging to $\mathscr{L}_{q,w}^o(\varphi)$ is given by

$$(13) \qquad V(x) = \varphi(x)\,(1+|\log(\varphi(x))|)^{-1/q}, \quad 2q > m.$$

In the following φ plays the role of a functional parameter in the estimates. Very often we will use $\varphi(x) = \gamma|x|^{-2}$ or, more generally,

$$(14) \qquad \varphi(x) = \Phi\left(\frac{x}{|x|}\right)|x|^{-2}, \quad \Phi \in L_{m/2}(S^{m-1}), \quad \Phi > 0.$$

Another example is given by the function

$$\varphi(x) = \left(\sum_{k=1}^{m}|x_k|^{\gamma_k}\right)^{-1}, \quad \gamma_k > 0, \quad 2\sum_k \gamma_k^{-1} = m.$$

Furthermore, we need the space $l_{p,w}(\mathbb{Z}^m)$ of sequences defined on the lattice $\mathbb{Z}^m$ and equipped with the quasi-norm

$$|b|_p^p := \sup_{t>0} t^p \operatorname{card}\{n \in \mathbb{Z}^m: |b_n| > t\}, \quad b = \{b_n\}_{n\in\mathbb{Z}^m}.$$

The separable subspace $l_{p,w}^o$ is characterized by the condition

$$\operatorname{card}\{n \in \mathbb{Z}^m: |b_n| > t\} = o(t^{-p}), \quad t \to 0.$$

Let us pass to the estimates supposing for simplicity that $V \geq 0$. In the opposite case we have to replace V by V_+. First we consider the case $a = 0$.

PROPOSITION 3. *If* $V \in \mathcal{L}_{q,w}(\varphi)$, $2q > m$, *then*

$$(15) \qquad D_q(V) \leq C(\varphi)|V|_{q,\varphi}^q, \quad 2q > m,$$

$$(16) \qquad V \in \mathcal{L}_{q,w}^o(\varphi) \Rightarrow \Delta_q(V) = 0.$$

We note that if $|\varphi|_q \leq 1$ the constant in the estimate (15) can be chosen depending only on m,q but independent of φ.

If a > 0 the estimates have a somewhat different form.

PROPOSITION 4. *If* $V \in L_{p,w}(\mathbb{R}^m)$, $2p > m$, *then*

$$(17) \qquad D_p(a,V) \leq C(a,m,p)|V|_p^p, \quad a > 0, \quad 2p > m,$$

$$V \in L_{p,w}^{\circ}(\mathbb{R}^m) \Rightarrow \Delta_p(a,V) = 0, \quad a > 0.$$

The estimate can be improved taking into account a decomposition of $\mathbb{R}^m$ into unit cubes. Let Q_n, $n \in \mathbb{Z}^m$, be the unit cube Q shifted by the integer-component vector n and let b(V) denote the sequence of real numbers given by

$$b(V) = \{b_n\}, \quad b_n = \|V\|_{L_{m/2}(Q_n)}, \quad n \in \mathbb{Z}^m.$$

PROPOSITION 5. *If* $b(V) \in l_{p,w}(\mathbb{Z}^m)$, $2p > m$, *then*

$$(18) \qquad D_p(a,V) \leq C(a,m,p)|b(V)|_p^p, \quad a > 0,$$

$$b(V) \in l_{p,w}^{\circ}(\mathbb{Z}^m) \Rightarrow \Delta_p(a,V) = 0, \quad a > 0.$$

We note that the constants contained in (17) and (18) grow as $a \to 0$.

6. Let us now compare the remarks (a)-(d) of Section 4 concerning potentials $V \in L_{m/2}$ with the contents of Section 5.

(a) If $V \notin L_{m/2}$, then the order of growth of $N(\alpha,V)$ is in general higher than that of $N(\alpha,a,V)$, $a > 0$. For instance, this is the case for the potential V considered in (13) provided $\varphi \in L_{m/2,w} \cap L_\infty$. Here we have $\Delta_p(a,V) = 0$, $a > 0$, $\forall p > m/2$. At the same time the estimate (15) certainly admits no improvement if $\varphi(x)$ has the form (14) for $|x| \geq 1$ and $\varphi(x) = 1$ for $|x| < 1$. However, the order of $N(\alpha,a,V)$ might not depend on $a \geq 0$ if the main singularities are situated in a bounded region. For instance, this is fulfilled if V has the form (13) for $\varphi(x) = |x|^{-2}$, $|x| \leq 1$ and $V(x) = 0$ for $|x| > 1$.

(b)-(c) The *rhs* of the estimate (15) includes a quasi-norm of a nonseparable class. Together with (16) this means that the functionals $\Delta_q(V)$, $\delta_q(v)$ are continuous with respect to the quasi-norm of the factor class $\mathcal{L}_{q,w}(\varphi)/\mathcal{L}^o_{q,w}(\varphi)$. In other words, these functionals depend only on the behaviour of V in a neighborhood of the main singularities. The remark remains true in application to Propositions 4 and 5.

(d) The quantities $\Delta_q(a,V)$, $\delta_q(a,V)$, $a > 0$, are affected in general by the whole region of the variation of the momentum. The picture is different for the quantities $\Delta_q(V)$, $\delta_q(V)$ if $\Delta_q(1,V) = 0$. Then for $a = 0$ we easily conclude from (5) and (6) that $\Delta_q(V)$ and $\delta_q(V)$ are determined by the integral operator (5) where the integration is restricted to an arbitrarily small neighborhood of the point $\xi = 0$.

7. We present now some asymptotic formulas confirming the remarks of section 6.

THEOREM 1. *If* $V \in L_{m/2,\text{loc}}(\mathbb{R}^m)$,

$$(19) \qquad V(x) \sim \Phi\left(\frac{x}{|x|}\right)|x|^\beta, \quad |x| \longrightarrow \infty,$$

where $-2 < \beta < 0$, $\Phi \in L_q(S^{m-1})$, $q = -m\beta^{-1}$, *then the asymptotics of* $N(\alpha,a,V)$ *for* $a > 0$ *is given by*

$$(20) \qquad \Delta_q(a,V) = \delta_q(a,V) = c(m,q)\, a^{(m/2)-q} \int_{S^{m-1}} \Phi_+^q \, dS,$$

$$c(m,q) = v_m (2\pi)^{-m} \int_0^1 (\rho^\beta - 1)^{m/2}\, \rho^{m-1}\, d\rho.$$

The asymptotics (20) confirms the sharpness of the estimate (17). Furthermore, we note that the quantity Δ_q of (20) depends explicitly on a. The density of the eigenvalues in a neighborhood of $\lambda = -a$ is proportional to $a^{(m/2)-q-1}$. This agrees with the fact that under the conditions of Theorem 1 $N(\alpha,V) = \infty$, $\forall\, \alpha > 0$. Let us present, in addition, two theorems ; one of them complements while the other one generalizes Theorem 1.

THEOREM 2. *If* $V \in L_{m/2,\,loc}(\mathbb{R}^m)$, $\Phi \in L_{m/2}(S^{m-1})$,

$$V(x) \sim \Phi\left(\frac{x}{|x|}\right) |x|^{-2}, \quad |x| \longrightarrow \infty,$$

then for a > 0 and $\alpha \longrightarrow \infty$ *the asymptotics*

$$(21) \qquad N(\alpha,a,V) \sim \frac{v_m}{2(2\pi)^m} \int_{S^{m-1}} \Phi_+^{m/2}\, dS\ (\alpha^{m/2}\, \log(\alpha))$$

is valid.

In (21) the asymptotic coefficient does not depend on a > 0 but the asymptotics is not uniform as $a \longrightarrow \infty$.

Let us now consider the operator

$$(22) \qquad -\Delta + \Psi\left(\frac{x}{|x|}\right) |x|^\gamma - \alpha V(x)$$

and let us denote the number of its negative eigenvalues by $\tilde{N}(\alpha,V)$.

THEOREM 3. *If* $\Psi > 0$, $\Psi + \Psi^{-1} \in L_\infty(S^{m-1})$ *and* $V \in L_{m/2,\text{loc}}(\mathbb{R}^m)$ *in (22) and the assumption (19) is satisfied, where* $-2 < \beta < \gamma < \infty$, *then for* $\alpha \longrightarrow \infty$ *we have*

$$(23) \qquad \tilde{N}(\alpha,V) \sim c(m,\beta,\gamma) \int_{S^{m-1}} \Phi_+^q \, \Psi^{-p} \, dS \; \alpha^q,$$

where $2p = m(\beta+2)(\gamma-\beta)^{-1}$, $2q = m(\gamma+2)(\gamma-\beta)^{-1}$,

$$c(m,\beta,\gamma) = (2\pi)^{-m} v_m \int_0^1 (\rho^\beta - \rho^\gamma)^{m/2} \rho^{m-1} \, d\rho.$$

If $\gamma = 0$, $\Psi = a$, *then (23) transforms to (20).*

Returning to the operator (1) we note that we are able to calculate the asymptotics of $N(\alpha,a,V)$ for $a = 0$ only for particular potentials.

THEOREM 4. *If* $V \in L_{m/2,\text{loc}}(\mathbb{R}^m)$,

$$(24) \qquad V(x) \sim |x|^{-2}(\log|x|)^{-1/q}, \quad |x| \longrightarrow \infty, \; 2q > m,$$

then for $\alpha \longrightarrow \infty$ *the asymptotics*

$$(25) \qquad \Delta_q(V) = \delta_q(V) = \frac{1}{2\sqrt{\pi}} \frac{\Gamma(q-\frac{1}{2})}{\Gamma(q)} \sum_j \left(\Lambda_j + \frac{(m-2)^2}{4}\right)^{(1/2)-q},$$

is valid where Λ_j *denote the eigenvalues (including their multiplicity) of the spherical part of the operator* $-\Delta$.

REMARK. (i) The same asymptotics can be obtained if $V(x) \sim |x-x_o|^{-2}|\log|x-x_o||^{-1/q}$, $x \longrightarrow x_o$, $2q > m$. If there are some points (including infinity with the asymptotics (24)), then the coefficient (25) is multiplied by the number of singularities.

(ii) If $V \in L_{m/2,\,loc}$ and instead of (24) the condition

$$V(x) \sim |x|^{-2}(\log|x|)^{-1/q}\,\Phi(\tfrac{x}{|x|}),\quad |x| \longrightarrow \infty,$$

is satisfied for $2q > m$, where $\Phi \in L_{m/2}(S^{m-1})$, then the asymptotics of $N(\alpha,V)$ has not been calculated. However, if $\Phi_+ \neq 0$, then in any case we have

$$0 < \delta_q(V) \leq \Delta_q(V) < +\infty.$$

This shows sharpness of the estimate (15) for potentials of the form (13). For the case under consideration the quantities $\Delta_q(V)$, $\delta_q(V)$ are continuous functionals of the argument Φ varying in the class $L_{m/2}(S^{m-1})$.

8. In conclusion we are going to formulate one asymptotical result for the operator

$$(26) \qquad -\Delta + P(x) - \alpha V(x)$$

where p is a periodic function and the perturbation V is regular. For simplicity we assume that the lower bound of the spectrum of the operator $-\Delta + p(x)$ equals zero. Restrictions for p are formulated implicitly: the positive periodic solution ω of the equation $-\Delta\omega + p\omega = 0$ should be continuous.

By $\hat{N}(\alpha,a,V)$ we denote the number of negative eigenvalues of the operator (26). The following theorem can easily be obtained from well known results but, apparently, it has not been formulated explicitly.

THEOREM 5. *If* $V \in L_{m/2}(\mathbb{R}^m)$, *then for any* $a \geq 0$ *the relations* (2) *and* (3), *where we have to replace* $N(\alpha,a,V)$ *by* $\hat{N}(\alpha,a,V)$, *are valid.*

Let us sketch proof of the theorem. Instead of (4) we have now to consider the quotient

$$(27) \qquad \int V|v|^2 \, dx \; \Big/ \; \int (|\nabla v|^2 + (p + a)|v|^2) \, dx.$$

The substitution $v = \omega u$ (compare to [2]) transforms (27) to the form

$$(28) \qquad \int V\omega^2|u|^2 \, dx \; \Big/ \; \int \omega^2(|\nabla u|^2 + a|u|^2) \, dx.$$

The function ω^2 is bounded above and bounded away from zero. Therefore, the relation (28) can be estimated by (4) and the estimate (2) can be extended to $\hat{N}(\alpha,a,V)$. Having the estimate it is sufficient to obtain the asymptotics for $V \in C_o^\infty(\mathbb{R}^m)$. In this case the problem leads to the so-called "Dirichlet-Neumann bracketing" in a ball containing supp(V). The spectral asymptotics for these problems (i.e., the quotient (28)) is well known - see, for instance, [6, 3] where substantially more general results are mentioned. In the expression for the corresponding asymptotic coefficient the term ω^2 is "reduced" and we arrive at the asymptotics (3) for $\hat{N}(\alpha,a,V)$.

REFERENCES

1   Reed,M.;Simon,B.: *Methods of Modern Mathematical Physics IV*
     Academic Press, New York, 1978.
2   Birman, M.S.: *Matem. Sbornik* 55 (1961), 125-174 (in
     Russian); English translation: *Am.Math.Soc.Trans.*53
     (1966),23-80.

3   Birman, M.S.; Solomyak, M.Z.: In *Schroedinger operators,
        standard and non-standard* (eds. P.Exner and P.Seba), World
        Scientific, Singapore 1989, pp 3-18.
4   Birman, M.S.; Solomyak, M,Z.: In Proceedings of the
        conference *Integral Equations and Inverse Problems*, Varna,
        1989, 18-24 Sept. (in print).
5   Cwikel,M.: *Ann.Math.*106 (1977), 93-100.
6   Birman, M.S.; Solomyak, M.Z.: *Trudy Mosk.Mat.Ob-va* 27
        (1972),3-52 (in Russian).

M.S.Birman   M.Z.Solomyak
Department of Physics   Department of Mathematics
Insitute of Physics   Leningrad State University
Leningrad State University   Petrodworez,
Petrodworez,   Leningrad 198 904, U.S.S.R.
Leningrad 198 904, U.S.S.R.

Operator Theory:
Advances and Applications, Vol. 46
© 1990 Birkhäuser Verlag Basel

DISCRETE SPECTRUM IN THE GAPS OF THE CONTINUOUS ONE IN THE LARGE-COUPLING-CONSTANT LIMIT[*)]

M.S.Birman

1. The present paper is related to the paper [1], however, it does not depend on it formally. When the gap in the spectrum of the unperturbed operator is semi-infinite (coincides with $(-\infty, 0)$) then the problem considered below corresponds to investigation of the function $N(\alpha, a, V)$ introduced in [1] for a >0 and a "regular" perturbation $V(x) \geq 0$. As we shall show in the following, the problem can be reduced in general to the application of a simple abstract theorem. If the "control point" (see Section 2) is situated on the boundary of a gap and V is allowed to be "nonregular", then substantially more special considerations are necessary. For the periodic Schroedinger operator this will be discussed in another paper. Here we restrict ourselves to some applications of the abstract theorem of Section 3.

[*)] Translated by the editors

2. Let $\mathcal{H}$ be a Hilbert space and let S_∞ be the class of compact operators in $\mathcal{H}$. If $T \in S_\infty$, then we denote by $\nu(.,T)$ the distribution function of the singular numbers of the operator T. We will write $T \in \Sigma_p$, $0 < p < \infty$, if

$$(1) \qquad \nu(s,T) = O(s^{-p}), \quad s \longrightarrow +0,$$

and $T \in \Sigma_p^o$, if in (1) we can write o instead of O.
Let A and V be self-adjoint operators in $\mathcal{H}$ such that the conditions $V > 0$, $\text{Dom}(W) \supset \text{Dom}(|A|^{1/2})$, where $W := V^{1/2}$, and

$$(2) \qquad W|A - iI|^{-1/2} \in S_\infty$$

are satisfied. We consider the self-adjoint operator

$$(3) \qquad A_\pm(t) = A \mp tV, \quad t > 0.$$

The sum (3) is defined at least in the form sense (in view of (2))[*]. Assume now that there is a point $\lambda = \bar{\lambda} \in \rho(A)$ where $\rho(A)$ denotes the resolvent set of the operator A. In some neighborhood of the point λ the spectrum of $A_\pm(t)$ is discrete for all $t > 0$. The eigenvalues of $A_+(t)$ (of $A_-(t)$) move to the left (to the right) when t increases. By $N_\pm(\alpha,A,V,\lambda)$, $\alpha > 0$, we denote the total number of eigenvalues of the operator families $A_\pm(t)$ passing through the "control point" λ when t increases from 0 to α. In terms of the power scale the behavior of the function $N_\pm(\alpha)$ can be characterized by the quantities

---

[*] In this case the semiboundedness of A is not necessary. This fact was pointed out by D.R.Yafaev. The author thanks D.R.Yafaev for helpful discussions.

$$(4) \qquad \Delta_q^{(\pm)}(A,V,\lambda) = \lim_{\alpha\to\infty} \sup \ \alpha^{-q} \ N_\pm(\alpha,A,V,\lambda), \quad 0 < q < \infty,$$

and

$$(5) \qquad \delta_q^{(\pm)}(A,V,\lambda) = \lim_{\alpha\to\infty} \inf \ \alpha^{-q} \ N_\pm(\alpha,A,V,\lambda), \quad 0 < q < \infty.$$

3. Let F be now a symmetric operator in $\mathcal{H}$ such that $\mathrm{Dom}(F) \supset \mathrm{Dom}(A)$ and the condition

$$(6) \qquad \|Fu\|^2 \le \varepsilon\|Au\|^2 + c(\varepsilon)\|u\|^2, \quad u \in \mathrm{Dom}(A), \ \forall \varepsilon > 0$$

is satisfied. Setting $B = A + F$ we have $B = B^*$ and $W|B-iI|^{-1/2} \in S_\infty$. We introduce the self-adjoint families $B_\pm(t) = B \mp tV$ of the type (3). From the resolvent identity for the pair A,B the following assertion on the steadiness of the functionals (4) and (5) is obtained.

THEOREM. *Let* $\lambda = \bar\lambda \in \rho(A)$ *and let the conditions*

$$(7) \qquad W|A - \lambda I|^{-1/2} \in \Sigma_{2q} \ , \quad W|A - \lambda I|^{-1/2} \in \Sigma_{2q}^o$$

be satisfied for some $q > 0$. *If* F *obeys* (6) *and* $B = A + F$, *then for every* $\mu = \bar\mu \in \rho(B)$ *we have*

$$(8) \qquad \Delta_q^{(\pm)}(A,V,\lambda) = \Delta_q^{(\pm)}(B,V,\mu), \quad \delta_q^{(\pm)}(A,V,\lambda) = \delta_q^{(\pm)}(B,V,\mu).$$

REMARKS. (i) The condition (6) is certainly satisfied if $\mathrm{Dom}(F) \supset \mathrm{Dom}(|A|^\rho)$, $0 \le \rho < 1$. It implies, in particular, that under the condition (7) the quantities (4) and (5) are independent on $\lambda \in \rho(A)$.

(ii) The theorem can be easily generalized if we understand the sum A+F in the form sense. Then we have to replace the condition (6) by the analogous one for forms. However, we will not go into details.

(iii) If A > 0, then it is convenient to set $\lambda = -a < 0$. To check the conditions (7) and to compute the functionals (4) and (5) is just the same sort of problem as in [1] (for "regular" perturbations).

4. Let $\mathcal{H} = L_2(\mathbb{R}^m)$ and $A = (-\Delta)^l$, where $l > 0$ is not necessarily an integer. Then $\mathrm{Dom}(A) = H^{2l}(\mathbb{R}^m)$. Furthermore, let F be a symmetric operator in $L_2(\mathbb{R}^m)$ and let $\mathrm{Dom}(F) \supset H^\rho(\mathbb{R}^m)$, $0 \leq \rho < 2l$. If V is a multiplication operator generated by the function $v(x) \geq 0$ obeying

$$(9) \qquad v \in L_q(\mathbb{R}^m), \quad q = \frac{m}{2l} > 1,$$

then for every $\mu = \bar{\mu} \in \rho(A + F)$ we have

$$(10) \qquad \Delta_q^{(+)}(A + F, V, \mu) = \delta_q^{(+)}(A + F, V, \mu) = (2\pi)^{-m} \omega_m \int v^q dx, \quad {}^{*)}$$

$$(11) \qquad \Delta_q^{(-)}(A + F, V, \mu) = 0.$$

To deduce (10) and (11) from the theorem we have to use the results available in [2] and [3] concerning estimates of the spectrum and of spectral asymptotics for variational problems.

If $2l \geq m$, then we have to replace (9) by a lattice type condition (see [4]). Let $\mathbb{R}^m$ be decomposed into a lattice of unit cubes Q_n, $n \in \mathbb{Z}^m$, and let

---

${}^{*)}$ Here ω_m denotes volume of the unit ball of $\mathbb{R}^m$.

$$(12) \quad \begin{cases} \sum_n \|v\|^q_{L_p(Q_n)} < \infty, \quad q = \dfrac{m}{2l}, \\[2em] p = 1 \text{ for } 2l > m \text{ and } \forall p > 1 \text{ for } 2l = m. \end{cases}$$

Then the relations (10) and (11) are still fulfilled.

The condition for F is satisfied if, for instance, F is a multiplication operator generated by a real function f where f obeys

$$(13) \quad \begin{cases} \sup_{n \in \mathbb{Z}^m} \int_{Q_n} |f|^p \, dx < \infty, \\[2em] p = 2 \text{ for } 4l > m \text{ and } p > m/2l \text{ for } 4l \leq m. \end{cases}$$

Another possibility : the operator F is generated by a formally self-adjoint differential expression of order $\rho < 2l$ and all coefficients have bounded derivatives up to order ρ.

5. Let us discuss in more detail the case when $l = 1$ and V and F are multiplication operators generated by functions f and v, respectively. This case was investigated in [5] and [6] under the assumptions that $f, v \in L_\infty(\mathbb{R}^m)$. In [5] it was shown that under the condition

$$(14) \qquad 0 \leq v(x) \leq C(1 + |x|)^{-\tau}, \quad \tau > 2,$$

the relation (10) is fulfilled (for $l = 1$ and $2q = m$). This result is obviously covered by the results of Section 4 : the conditions (9) and (12) (for $l = 1$) are more general than the condition (14). Moreover, for $m \geq 3$ the condition (14) seems to

be an unnecessary restriction when compared to the "necessary"
condition $v \in L_{m/2}$. Likewise, for f the condition (13) is more
general than the condition $f \in L_\infty$.

In [6] the more complicated problem of the behaviour of
$N_-(\alpha)$ as $\alpha \longrightarrow \infty$ is considered. Assuming that $v(x) \sim c|x|^{-\beta}$ as
$|x| \longrightarrow \infty$, $0 < \beta < \infty$, the limit $\Delta_p^{(-)} = \delta_p^{(-)} > 0$, $p\beta = m$, is
calculated. In comparison with these results the relation (11) (l
= 1, 2q = m) looks coarser, however, it gives some additional
information (for $\beta > 2$) in view of the less restrictive
conditions for v.

6. The results of Section 4 allow to handle the problem
of a quantum particle in magnetic and electric fields. Let $m \geq 2$
and let $\vec{f}$ be a real vector function in $\mathbb{R}^m$ such that $\vec{f}, \mathrm{div}(\vec{f}) \in$
$L_\infty(\mathbb{R}^m)$. A periodic vector function $\vec{f}$ can be regarded as a
typical example. We set

$$B = (-i\nabla - \vec{f})^2 = A + F, \quad F = i(\vec{f}\nabla + \nabla\vec{f}) + |\vec{f}|^2,$$

and assume that V is a multiplication operator induced by $v \geq 0$.
If the conditions (9) for $m \geq 3$, l = 1 and (12) for m = 2, l = 1
are satisfied, then for any $\mu \in \rho(B)$ (in particular for $\mu < 0$)
the relations (10) and (11) are fulfilled for 2q = m . We note
that for $\mu < 0$ similar problems were discussed in [7], however,
in an substantially different case where the magnetic field is
assumed to be homogeneous.

7. Now we are going to discuss the example when A is a
differential operator with non-constant coefficients. To make the
formulations simple we consider only the periodic operator of
second order for $m \geq 3$, i.e. we set

$$B = A + F = -\mathrm{div}(a\,\mathrm{grad}) + f$$

where the positive definite matrix function a and the function f(x) are periodic. Denoting by Q the elementary cube of the lattice of periods and assuming $a \in C^1(\bar{Q})$, $f \in L_p(Q)$, where p = 2 for m = 3 and 2p > m for m ≥ 4, and $0 \leq v(x) \in L_{m/2}(\mathbb{R}^m)$ then for any $\mu \in \rho(B)$ we have

$$\Delta^{(+)}_{m/2}(B,V,\mu) = \delta^{(+)}_{m/2}(B,V,\mu) = (2\pi)^{-m} \omega_m \int \frac{v^{m/2}}{(\det a)^{1/2}} \, dx,$$

$$\Delta^{(-)}_{m/2}(B,V,\mu) = 0.$$

The condition $a \in C^1(\bar{Q})$ can be replaced by the condition $a+a^{-1} \in L_\infty(Q)$. This generalization is based on Remark (ii). In order to check the condition (7) we have to use the spectral estimates and asymptotics of [2,3] for variational problems with non-smooth main coefficient.

8. The number of examples can be extended easily. In particular, under the conditions of Section 4 it is possible to consider V as sign-definite differential operator (of order < 21) with non-constant coefficients. Similarly, it is possible to investigate operators of boundary problems in $L_2(\Omega)$ where $\Omega \subset \mathbb{R}^m$ is an unbounded region, etc..

9. In conclusion we are going to explain shortly how to check the conditions (7) under the assumptions of Section 4. For simplicity we restrict ourselves to the case 21 < m. As it was pointed out it is possible to set $\lambda = -1$ in (7). We consider the spectral problem for the quotient of quadratic forms

$$(15) \qquad \int v|u|^2 dx \Big/ \|u\|^2_{H^\rho(\mathbb{R}^m)}, \qquad 2\rho < m.$$

Denote by T_ρ the operator generated by (15) and by $n_\rho(s)$ the distribution function of its spectrum. The well-known [2,3] estimate for $n_\rho(s)$ holds:

$$(16) \qquad n_\rho(s) \leq C(m,\rho) \; s^{-\frac{m}{2\rho}} \int v^{\frac{m}{2\rho}} \, dx, \quad 2\rho < m.$$

Since $n_1(s) = N_+(s^{-1},A,V,-1)$ (see also (6) from [1]) it is easy to see that for $\rho = 1$, (16) leads directly to the relation $W(A+I)^{-1/2} \in \Sigma_{m/1}$.

The second condition of (7) can also be satisfied for q = m/1 if there exists a number σ such that

$$(17) \qquad W(A + I)^{-\sigma} \in \Sigma^o_{m/1}, \quad 1/2 < \sigma \leq 1.$$

In order to verify (17) we use interpolation methods. Let ρ obey the condition $1 < \rho \leq 21$, $2\rho < m$. If in addition to (16) we take into account the boundedness of T_ρ for $v \in L_\infty(\mathbb{R}^m)$, then real-interpolation method implies that under the condition $v \in L_p(\mathbb{R}^m)$, $m/2\rho < p < \infty$ the operator T_ρ belongs to S_p. In particular, for p = m/21 we obtain

$$(18) \qquad T_\rho \in S_{m/21} \subset \Sigma^o_{m/21}.$$

It remains to note that (18) is equivalent to (17) if we set $\sigma = \rho/21$.

Analogously we handle the case $21 \geq m$ but here we have to interpolate classes with quasi-norms of the "lattice" type.

REFERENCES

[1] Birman, M.S.; Solomyak, M.Z.: *Negative discrete spectrum of the Schroedinger operator with large coupling constant: quantitative discussion*, in this volume.
[2] Rosenblum, G.V.: *Izvest. VUZ'ov (Matematika)*, 1976, N.1, pp 75-86 (in Russian).
[3] Birman, M.S.; Solomyak,M.Z.: *Quantitative analysis in Sobolev embedding theorems and application to spectral theory*, Am. Math. Soc. Translation, 2.ser., vol.114 (1980).

[4] Birman, M.S.; Borzov, V.V.: *Problemy Matem. Fiziki*, Leningrad State University Publ., issue 5 (1971), pp. 24-38 (in Russian).
[5] Hempel,R.: *J. reine angew. Math.* 399 (1989), pp. 38-59.
[6] Alama, S.; Deift, P.; Hempel, R.: *Commun. Math. Phys.* 121 (1989), pp. 291-321.
[7] Raikov, G.: In Proceedings of the Conf. on *Integral Equations and Inverse Problems*, Varna, 1989, 18-24 Sept. (in print).

Department of Mathematical Physics
Leningrad State University
198904 Leningrad - St.Peterhof
USSR

Operator Theory:
Advances and Applications, Vol. 46
© 1990 Birkhäuser Verlag Basel

ON THE ASYMPTOPTICS OF DISCRETE SPECTRUM FOR   THE SCHROEDINGER OPERATOR IN ELECTRIC AND HOMOGENEOUS MAGNETIC FIELDS

A.V. Sobolev

One studies the asymptotics of bound states below the bottom of essential spectrum for the Schroedinger operator in a homogeneous magnetic and a decreasing electric fields. The electric potential is not assumed to be non-positive. The potential integrated along the direction of the magnetic field is supposed to have a power-like behaviour at infinity. The asymptotics of bound states is shown to be of a power-like character, and its main term is evaluated.

Let us consider a Schroedinger operator H in a homogeneous magnetic field and a decreasing electric field in $L^2(\mathbb{R}^3)$. Under a suitable choice of units the operator H can be written in cylindrical coordinates $x=(\rho,\phi,z)$ (the magnetic field is directed along the z-axis) in the form

$$H = -\Delta - 2i\partial/\partial\phi + \rho^2 + V(\rho,\phi,z).$$

Its essential spectrum coincides with the halfline $[2,\infty)$ [1].

Conditions on the potential given below guarantee infinitude of the discrete spectrum of H. Denote by $N(\mu,H)$, $\mu>0$, the number of eigenvalues of H (counting multiplicity) less than $2-\mu^2$. Our aim is to study the asymptotics of the function $N(\mu,H)$ as $\mu \to 0$. Its principal term is determined by the potential V integrated along the direction of magnetic field only [2-5]:

$$v(\rho,\phi) = \int_{-\infty}^{\infty} V(\rho,\phi,z)\,dz.$$

We assume the negative part $v_- = (|v|-v)/2$ of v to be an asymptotically homogeneous function:

$$v_-(\rho,\phi) = \Phi(\phi)\rho^{-\alpha} + o(\rho^{-\alpha}), \quad \alpha>0, \quad \rho \quad \infty, \tag{1}$$

Our main result is the following

THEOREM 1. *Let* V *satisfy the estimates*

$$\int_{-\infty}^{\infty} |V(\rho,\phi,z)|\,dz \leq C(1+\rho)^{-\beta},$$

$$\int_{-\infty}^{\infty} |z||V(\rho,\phi,z)|\,dz \leq C(1+\rho)^{-\gamma},$$

with some $\beta>0$, $0<\gamma\leq\beta$. *Let* v_- *have the asymptotics* (1) *with some* $\alpha>0$ *and* v *obey the bound*

$$|v(\rho,\phi)| \leq C(1+\rho)^{-\alpha}.$$

If $\beta\leq\alpha<\beta+\gamma$ *then the relation*

$$\lim_{\mu \to 0} \mu^{2/\alpha} \, N(\mu,H) = (2\pi)^{-1} \int_0^{2\pi} \{\Phi(\phi)\}^{2/\alpha} d\phi. \qquad (2)$$

holds.

The asymptotics of $N(\mu,H)$ was obtained for the first time in [2] for non-positive potentials symmetric with respect to the direction of magnetic field without the assumption (1) For the potentials $V \leq 0$ obeying (1) but not necessarily symmetric, the relation (2) was justified in [3] (see also [5]). The asymptotics (2) without the condition $V \leq 0$ was proven in [4] for the particular case $\alpha = \beta$. Thus Theorem 1 extends the last named result to all $\alpha \in [\beta, \beta+\gamma)$. Notice that for nonpositive functions V the equality $\alpha = \beta$ is automatically fulfilled, so the case $\alpha > \beta$ can be realized only if V changes sign.

Let $H_0 = -\Delta - 2i\partial/\partial\phi + \rho^2$ be the free Hamiltonian, $R_0 = (H_0 - 2 + \mu^2)^{-1}$ be its resolvent. For an arbitrary compact operator K denote by $n_+(\lambda,K)$, $\lambda > 0$, its distribution functions, i.e. the number of positive (respectively, negative) eigenvalues greater (respectively, less) than $\pm\lambda$.

The proof of Theorem 1 essentially follows the method of the papers [3,4]. We proceed from the following elementary relation

$$N(\mu,H) = n_-(1, R_0^{1/2}(\mu) V R_0^{1/2}(\mu)), \qquad (3)$$

which is usually referred to as Birman-Schwinger principle. Further argument is divided into two steps. The first one is to prove that the asymptotics of the function in the *rhs* of (3) coincides with the spectral asymptotics of the operator T acting in $L^2(\mathbb{R}^2)$, whose integral kernel is

$$(2\pi)^{-1}v_-^{1/2}(\rho,\phi)\ \exp\left\{-(\rho^2+\rho'^2)/2 + \rho\rho'e^{-i(\phi-\phi')}\right\}\ v_-^{1/2}(\rho',\phi').$$

More precisely, we establish that

$$\lim_{\mu\to0}\mu^{2/\alpha}n_-(1,R_0^{1/2}(\mu)VR_0^{1/2}(\mu)) = \lim_{\mu\to0}\mu^{2/\alpha}n_+(\mu,T). \tag{4}$$

To that end we derive an explicit representation for the resolvent $R_0(\mu)$ in terms of the eigenfunctions for H_0 and then retain the leading order (as $\mu \to 0$) only. It is exactly here, where the condition $\alpha\in[\beta,\beta+\gamma)$ plays role.

The most difficult part of the proof is to check that the *rhs* of (4) equals that of (2). Though the kernel of T has an explicit form, there are presumably no results in the literature which could have been applied directly to evaluate its spectral asymptotics. Nevertheless the operator T can be studied on the basis of general ideas developed earlier (see e.g. [6]) for pseudo-differential operators. A detailed analysis of the spectral asymptotics for T is given in [3].

ACKNOWLEDGMENT

The author is grateful to the organizers for their kind invitation to participate in the Dubna workshop.

REFERENCES

1   J.Avron, I.Herbst, B.Simon, *Schroedinger operators with magnetic fields,1,General interactions,* Duke Math.J. **45** (1978), 847-883.

2   S.N.Solnyshkin, *Asymptotics of the energy of bound states of the Schroedinger operator in the presence of electric and homogeneous magnetic fields,* Selecta Math. Soviet.5 (1986), 297-306.

3   A.V.Sobolev, *On the asymptotics for energy levels of a quantum particle in homogeneous magnetic field, perturbed by a decreasing electric field*, 1. J.Soviet Math. **35** (1986), 2201-2211.

4   A.V.Sobolev, *On the asymptotics for energy levels of a quantum particle in homogeneous magnetic field, perturbed by a decreasing electric field*, 2. Probl.Math.Phys.**11** (1986), 232-248.

5   H.Tamura. *Asymptotic distribution of eigenvalues for Schroedinger operators with homogeneous magnetic fields*, Osaka J.Math. **25** (1988), 633-647.

6   M.Sh.Birman, M.Z.Solomjak. *Asymptotic behaviour of the spectrum of weakly polar integral operators*, Izv.Akad.Nauk SSSR, Ser.Math. **34** (1970), 1142-1158.

Leningrad Branch of Steklov Mathematical Institute
Fontanka 27
191011 Leningrad, USSR

Operator Theory:
Advances and Applications, Vol. 46
© 1990 Birkhäuser Verlag Basel

ASYMPTOTICS OF THE DISCRETE SPECTRUM OF HAMILTONIANS OF QUANTUM SYSTEMS WITH A HOMOGENEOUS MAGNETIC FIELD

S.A.Vugal'ter, G.M.Zhislin

We consider Hamiltonians of systems of N identical quantum particles in an external homogeneous magnetic field combined with the potential field of a fixed centre. Double-sided estimates are obtained for the number $N^\alpha(\mu^\alpha+\lambda)$ of the eigenvalues of a given symmetry (taking multiplicity into account) which are less than $\mu^\alpha+\lambda$, using similar estimates of effective one-particle operators; here μ^α is the bottom of the essential spectrum of the symmetry α , $\lambda < 0$. Using these estimates we have got the asymptotics of $N^\alpha(\mu^\alpha+\lambda)$ as $\lambda \to -0$.

INTRODUCTION

In this paper we investigate the discrete spectrum of a given permutational symmetry for the quantum system of N identical particles in an external homogeneous magnetic field combined with the potential field of a fixed centre.

Such systems are, for example, atoms with a fixed nucleus with a homogeneous magnetic field, their negative and positive ions. For Hamiltonians of the considered systems, the asymptotic estimates are obtained for the number of eigenvalues (with the account of multiplicity) which are less than $\mu^\alpha+\lambda$, where μ^α is the bottom of the essential spectrum of the sym - metry α , $\lambda < 0$. These estimates are expressed by similar ones for effective one-particle operators. The method of the proof of the main results modifies the method, which has been

proposed by the authors for Hamiltonians of quantum systems when
the external magnetic field is absent [1].

DEFINITIONS

Let $Z = (1,2,\ldots N)$ be the system of N identical
particles, m be the mass of a particle, $r_i = (x_i, y_i, z_i)$ be
the position of the i-th particle, $\{0\}$ be the index of an infinitely heavy particle, which is situated at the point $(0,0,0)$,
$Z_0 = (0,1,\ldots, N)$. Hamiltonian of the system Z_0 with a homogeneous magnetic field has the form

$$\mathcal{H} = \sum_{j=1}^{N} m^{-1}(i\nabla_j + A_j)^2 + \sum_{i,j;\,i<j}^{1,N} V(|r_{ij}|) + \sum_{i=1}^{N} V_0(|r_i|) \qquad (2.1)$$

where $r_{ij} = r_i - r_j$,

$$A_j = \left\{ -\frac{H}{2} y_j, \; \frac{H}{2} x_j, \; 0 \right\}, \; H > 0 \qquad \text{is the constant.}$$

What concerns the potentials we suppose that

$$V_0(|r_1|), V(|r_1|) \subset C^{\infty}(R^3 \setminus \{0\}) \cap \mathcal{L}_{2,\ell oc}(R^3),$$

$$\exists c > 0, \; e \neq 0, \; Z_0 \neq 0, \; \mathscr{X} > 0 \qquad \text{for which}$$

$$V(|r_1|) = e|r_1|^{-\mathscr{X}}, \; V_0(|r_1|) = Z_0 e|r_1|^{-\mathscr{X}} \; if \; |r_1| > c$$

The operator $\mathcal{H}$ with the domain C_0^{∞} is essentialy self-
adjoint in $\mathcal{L}_2(R^{3N})$. We extend it to a self-adjoint operator
retaining the previous notation.

Let S be the group of permutations of the identi-
cal particles, α be the types of irreducible representations
of S in $\mathcal{L}_2(R^{3N})$ by the operators T_g : $T_g \varphi(r_1,\ldots,r_N) =
= \varphi(g^{-1}r_1,\ldots,g^{-1}r_N), P^{\alpha}$ be the projection in $\mathcal{L}_2(R^{3N})$ on the space
of the functions of symmetry $\alpha, \mathcal{H}^{\alpha} = \mathcal{H} P^{\alpha}, D(\mathcal{H}^{\alpha})$ be
the domain of operator $\mathcal{H}^{\alpha}$. In the following we shall in-
vestigate the discrete spectrum of $\mathcal{H}^{\alpha}$. We denote by $Z_2 =
(C_1, C_2)$ an arbitrary decomposition of the system Z_0 into
two nonempty nonintersecting subsystems C_j. We suppose that

the particle $\{0\}$ is always in C_1 and write $C_1^0 = C_1 \setminus \{0\}$. Furthermore, let $n(C_1^0)$, $n(C_2)$ be the number of the particles in C_1^0 and in C_2, respectively; $S[C_i]$ be the group of the permutations of the identical particles in C_i ; $S(z_2) =$

$$= S[C_1] \times S[C_2], \; m[C_2] = \sum_{i \in C_2} m_i \; ; \; \zeta \equiv \zeta(z_2) = \zeta[C_2] = (\zeta_{z_2 x}, \zeta_{z_4}, \zeta_{z_2})=$$
$$= m^{-1}[C_2] \sum_{i \in C_2} m_i \, r_i$$

be the centre-of-mass position of the subsystem C_2; $q_t = (q_{t,x}, q_{t,y}, q_{t,z})$, $q_t = r_t$ for $t \in C_1^0$, $q_t = r_t - \zeta$ for $t \in C_2$; $q = (q_1, \ldots, q_N)$, $R_c(z_2) = \{\zeta\}$, $R_0(z_2) = \{q\}$. Evidently the space $R_c(z_2)$ corresponds to the centre-of-mass motion of subsystem C_2, $R_0(z_2)$ — to the relative motion of the particle of the compound system $z_2 = (C_1, C_2)$, and $R^{3N} = R_c(z_2) \oplus R_0(z_2)$. The operator $\mathcal{H}$ can be written as

$$\mathcal{H} = \mathcal{H}(z_2) + m[C_2]^{-1}(i\nabla_\zeta + A_\zeta^c)^2 + I(z_2) \tag{2.2}$$

where

$$\mathcal{H}(z_2) = \mathcal{H}[C_1] + \mathcal{H}_0[C_2],$$

$$\mathcal{H}[C_1] = \sum_{j \in C_1^0} m^{-1}(i\nabla_j + A_j)^2 +$$

$$+ \sum_{\kappa, p \in C_1^0 ; \kappa < p} V(|r_{\kappa p}|) + \sum_{j \in C_1^0} V_0(|r_j|),$$

$$\mathcal{H}_0[C_2] = \sum_{j \in C_2} m^{-1}(i[\nabla_{q_j} - n[C_2]^{-1} \sum_{\kappa \in C_2} \nabla_{q_\kappa}] + A_j^0)^2 +$$

$$+ \sum_{j, \kappa \in C_2, j < \kappa} V(|r_{j\kappa}|),$$

$$A_j^0 = \left\{ -\frac{H}{2} q_{j,y} , \frac{H}{2} q_{j,x} , 0 \right\},$$

$$A_\zeta^c = \left\{ -\frac{H}{2} n[C_2] \zeta_y , \frac{H}{2} n[C_2] \zeta_x , 0 \right\}$$

$$I(z_2) = \sum_{j \in C_2} V_0(|r_j|) + \sum_{i \in C_1^0, j \in C_2} V(|r_{ij}|).$$

Obviously, the operator $\mathcal{H}_0(z_2)$ affects only the variables of $R_0(z_2)$

      Let $\alpha'(z_2)$ be the type of irreducible representation of the group $S(z_2)$ in $\mathcal{L}_2(R_0(z_2))$, $P^{\alpha'}$ be the projection in $\mathcal{L}_2(R_0(z_2))$ on the subspace of the functions of symmetry α', $m_{\alpha'}^\alpha$ be the multiplicity of the representation α' in the representation α after the restriction of the latter from S to $S(z_2)$. Notice that if the particles $1,2,\ldots,N$ are electrons and only the types α permitted by Pauli principle are considered then $m_{\alpha'}^\alpha \leqslant 1$.

      We shall write $\alpha' \prec \alpha$ if $m_{\alpha'}^\alpha \geqslant 1$. Let

$$\mathcal{H}(\alpha; z_2) = \sum_{\alpha' \prec \alpha} \mathcal{H}(z_2) \cdot P^{\alpha'},$$

$$\nu^\alpha = \min_{z_2} \inf \mathcal{H}(\alpha; z_2), \quad \mu^\alpha = \nu^\alpha + m^{-1}H.$$

Similarly to the proof of HVZ-theorem for the Hamiltonians of systems without the magnetic field, one can show that the essential spectrum of operator $\mathcal{H}^\alpha$ coincides with the half-line $[\mu^\alpha, +\infty)$. Let

$$0(\alpha) = \left\{ z_2 \mid \inf \mathcal{H}(\alpha; z_2) = \nu^\alpha \right\}$$

We suppose everywhere that the following assumptions are valid:

A) All $z_2 \in 0(\alpha)$ may be constructed from one decomposition by permutations of the identical particles.

B) The number ν_α is the point of discrete spectrum of the operator $\mathcal{H}(\alpha, z_2)$ for $z_2 \in 0(\alpha)$.

We denote by $\mathcal{W}(\alpha; z_2)$ the eigenspace of the operator $\mathcal{H}(\alpha; z_2)$ which corresponds to the eigenvalue ν_α .

In addition we suppose that the following condition is fulfilled:

C) There is a unique $\alpha'(Z_2)$ such that

$$\rho^{\alpha'(Z_2)} \, W(\alpha; Z_2) = W(\alpha; Z_2),$$

$$\dim W(\alpha; Z_2) = \dim \alpha'(Z_2)$$

We denote by $\varphi_j(q)$, $j = 1, 2, \ldots, t$, the orthonormal basis in $W(\alpha; Z_2)$ that coincides with the canonical basis of the representation α' . Let $N^\alpha(\mu^\alpha + \lambda)$ be the number of the ei-genvalues (with account of multiplicity) of the operator $\mathcal{H}^\alpha$ which are less than $\mu^\alpha + \lambda$; $N_1(\lambda, V)$ be the number of the eigenvalues (with account of multiplicity) which are less than $m^{-1} H + \lambda$ for the one-particle operator

$$h = m[C_2]^{-1}(i\nabla_{\mathfrak{z}} + A_{\mathfrak{z}}^c)^2 + V(\mathfrak{z}). \tag{2.3}$$

THE RESULTS

Let us first consider the case when in the decomposition $Z_2 \in 0(\alpha)$ the subsystem C_1 is "charged", that is $n[C_1^0]^2 + Z_0 \neq 0$. Let $Q = en[C_2](n[C_1] + Z_0)$

THEOREM 1. Let $Q \neq 0$. Then there are $a > 0$, $a_1 > 0$ and $\beta > 0$ such that for each $\lambda < 0$

$$m_{\alpha'}^\alpha(Z_2) \dim \alpha \, N_1(\lambda, V_1) - a \leqslant N^\alpha(\mu^\alpha + \lambda) \leqslant$$

$$\leqslant m_{\alpha'}^\alpha(Z_2) \dim \alpha \, N_1(\lambda, V_2) + a,$$

where $V_{1(2)} = Q|\mathfrak{z}|^{-\gamma} \underset{(-)}{+} a_1(1 + |\mathfrak{z}|)^{-\gamma - 1}$ if $|\mathfrak{z}| \geqslant \beta$,

$$V_i(\mathfrak{z}) = 0 \quad \text{if } |\mathfrak{z}| < \beta.$$

REMARKS. 1. Theorem 1 remains valid when the poten-tials $V(r_1)$ and $V_0(r_1)$ in (2.1) are not spherically symmetric.

2. Theorem 1 can be generalized for the case when the assumptions A and C are relaxed.

Let us apply Theorem 1 to the case $Q > 0$. In this situation the numbers $N_1(\lambda, V_1)$ and $N_1(\lambda, V_2)$ will be bounded uni-formly for all $\lambda < 0$ and from Theorem 1 we get the following assertion.

THEOREM 2. Let $Q > 0$. Then the discrete spectrum of the operator $\mathcal{H}^\alpha$ is finite.

Now we shall consider the case $Q < 0$. From Theorem 1 and the asymptotics of the number $N_1(\lambda, V)$ for $\gamma < 2$ (deduced by Ivrii [2]) and for $\gamma > 2$ (cf. Sobolev [3]) we get the asymptotics for $N^\alpha(\mu^\alpha + \lambda)$.

THEOREM 3. Let $Q < 0$, $\gamma < 2$. Then

$$N^\alpha(\mu^\alpha + \lambda) = |\lambda|^{\frac{1}{2} - \frac{3}{\gamma}} \cdot H \cdot m^{-1} \cdot m [C_2]^{3/2} \cdot Q^{3/\gamma} \cdot \pi^{-1} \cdot$$
$$\dim \alpha \cdot m_{\alpha'}^\alpha \cdot \mathcal{I}_1 + O\left(|\lambda|^{-\frac{2}{\gamma}}\right)$$

where $\mathcal{I}_1 = \int_0^1 (\tau^{-\gamma} - 1)^{1/2} \tau^2 \, d\tau$.

THEOREM 4. Let $Q < 0$, $\gamma > 2$. Then

$$\lim_{\lambda \to -0} N^\alpha(\mu^\alpha + \lambda) \cdot |\lambda|^{\frac{1}{\gamma - 1}} = H \cdot m^{-1} m [C_2]^{\frac{\gamma}{\gamma - 1}} \cdot Q^{\frac{2}{\gamma - 1}} \, 2^{\frac{1 + \gamma}{1 - \gamma}}$$
$$\cdot \dim \alpha \cdot m_{\alpha'}^\alpha \cdot \mathcal{I}_2$$

where $\mathcal{I}_2 = \int_0^\infty (1 + \tau^2)^{-\frac{\gamma}{2}} \, d\tau$.

Let us now consider the more complicated case $Q = 0$. Under this condition the asymptotics of $N^\alpha(\mu^\alpha + \lambda)$ are defined by effective interactions of a multipole type between the subsystems C_1 and $C_2 \in Z_2 \in D(\alpha)$. Let

$$K_\tau^2 = \sum_{n \in C_1^0, \, j \in C_2} \int (q_{n\tau} - q_{j\tau})^2 |\varphi_p|^2 \, d\tau \qquad \tau = x \, ; \, y \, ; \, z \, .$$

One can demonstrate that $K_x^2 = K_y^2$ and the numbers K_τ^2 do not depend on the choice of the function φ_p from the canonical basis in $\mathcal{W}(\alpha \, ; \, Z_2)$. We write $\sin \theta_{\zeta} = \zeta_z \cdot |\zeta|^{-1}$

THEOREM 5. Let $Q = 0$. Then there are $a > 0$, $a_1 > 0$, $\beta > 0$, such that

$$m_{\alpha'}^\alpha \, \dim \alpha \, N_1 \, (\lambda, V_1) - a \leqslant N^\alpha (\mu^\alpha + \lambda) \leqslant$$

$$\leqslant m_{\alpha'}^\alpha \cdot \dim \alpha \, N_1 \, (\lambda, V_2) + a \, ,$$

where

$$V_{1(2)} = e \, \gamma \, |\zeta|^{-\gamma - 2} \left\{ K_x^2 \left[2 - (\gamma + 2) \cos^2 \theta_{\zeta} \right] + \right.$$

$$\left. + K_z^2 \left[1 - (\gamma + 2) \cdot \sin^2 \theta_{\zeta} \right] \right\} \underset{(-)}{\overset{+}{}} a_1 |\zeta|^{-\omega} \text{ if } |\zeta| \geqslant \beta \, ,$$

$$V_i(\zeta) = 0 \qquad \text{if } |\zeta| < \beta \, ; \quad \omega = \min \left\{ 2\gamma + 2 \, , \, 4 + \gamma \right\} \, .$$

The effective potentials $V_i(\zeta)$ decrease more rapidly than $|\zeta|^{-2}$, and consequently [3] the asymptotics of the numbers $N_1(\lambda, V_i)$ are defined by the integral of the main part of the potential along the direction of the field (along the Z - axis). This integral equals:

$$I_0 = e \, K_x^2 \, J_1 \cdot (1 - \gamma)(\zeta_x^2 + \zeta_y^2)^{-\frac{\gamma + 1}{2}} \, ,$$

where $J_1 = 2 \gamma \int^{\pi/2} \cos^\gamma x \, d x$.

If $(1 - \gamma) e < 0$, then to get $N_1(\lambda, V_i)$ we have to apply the results of [3] and the asymptotic of $N^\alpha (\mu^\alpha + \lambda)$ follows from Theorem 5.

THEOREM 6. Let $Q=0$ and $(1-\gamma)e<0$. Then

$$\lim_{\lambda \to -0} N^\alpha(\mu^\alpha+\lambda)\cdot|\lambda|^{\frac{1}{\gamma+1}} = \dim \alpha \cdot m^\alpha_{\alpha',n}[C_2]^{\frac{\gamma}{\gamma+1}} .$$

$$\cdot(K^2_x)^{\frac{2}{\gamma+1}}\cdot H\cdot|e|^{\frac{2}{\gamma+1}}m^{-\frac{1}{\gamma+1}}a_0(\gamma),$$

where $a_0(\gamma)=J_1^{\frac{2}{\gamma+1}}\cdot 2^{\frac{\gamma+3}{\gamma+1}}\cdot|(1-\gamma)|^{\frac{2}{\gamma+1}}$.

We remark that in the considered case the asymptotics of $N^\alpha(\mu^\alpha+\lambda)$ are defined by the multipole moment K^2_x in the plane orthogonal to the field direction and does not depend on K^2_z . Let us now suppose that $e(1-\gamma)>0$. Under this condition we get a positive value after integrating U_j along the Z-axis. In this case, we conjecture that the discrete spectrum of operator $\mathcal{H}^\alpha$ is finite, but we have been able to prove this assertion only under an extra condition.

THEOREM 7. Let $Q=0$, $e(1-\gamma)>0$ and

$$\frac{1}{2}(\gamma+1)< K^2_x(K^2_z)^{-1}< \gamma^{-1} \quad \text{or} \quad \gamma^{-1}< K^2_x(K^2_z)^{-1}< \frac{1}{2}(\gamma+1)$$

Then the discrete spectrum of $\mathcal{H}^\alpha$ is finite.

When $\gamma=1$ (Coulomb potentials) we have $I_0=0$ and Theorem 5 does not permit to find the asymptotic behaviour of $N^\alpha(\mu^\alpha+\lambda)$ That is why the case $\gamma=1$ requires special consideration. Let $\tau=x$ or $\tau=y$ or $\tau=z$,

$$\Phi_{p,\tau}=\sum_{\kappa\in C^o_1, j\in C_2}(q_{\kappa\tau}-q_{j\tau})\varphi_p , \eta^2_\tau = ((\mathcal{H}(z_2)-\nu^\alpha)^{-1}\Phi_{p,\tau},\Phi_{p,\tau})$$

One can prove that $\Phi_{p,\tau}\perp \mathcal{W}(\alpha;z_2)$ and therefore these functions belong to the domain of the operator $(\mathcal{H}(z_2)-\nu^\alpha)^{-1}$ and $\eta^2_\tau>0$. Moreover $\eta^2_x=\eta_y$ and η^2_τ does not depend on p .

THEOREM 8. Let $Q=0$, $\gamma=1$. Then for arbitrary

$\varepsilon > 0$, some positive constants $a = a(\varepsilon)$, $b = b(\varepsilon)$, $\lambda_0(\varepsilon)$ and all $\lambda_0(\varepsilon) < \lambda < 0$, we have

$$m_{\alpha'}^{\alpha} \dim\alpha \, N_1(\lambda, \tilde{v}_1) - a \leqslant N^{\alpha}(\mu^{\alpha} + \lambda) \leqslant$$

$$\leqslant m_{\alpha}^{\alpha} \, \dim\alpha \, N_1(\lambda, \tilde{v}_2) + a ,$$

where

$$v_{1(2)}(\mathfrak{z}) = e \, |\mathfrak{z}|^{-3} \left\{ K_x^2 [2 - 3\cos^2\theta_{\mathfrak{z}}] + K_{\mathfrak{z}}^2 [1 - 3\sin^2\theta_{\mathfrak{z}}] \right\} -$$

$$- (1(\mp)\varepsilon) e^2 |\mathfrak{z}|^{-4} \left\{ \eta_x^2 \cos^2\theta_{\mathfrak{z}} + \eta_{\mathfrak{z}}^2 \sin^2\theta_{\mathfrak{z}} \right\}$$

if $|\mathfrak{z}| > b$; $\tilde{v}_i(\mathfrak{z}) = 0$ if $|\mathfrak{z}| \leqslant b$.

Now the asymptotics for $N_1(\lambda, \tilde{v}_1)$, $N_1(\lambda, \tilde{v}_2)$ are unknown, but we expect that the main terms of the asymptotics of $N_1(\lambda, \tilde{v}_1)$ and $N_1(\lambda, \tilde{v}_2)$ coincide.

PROOF OF THEOREM 1.

Proofs of Theorems 1,5,8 are based on the construction of subspaces $\mathcal{M}_i(\lambda) \subset D(\mathcal{H}^{\alpha}) \, i = 1,2$, such that for all $\varphi \in \mathcal{M}_1(\lambda)$, $\lambda < 0$, we have

$$(\mathcal{H}^{\alpha} \varphi, \varphi) < (\mu^{\alpha} + \lambda) \, \| \varphi \|^2 \tag{4.1}$$

and for all $\varphi \in D(\mathcal{H}^{\alpha})$, $\varphi \perp \mathcal{M}_2(\lambda)$

$$(\mathcal{H}^{\alpha} \varphi, \varphi) > (\mu^{\alpha} + \lambda) \, \| \varphi \|^2 \tag{4.2}$$

It follows from (4.1), (4.2) that $\dim \mathcal{M}_1(\lambda) \leqslant N^{\alpha}(\mu^{\alpha} + \lambda) \leqslant \dim \mathcal{M}_2(\lambda)$. The dimensions of the subspaces $\mathcal{M}_i(\lambda)$ are estimated by the dimensions of the corresponding subspaces for one-particles effective operators.

To demonstrate our methods, we shall prove Theorem 1. Let

$$K(\mathfrak{z}_2 ; \delta) = \left\{ r \, | \, r \in R^{3N}, \, |\mathfrak{z}_{\mathfrak{z}_2}| > (1 - \delta) |r| \right\},$$

$$u_{\mathfrak{z}_2} = u_{\mathfrak{z}_2}(|\mathfrak{z}_{\mathfrak{z}_2}| \cdot |r|^{-1}) u_{\mathfrak{z}_2} \in C^2(R^{3N}), \, u_{\mathfrak{z}_2} = 1 \text{ for } r \in K(\mathfrak{z}_2 ; \delta),$$

$$u_{z_2} = 0 \quad \text{for} \quad r \notin K(z_2;\beta), \ \beta > \delta > 0, \ |\nabla u_{z_2}|^2 (1 - u_{z_2}^2)^{-1} \leqslant \text{const}.$$

As the space $\mathcal{M}_1$ we take the subspace of the functions φ of the following form

$$\varphi = P^{\alpha} \sum_{i,p} \sum_{g \in S} C_{ip}(g) \, T_g \, \psi_i(q_{z_2}) \, f_p(z_{z_2}) \, u_{z_2} \qquad (4.3)$$

where C_{ip} are the arbitrary numbers, $\psi_i \in W(\alpha; z_2)$; $f_p(z)$ $p = 1, \ldots, P$ are the orthonormal eigenfunctions which correspond to the eigenvalues $\lambda_p \leqslant m^{-1} H + \lambda$ of the operator

$$h_K(\beta) = m[C_2]^{-1}(i\nabla_z + A_z^c)^2 + \mathcal{V}_K(z) \quad K = 1 \qquad (4.4)$$

in the region $|z| \geqslant \beta$ with a zero boundary condition for functions at $|z| = \beta$; here $\beta > 0$ is some constant and are defined in the theorem. According to Lemmas 5.2, 5.3

$$\dim \mathcal{M}_1(\lambda) \geqslant m_{\alpha}^{\alpha}, \ \dim \alpha \, N_1(\lambda, v_1) - \text{const}.$$

If (4.1) is valid then $N^{\alpha}(\mu^{\alpha} + \lambda) \geqslant \dim \mathcal{M}_1(\lambda)$. Consequently to estimate $N^{\alpha}(\mu^{\alpha} + \lambda)$ from below, it is sufficient to prove (4.1). Let $\varphi \in \mathcal{M}_1(\lambda)$. It is clear that for small $\beta > 0$ one has

$$(\mathcal{H}^{\alpha}\varphi, \varphi) = \sum_{z_2 \in 0(\alpha)} (\mathcal{H}^{\alpha}\varphi_{z_2}, \varphi_{z_2}), \ \|\varphi\|^2 = \sum_{z_2} \|\varphi_{z_2}\|^2, \qquad (4.5)$$

where

$$\varphi_{z_2} = \sum_{i,p} a_{ip} \, \psi_i \, f_p \, u_{z_2}$$

and a_{ip} are some numbers which depend on α. We write

$$F_i = \sum_p a_{ip} \, f_p.$$

Then $\psi_{z_2} = \sum_i \varphi_i F_i u_{z_2}$ and according to (2.2)

$$(\mathcal{H}^\alpha \psi_{z_2}, \psi_{z_2}) = (\mathcal{H}(z_2)\psi_{z_2}, \psi_{z_2}) + (I(z_2)\psi_{z_2}, \psi_{z_2}) \quad (4.6)$$

$$+ m[C_2]^{-1}((i\nabla_{\mathfrak{z}} + A_{\mathfrak{z}}^c)^2 \psi_{z_2}, \psi_{z_2}).$$

It is easy to demonstrate that

$$\left(\mathcal{H}(z_2)\psi_{z_2}, \psi_{z_2}\right) \leq \left(\mathcal{H}(z_2)\sum_{i=1}^t \varphi_i F_i, \sum_{i=1}^t \varphi_i F_i\right) - \quad (4.7)$$

$$- \left(\mathcal{H}(z_2)\sum_{i=1}^t \varphi_i F_i v_{z_2}, \sum_{i=1}^t \varphi_i F_i v_{z_2}\right) +$$

$$+ C_0 \sum_{\kappa \in C_2} \int \left(|\nabla_{g\kappa} u_{z_2}|^2 + |\nabla_{q\kappa} v_{z_2}|^2\right) \left|\sum_{i=1}^t \varphi_i F_i\right|^2 d\Omega,$$

where $v_{z_2} = (1 - u_{z_2}^2)^{\frac{1}{2}}$ and C_0 is some constant.
Since operator $\mathcal{H}(z_2)$ is semibounded from below, we have

$$- \left(\mathcal{H}(z_2)\sum_{i=1}^t \varphi_i F_i v_{z_2}, \sum_{i=1}^t \varphi_i F_i v_{z_2}\right) \leq C \left\|\sum_{i=1}^t \varphi_i F_i v_{z_2}\right\|^2 \quad (4.8)$$

The functions v_{z_2}, $\nabla_{q_\kappa} u_{z_2}, \nabla_{q_\kappa} v_{z_2}$ can be non-zero only for $(1-\delta)|q_{z_2}| > \delta|\mathfrak{z}_{z_2}|$. Consequently it follows from Lemma 5.1 that

$$C_0 \sum_{\kappa \in C_2} \int \left(|\nabla_{q\kappa} u_{z_2}|^2 + |\nabla_{q\kappa} v_{z_2}|^2\right) \left|\sum_{i=1}^t \varphi_i F_i\right|^2 d\Omega + \quad (4.9)$$

$$+ C \left\|\sum_{i=1}^t \varphi_i F v_{z_2}\right\|^2 \leq C_1 \sum_{i=1}^t \left\| F_i |\mathfrak{z}|^{-\gamma-2}\right\|^2.$$

Therefore

$$(\mathcal{H}(z_2)\psi_{z_2}, \psi_{z_2}) \leq \sum_{i=1}^t \left(\nu^\alpha \|F_i\|^2 + C_1 \| F_i |\mathfrak{z}|^{-\gamma-2}\|^2\right) \leq \quad (4.10)$$

$$\leq \nu^\alpha \|\psi_{z_2}\|^2 + 2C_1 \sum_{i=1}^t \| F_i |\mathfrak{z}|^{-\gamma-2}\|^2.$$

Now we estimate the expression $m[c_2]^{-1}((i\nabla_{\mathfrak{z}}+A_{\mathfrak{z}}^C)^2\psi_{z_2},\psi_{z_2})$. Evidently,

$$((i\nabla_{\mathfrak{z}}+A_{\mathfrak{z}}^C)^2\varphi_i F_i u_{z_2},\varphi_i F_i u_{z_2})= \tag{4.11}$$

$$=\int(|\nabla_{\mathfrak{z}}u_{z_2}|^2+|\nabla_{\mathfrak{z}}v_{z_2}|^2)|\varphi_i F_i|^2\,d\Omega+$$

$$+((i\nabla_{\mathfrak{z}}+A_{\mathfrak{z}}^C)^2\varphi_i F_i,\varphi_i F_i)-((i\nabla_{\mathfrak{z}}+A_{\mathfrak{z}}^C)\varphi_i F_i v_{z_2},\varphi_i F_i v_{z_2})$$

Using (4.11) in analogy with (4.8)-(4.10), we get

$$((i\nabla_{\mathfrak{z}}+A_{\mathfrak{z}}^C)^2\psi_{z_2},\psi_{z_2})\le\sum_{i=1}^{t}((i\nabla_{\mathfrak{z}}+A_{\mathfrak{z}}^c)^2 F_i F_i)+ \tag{4.12}$$

$$+2C\|F_i|\mathfrak{z}|^{-\gamma-2}\|^2.$$

From the relations (4.4), (4.6), (4.10), (4.12) it follows that to prove (4.1), it is sufficient to estimate the expression $(I(z_2)\psi_{z_2},\psi_{z_2})$.

For a sufficiently small β , $r\in K(z_2,\beta)$ and $K\in C_1^0$, $n\in C_2$ we have

$$\|q_K+\mathfrak{z}z_2-q_n|^{-\gamma}-|\mathfrak{z}z_2|^{-\gamma}\le C(|q_K|+|q_n|)|\mathfrak{z}z_2|^{-\gamma-1}, \tag{4.13}$$

$$\|\mathfrak{z}z_2-q_n|^{-\gamma}-|\mathfrak{z}z_2|^{-\gamma}\|\le C|q_n\|\mathfrak{z}z_2|^{-\gamma-1}.$$

That is why for $r\in K(z_2,\beta)$ the inequality

$$((I(z_2)-Q|\mathfrak{z}|^{-\gamma})\varphi_i F_i u_{z_2},\varphi_i F_i u_{z_2})\le C\|F_i|\mathfrak{z}|^{-\frac{\gamma+1}{2}}\|^2$$

holds, and hence

$$(I(z_2)\psi_{z_2},\psi_{z_2})\le\sum_{i=1}^{t}\Big\{Q(z_2)(F_i,|\mathfrak{z}|^{-\gamma}F_i)+ \tag{4.14}$$

$$+2C(F_i,|\mathfrak{z}|^{-\gamma-1}F_i)\Big\}.$$

The relations (4.6), (4.10), (4.12), (4.13) demonstrate the validity of (4.1), because F_i is the sum of eigenfunctions of the operator h_1, which correspond to the eigenvalues $\varepsilon_i \leqslant \bar{m} H + \lambda'_d$.

Now let us begin to construct subspace $M_2(\lambda)$. Similarly to case $H = 0$ [1,4] we can find a subspace M'_2, $\dim M'_2 < +\infty$, such that for $\varphi \perp M'_2$ one has

$$(\mathcal{H}^\alpha \varphi, \varphi) \geqslant \sum_{z_2 \in 0(\alpha)} \left\{ (\mathcal{H}^\alpha \varphi_{z_2}, \varphi_{z_2}) - \varepsilon \| \varphi_{z_2} |z|^{-\gamma-2} \|^2 - \tag{4.15}$$

$$- C(\varepsilon) \| \varphi_{z_2} \cdot \chi(z_2) \cdot |z|^{-1} \|^2 \right\} + \mu^\alpha \left(\| \varphi \|^2 - \sum_{z_2} \| \varphi_{z_2} \|^2 \right),$$

where $\chi(z_2)$ is the characteristic function of the region

$K(z_2; \rho) \backslash K(z_2; \delta)$; $\varphi_{z_2} = \varphi \cdot u_{z_2} \omega_\beta(z)$, where $\omega_\beta(z) \in C^2$ with $\omega_\beta(z) = 1$ if $|z| \geqslant 2\beta$ and $\omega_\beta(z) = 0$ if $|z| < \beta$. We can take here $\varepsilon > 0$ arbitrarily small but then $C(\varepsilon)$ will be large. Let $M''_2(\lambda)$ consist of the functions φ of the type (4.3), where f_p are the eigenfunctions of the operator (4.4) with $K = 2$, and $M_2(\lambda) = M'_2(\lambda) \cup M''_2$. By virtue of Lemmas 5.2, 5.3 we get $\dim M_2(\lambda) \leqslant m^\alpha_{\alpha'} \dim N_1(\lambda, v_2) + \text{const}$. Let $\varphi \perp M_2(\lambda)$. We shall prove (4.2). We write φ_{z_2} as

$$\varphi_{z_2} = \sum_{i=1}^t \varphi_i(q_{z_2}) F_i(z_{z_2}) + g , \tag{4.16}$$

where $F_i(z_{z_2}) = (\varphi_{z_2}, \varphi_i(q_{z_2}))_{R_0(z_2)}, g \perp \varphi_i, i = 1, \ldots, t$. It is clear that

$$\varepsilon \| \varphi_{z_2} |z_{z_2}|^{-\gamma-2} \|^2 \leqslant \varepsilon \| g \|^2 + \varepsilon \sum_{i=1}^t \| F_i |z|^{-\gamma-2} \|^2. \tag{4.17}$$

According to Lemma 5.1

$$C \| \varphi_{z_2} \cdot \chi(z_2) |z_{z_2}|^{-1} \|^2 \leqslant \varepsilon \| g \|^2 + C_1 \sum_{i=1}^t \| F_i |z_{z_2}|^{-\gamma-2} \|^2. \tag{4.18}$$

Let us estimate the expression $(\mathcal{H}^\alpha \varphi_{z_2}, \varphi_{z_2})$. Let $\eta > 0$ be the distance between ν^α and the next point of the spectrum $\mathcal{H}^\alpha(z_2)$.

Evidently,

$$(\mathcal{H}(Z_2)\Psi_{Z_2}, \Psi_{Z_2}) \geq \nu^{\alpha} \|\Psi_{Z_2}\|^2 + \eta \|g\|^2. \tag{4.19}$$

In addition, one has

$$m[C_2]^{-1}((i\nabla_{\mathfrak{z}} + A_{\mathfrak{z}}^c)^2 \Psi_{Z_2}, \Psi_{Z_2}) = \tag{4.20}$$

$$= m[C_2]^{-1}((i\nabla_{\mathfrak{z}} + A_{\mathfrak{z}}^c)^2 g, g) + m[C_2]^{-1}\sum_{i=1}^{t}((i\nabla_{\mathfrak{z}} + A_{\mathfrak{z}}^c)^2 F_i, F_i) \geq$$

$$\geq m^{-1}H\|g\|^2 + m[C_2]^{-1}\sum_{i=1}^{t}((i\nabla_{\mathfrak{z}} + A_{\mathfrak{z}}^c)^2 F_i, F_i).$$

We denote by $\tilde{\chi}$ the characteristic function of the region Supp Ψ_{Z_2}. Then

$$(I(Z_2)\Psi_{Z_2}, \Psi_{Z_2}) = (I(Z_2)\tilde{\chi}\Psi_{Z_2}, \Psi_{Z_2}) = \tag{4.21}$$

$$= (I(Z_2)\tilde{\chi} g, g) + 2\,\mathrm{Re}\,(I(Z_2)\tilde{\chi}\sum_{i=1}^{t}\Psi_i F_i, g) + (I(Z_2)\sum_{i=1}^{t}\Psi_i F_i, \sum_{i=1}^{t}\Psi_i F_i).$$

It is easy to see that

$$(I(Z_2)\tilde{\chi} g, g) < \varepsilon\|g\|^2. \tag{4.22}$$

Moreover, similarly to (4.14) one gets

$$(I(Z_2)\tilde{\chi}\Psi_i F_i, \Psi_i F_i) \geq Q(F_i, F_i |\mathfrak{z}|^{-\gamma}) - C(F_i, F_i |\mathfrak{z}|^{-\gamma-1}), \tag{4.23}$$

$$(I(Z_2)\tilde{\chi}\Psi_i F_i, g) = Q(|\mathfrak{z}|^{-\gamma}\Psi_i F_i, g) - \tag{4.24}$$

$$- Q(|\mathfrak{z}|^{-\gamma}(1-\tilde{\chi})\Psi_i F_i, g) + ((I(Z_2) - Q|\mathfrak{z}|^{-\gamma})\tilde{\chi}\Psi_i F_i, g).$$

According to Lemma 5.1, one has

$$\left|((I(Z_2) - Q|\mathfrak{z}|^{-\gamma})(1-\tilde{\chi})\Psi_i F_i, g)\right| \leq \varepsilon\|g\|^2 + C\|F_i|\mathfrak{z}|^{-\gamma-1}\|^2$$

and (4.13) yields

$$\left| \left(\left(I(z_2) - Q |z|^{-\gamma} \tilde{\chi} \, \psi_i F_i, \, g \right) \right) \right| \leq \varepsilon \| g \|^2 + C \| F_i \, |z|^{-\frac{\gamma+1}{2}} \|^2 \tag{4.25}$$

It follows from the relations (4.5), (4.17)-(4.25) that (4.2) is valid if $\delta > 0$ is taken sufficiently large and ε sufficiently small, so theorem 1 is proved.

The main idea of the proofs of Theorem 5,7,8 is close to the above described one. The main difference is that constructing the subspace $\mathcal{M}_1(\lambda)$ under the assumptions of Theorem 8, we take the functions

$$\varphi = P^{\alpha} \sum_{i,p} \sum_{g \in S} C_{ip}(g) T_g \left\{ \psi_i(q_{z_2}) u_{z_2}^+ + (\psi_{i1} \partial_x + \psi_{i2} \partial_y + \psi_{i3} \partial_z) |z|^{-3} \right\} f_p(z)$$

instead of (4.3), where $\psi_{iK}(q_{z_2})$ are some finite functions, f_p is the eigenfunctions of the operator $h_1(\delta)$ given by (4.4) with the potential $U_1 = \tilde{U}_1$ specified in Theorem 8.

LEMMAS.

In this section we prove Lemmas 5.1-5.3 which we have used in the proof presented above.

LEMMA 5.1. Let α be an arbitrary type of the permutational symmetry and $S = (1 + |r|^2)^{\frac{1}{2}}$. Then $S^K \varphi \in \mathcal{L}_2(R^{3N})$, $K = 1, 2, \ldots$, for every eigenfunction φ of the discrete spectrum of the operator $\mathcal{H} \cdot P^{\alpha}$.

LEMMA 5.2. Let the functions $\psi_i(q_{z_2})$ $i = 1, \ldots, t$ form a canonical basis of the irreducible representation of the type α', and $f_p(z), p = 1, \ldots, \bar{p}$, be the orthonormal eigenfunctions of the operator $h_K(\delta)$ (see (4.4)), $u_{z_2}(r)$ be some functions such that $u_{z_2}(r) = 1$ if $r \in K(z_2; \delta)$ and $u_{z_2}(r) = 0$ for $r \in K(z_2, \beta)$. Let farthemore $\beta > 0$ be an arbitrary number, and $\beta > \delta > 0$ be a small number,

$$F = \left\{ \varphi \mid \varphi = \sum_{i,p} \sum_{g \in S} C_{ij}(g) T_g \psi_i(q_{z_2}) f_p(z_{z_2}) u_{z_2} \right\}.$$

Then

$$0 \leqslant m^{\alpha}_{\alpha'},\ \bar{p}\ \dim\alpha - \dim P^{\alpha}F \leqslant C,$$

where C does not depend on $\bar{P}$.

LEMMA 5.3. Let $\beta > 0$, $\gamma_0 > 0$, $C > 0$ be constants, $\theta(x) \equiv 0$ for $x < 0$ while $\theta(x) = 1$ for $x \geqslant 0$. Let $\bar{v}_1(\mathfrak{z}) \in \mathcal{L}_{2,loc}$ such that $\bar{v}_1(\mathfrak{z}) \to 0$ if $|\mathfrak{z}| \to \infty$,

$$\bar{v}_2(\mathfrak{z}) = \bar{v}_1(\mathfrak{z}) + C\theta(|\mathfrak{z}| - \beta)\cdot|\mathfrak{z}|^{-\gamma_0} \text{ and } h_K(\beta) =$$
$$= m[C_2]^{-1}(i\nabla_{\mathfrak{z}} + A^c_{\mathfrak{z}})^2 + \bar{v}_K.$$
Then for an arbitrary $\gamma_0 > 0$ one can find such $C > 0$ that for all $\lambda < 0$,

$$N_1(\lambda, \bar{v}_1) \geqslant N_{1,\beta}(\lambda, \bar{v}_1) \geqslant N_1(\lambda, \bar{v}_2) - const,$$

where $N_1(\lambda, v)$ is defined in Section 3, $N_{1,\beta}(\lambda, v)$ is defined similarly to $N_1(\lambda, v)$, but for the operator $h_K(\beta)$

PROOF OF LEMMA 5.1. We shall prove Lemma 5.1 according to scheme [5,6]. Let

$$D_n = \bigcap_{k \leqslant n,\, m \leqslant n-k} D(s^k \mathcal{H}^m),\ \|u\|_n = \sup_{k \leqslant n,\, m \leqslant n-k} \|s^k \mathcal{H}^m u\|.$$

First we establish that
A) for every n the set D_n is invariant with respect to the unitary transformations $e^{-i\mathcal{H}t}$, and if $u \in D_n$ then the function $e^{-i\mathcal{H}t}$ is continuous in t with respect to the norm $\|\cdot\|$ and $\|e^{-i\mathcal{H}t}u\| \leqslant C_n(1+|t|^n)\|u\|_n$ for some $C_n > 0$.
To prove A) we can use the methods of [5] but with the following alterations. In Theorem 1 and in Lemma 1 of the paper [5] we must take K and n as integers, to substitute s^K for the factor x^K, and to substitute the operators

$$\mathcal{H}, \; T(A) = \sum_{j=1}^{N} \nabla_j^2, \; \nabla_j(A) = i\nabla_j + A_j$$

for H, H_o and $P_j = i\frac{\partial}{\partial x_j}$, respectively; in the assertion C) of Theorem 1 instead of $2i\sum_{j=1}^{N} P_j x_{,j}^n + x_{,jj}^n$ we write $-2i\sum_{j=1}^{N}\nabla_j(A)\cdot\nabla_j S^n + \Delta S^n$; where the $\cdot$ let means the inner product in R^3. Then in Lemma 1 of the paper [5] we get

$$(u, S^n \mathcal{H}v) = (u, \mathcal{H}S^n v) - 2i(u, \nabla S^n \cdot \nabla(A)v) + (u, \upsilon S^n)$$

instead of (8). Using the induction assumption we obtain $u \in D(T(A)S^k)$ for $K < n$ and

$$\|u\|_K^1 = \sup_{m \leqslant K}(\|S^m u\| + \|T(A)S^m\|) = \text{const}(\|S^n u\| + \|S^n \mathcal{H}u\|) < +\infty$$

It is clear that $\nabla S^n \cdot \nabla(A) = S^{n-1} f \cdot \nabla(A)$ where $f = (f_1, ..., f_N)$, $f_i \in C^\infty$ are bounded in R^{3N} together with their derivatives. Hence $u \in D(\nabla(A)\cdot f \cdot S^{n-1})$, and instead of (9), (10) from [5] we get

$$\|\nabla(A)\cdot f \cdot S^{n-1} u\| \leqslant \|f\cdot\nabla(A)S^{n-1}u\| + \|S^{n-1}u \, \text{div} f\| \leqslant \quad (5.1)$$

$$\leqslant \text{const}(\|S^{n-1}u\| + \|T(A)S^{n-1}u\|) \leqslant \text{const}\|u\|_n^{(4)},$$

$$(S^n u, \mathcal{H}v) = (S^n \mathcal{H}u, v) + 2i(\nabla(A)\cdot f u, v) + \quad (5.2)$$

$$+ (\Delta S^n u, v).$$

To prove the assertion (e) of Lemma 1 in [5] after the formula (13) we must use the factor S instead of ($i + x_1$). Finally, in the proof of Theorem 1 of [5] after the formula (13) we must not estimate the terms separatly, but the value

$\|\nabla(A)\cdot\nabla S^n v\|$ and to do it in the same way as in (5.1). The remaining part of the proof of A) follows to $[5]$.

To extend the results to the operator $\mathcal{H}P^\alpha$ it is sufficient to put $D_n^\alpha = P^\alpha D_n$ and to take into account that

$$D_n^\alpha = \bigcap_{j\leqslant n, K\leqslant n-j} D(S^j \mathcal{H}^K P^\alpha) \subseteq D_n .$$

Using that in combination of the invariance with respect to the permutations from S in the proof of $[5]$, the assertion A) is checked with D_n^α instead of D_n. Now the validity of Lemma 5.1 follows from A) and Theorem 4 in $[6]$ applied to the operator $\mathcal{H}P^\alpha$.

REMARK. It is easy to generalize Lemma 5.1 for the operator $\mathcal{H}(Z_2)P^{\alpha'}$

PROOF OF LEMMA 5.2. Since the functions $\varphi_i(q_{z_2})$ are mutually orthogonal, there is $d_0 > 0$ such that the functions $\varphi_i(q_{z_2})$ are linearly independent in the region $|q_{z_2}| \leqslant d_0$. We shall suppose that all the functions $f_p(z)$ in Lemma 5.2 correspond to such eigenvalues of the operator $h_K(\beta)$ that their eigenspaces do not contain the functions $f(z)$ with $f(z) \equiv 0$ for $|z| \leqslant d_0$; under this assumption we shall prove the equality

$$\dim F = m_{\alpha'}^\alpha \dim a \cdot \bar{p} . \tag{5.3}$$

Then the assertion of Lemma 5.2 follows from (5.3) because the relation $f_{p_i}(z)=0$ for $|z| \geqslant d_0$ may be valid for the eigenfunctions $f_{p_i}(z)$ of the operator h_K which correspond to not more than a finite number of eigenvalues λ_{p_i}. In fact these functions $f_{p_i}(z)$ are eigenfunctions of the operators $h_K(\beta)$ and at the same time of operator h_K in the region $\beta < |z| \leqslant d_0$, and consequently, $\lambda_{p_i} \to \infty$ if $i \to \infty$, but it is impossible since $\lambda_{p_i} \to \bar{m}'H$ if $i \to \infty$.

Proof of (5.3) is analogous to the proof of Corollary 1 to Lemma 4.4 in [1] . But since we do not assume here that the considered functions are analytical, one must modify the proof of their linear independence appearing in [1] in the cones $K(z_2 ; \delta)$

In section 4.10 of [1] the functions $\psi_j(q_{z_{2i}}) f_p(z_{2i}) u_{z_{2i}}$ form a linearly independent system, because the equality $\sum_{ji} c_{ji} \psi_j(q_{z_{2i}}) f_p(z_{2i}) = 0$ yields the relation $\sum_j c_{ji} \psi_j(q_{z_{2i}}) = 0$ for all $q_{z_{2i}}$, and hence $c_{ji} = 0, j = 1, \ldots, t$ for each i ; here z_{2i} runs over all decompositions which one can get from z_2 by permutations $g \in S$ and p is fixed.

In section 4.11 of [1] with $r = 1$ the functions W_p^K , $p = 1, \ldots, \bar{p}$; $K = 1, \ldots, K_1(p)$ (which for every p form a basis in the space

$$P^{\alpha} \left\{ \sum_{ji} \sum_{g \in S} c_{ji}(g) T_g \psi_j(q_{z_{2i}}) f_p(z_{2i}) u_{z_{2i}} \right\}$$

are linearly independent. Actually, let $\sum_{p,K} d_p^K W_p^K = 0$. Then for arbitrary i and $|q_{z_{2i}}| \leqslant d_0$ one has

$$\sum_p \left(\sum_j \psi_j(q_{z_{2i}}) \sum_K c_{ji}^K d_p^K \right) f_p(z_{2i}) = 0 \qquad (5.4)$$

when $|z_{2i}| > (1 - \delta)|q_{z_{2i}}|$, and if the functions $f_p(z)$ are linearly independent in the region $|z| \geqslant d_0$ then by virtue of (5.4),

$$\sum_j \left(\sum_K c_{ji}^K d_p^K \right) \psi_j(q_{z_{2i}}) = 0 \text{ for } |q_{z_{2i}}| \leqslant d_0, p = 1, \ldots, \bar{p} ;$$

hence

$$\sum_K c_{ji}^K d_p^K = 0 \qquad p = 1, \ldots, \bar{p}. \qquad (5.5)$$

From (5.5) in analogy with [1] we get $d_p^K = 0, p = 1, \ldots, \bar{p},$

$K = 1, \ldots, K_1(\rho)$, i.e., the functions W_ρ^K form a linearly independent system.

Now we shall prove linear independence of the functions $f_\rho(\mathfrak{z})$ in the region $|\mathfrak{z}| \geqslant d_o$, that is, we shall prove that the equality

$$\Phi(\mathfrak{z}) \equiv \sum_{\rho=1}^{\bar{\rho}} C_\rho \, f_\rho(\mathfrak{z}) \equiv 0 \qquad \text{for } |\mathfrak{z}| \geqslant d_o$$

is justified only if $C_\rho = 0, \rho = 1, \ldots, \bar{\rho}$.
We write $\Phi(\mathfrak{z})$ as $\Phi(\mathfrak{z}) \equiv \sum_{m=1}^{\bar{m}} \tilde{C}_m \, \tilde{f}_m(\mathfrak{z})$ where the functions $\tilde{f}_m(\mathfrak{z})$ correspond to different eigenvalues $\tilde{\lambda}_m$ of the operator $h_K(\beta)$ given by (4.4) and $\sum_m |\tilde{C}_m| > 0$ if $\sum_\rho |C_\rho| > 0$. Evidently,

$$(h_K(\beta))^S \Phi(\mathfrak{z}) = \sum_{m=1}^{\bar{m}} \tilde{C}_m (\tilde{\lambda}_m)^S \tilde{f}_m(\mathfrak{z}) \quad S = 1, \ldots, \bar{m}. \tag{5.6}$$

If $\Phi(\mathfrak{z}) \equiv 0$ for $|\mathfrak{z}| \geqslant d_o$, then from (5.6) one gets

$$\sum_{m=1}^{\bar{m}} \tilde{C}_m (\tilde{\lambda}_m)^S \tilde{f}_m(\mathfrak{z}) \equiv 0$$

and hence $\tilde{C}_m \tilde{f}_m(\mathfrak{z}) \equiv 0, m = 1, \ldots, \bar{m}$ when $|\mathfrak{z}| \geqslant d_o$.
We have $\sum_\rho |C_\rho| > 0$ under the condition $\sum_m |\tilde{C}_m| > 0$ and $\tilde{C}_{m_o} \neq 0$ for some m_o; hence $\tilde{f}_{m_o}(\mathfrak{z}) \equiv 0$ if $|\mathfrak{z}| \geqslant d_o$. But this equality contradicts to our assumption about $f_\rho(\mathfrak{z})$, and consequently, $C_\rho = 0$, $\rho = 1, \ldots, \bar{\rho}$. Lemma 5.2 is proved.

PROOF OF LEMMA 5.3. It is clear that $N_1(\lambda, \bar{U}_1) \geqslant N_{1,\beta}(\lambda, \bar{U}_1)$. One can prove the inequality $N_{1,\beta}(\lambda, \bar{U}_1) \geqslant N_1(\lambda, \bar{U}_2) - const$ similarly to Lemma 4.1 of [1]. But instead of usings analyticity of the eigenfunctions we must employ practically the same argument as at the end of proof of Lemma 5.2 where we have established linear independence of functions $f_\rho(\mathfrak{z})$ for $|\mathfrak{z}| \geqslant d_o$

REMARK. When the paper was ready for typing, the authors got to know about the paper by Tamura [7] where, with exception of symmetry, for the case $Q \neq 0$, the results close to ours were obtained.

REFERENCES

1. Vugal'ter S.A., Zhislin G.M. On the asymptotics of a discrete spectrum of the given symmetry for many-particles Hamil - tonians. Trans.Moscow Math.Soc. (to appear).
2. Ivrii V. Precise Spectral Asymptotics for Schrödinger and Dirac Operators with Strong Magnetic Field; Schrödinger Operators, Standard and Non-Standard, World Scientific, 1989, p.29-40.
3. Sobolev A.V. On the asymptotics of the energy levels of a quantum particle in a homogeneous magnetic field perturbed by the decreasing electric field. II. Problemy matem.fiziki, N 11 (1986), p.232-261, Leningrad University, Leningrad (in Russian).
4. Zhislin G.M. The finiteness of a discrete spectrum in the quantum problem of N -particles. Teor.mat.fiz., 21 (1974), p.60-73 (in Russian).
5. Hunziker W. On the space-time behaviour of Schrödinger wave functions. J.Math.Phys. 7 (1966), p. 300-304.
6. Hunziker W. On the spectra of Schrödinger multiparticle Hamiltonians. Helvet Phys.Acta 39 (1966), p.451-462.
7. Tamura H. Asymptotic distribution of eigenvalues for Schrödinger operators with homogeneous magnetic fields. II. Osaka Journ.of Math. 26 (1989), p.119-137.

Institute of Applied Physics, Academy of Sciences of the USSR
Uljanov Street 46, 603600 Gorky, USSR

Radiophysical Research Institute, Lyadov Street 25/14, 603600
Gorky,USSR

Operator Theory:
Advances and Applications, Vol. 46
© 1990 Birkhäuser Verlag Basel

ASYMPTOTICS OF EIGENVALUES FOR MANY-PARTICLE HAMILTONIANS AT SYMMETRY SUBSPACES

S.A.Vugal'ter

The asymptotic formulae of the eigenvalues for the states with a given orbital momentum L and a spin at $L \to \infty$ are obtained for many-particle Hamiltonians with a Coulomb interaction. As a result the asymptotic of the eigenvalues for the helium atom are given.

In this report we investigate the asymptotical behaviour of the eigenvalues for many-particle Schrödinger operations on the subspaces of the functions having a given permutational symmetry α and orbital momentum L when $L \to \infty$.

We consider systems with Coulomb interaction for which the threshold of the essential spectrum is due to the decompositions of the initial system into two subsystems with charges of the opposite sign.

Let $Z_1 = (1, \ldots, N)$ be an N-particles system, m_i, q_i, $z_i = (x_i, y_i, z_i)$ be the mass, charge and position of the i-th particle, respectively, $z_{ij} = z_i - z_j$, $(z_1, \ldots, z_N) \in R^{3N}$. Let us introduce the inner product $(z, \tilde{z})_1 = \sum_i m_i (z_i, \tilde{z}_i)_{R^3}$ in R^{3N}. After separation of the centre-of-mass motion the Hamiltonian of the system Z_1 has the form

$$H = -\frac{1}{2}\Delta_0 + \sum_{i<j}^{1,N} q_i q_j |z_{ij}|^{-1},$$

where Δ_0 is the Laplace operator on $R_0 = \{z \,|\, z \in R^{3N}, \sum_{i=1}^{N} m_i z_i = 0\}$.

Let $Z_2 = (C_1, C_2)$ be the decomposition of Z_1 into two non-empty subsystems

$$M[C_j] = \sum_{i \in C_j} m_i \quad, \quad z_{C_j} = \left(\sum_{i \in C_j} m_i z_i\right) \cdot M[C_j]^{-1},$$

$$\zeta(Z_2) = z_{C_2} - z_{C_1} \quad, \quad Q(Z_2) = -\left(\sum_{i \in C_1} q_i\right)\left(\sum_{j \in C_2} q_j\right),$$

$$M(Z_2) = M[C_1] \cdot M[C_2] \cdot (M[C_1] + M[C_2])^{-1},$$

$$R_0(Z_2) = \{z \,|\, z \in R_0, \sum_{i \in C_\kappa} m_i z_i = 0 \quad \kappa = 1,2\},$$

$$R_c(Z_2) = R_0 \ominus R_0(Z_2).$$

It is clear that

$$H = H(Z_2) - \frac{1}{2}\Delta_c(Z_2) + I(Z_2),$$

where $H(Z_2) = -\frac{1}{2}\Delta_0(Z_2) + \sum_{\kappa=1}^{2} \sum_{(i,j) \in C_\kappa} q_i q_j |z_{ij}|^{-1}$, furthermore, $\Delta_0(Z_2)$ and $\Delta_c(Z_2)$ are the Laplace operators on $R_0(Z_2)$ and $R_c(Z_2)$, respectively, $I(Z_2) = \sum_{i \in C_1, j \in C_2} q_i q_j |z_{ij}|^{-1}$.

Let S be the group of the permutations of identical particles of the system Z_1, and suppose that α is the type of irreducible representation of S. In addition we assume that the identical particles are electrons and we consider only such symmetry types α which are permitted by Pauli principle.

Let L be the weight of the irreducible representation of the group $SO(3)$, $\sigma = (\alpha, L)$ the type of the irreducible representation of the group $G = S \times SO(3)$, and let P^σ be the projection in $\mathcal{L}_2(R_0)$ onto the subspace of the functions of the symmetry $\mathfrak{G}$, $\quad H^{\mathfrak{G}} = P^{\mathfrak{G}} H$.

We denote by $S(C_i)$ the subgroup of S which consists of all permutations from S acting only on the particles from C_i; by $S(Z_2)$ we denote the subgroup of S which are generated by $S(C_1) \times S(C_2)$ and by the transposition $C_1 \leftrightarrow C_2$ if the subsystems C_1, C_2 are identical.

We denote by $\alpha'(Z_2)$ and $\mathfrak{G}'(Z_2)$ respectively the types of the irreducible representations of the group $S(Z_2)$ and $G(Z_2) = S(Z_2) \times SO(3)$. Of course, $\mathfrak{G}' = (\alpha', \ell')$. Let $P^{\mathfrak{G}'}$ be the projection in $\mathcal{L}_2(R_0(Z_2))$ onto the subspace of functions of the symmetry $\mathfrak{G}'$. We write $\mathfrak{G}'(Z_2) \prec \mathfrak{G}$ if the restriction of the representation α of the group S to the subgroup $S(Z_2)$ contains the representation α' .

Let
$$H(\mathfrak{G}; Z_2) = \sum_{\mathfrak{G}' \prec \mathfrak{G}} P^{\mathfrak{G}'} H(Z_2) \quad , \quad \mu^{\mathfrak{G}} = \min_{Z_2} \inf H(\mathfrak{G}; Z_2),$$

$$O(\mathfrak{G}) = \{ Z_2 \mid \inf H(\mathfrak{G}; Z_2) = \mu^{\mathfrak{G}} \} .$$

We know that $\mu^{\mathfrak{G}}$ is the threshold of the essential spectrum of $H^{\mathfrak{G}}$.

In this paper we consider only such systems Z_1 for which $Q(Z_2^0) > 0$ in case that $Z_2^0 \in O(\mathfrak{G})$. For those systems $\mu^{\mathfrak{G}}$ is a discrete eigenvalue of $H(\mathfrak{G}; Z_2)$. Let $W(\mathfrak{G}; Z_2)$ be the eigenspace of $H(\mathfrak{G}; Z_2)$ which corresponds to $\mu^{\mathfrak{G}}$. We denote by $\mathcal{x}_1^{\mathfrak{G}}$ the smallest eigenvalue of $H^{\mathfrak{G}}$ and by $A_1^{\mathfrak{G}}$ the eigenspace of $\mathcal{x}_1^{\mathfrak{G}}$. For simplicity we adopt in the following the assumptions:

1. There is such $Z_2^0 \in O(\mathfrak{G})$ that all $Z_2 \in O(\mathfrak{G})$ may be constructed from Z_2^0 by permutations from S .

2. The subspace $W(\mathfrak{G}; Z_2^0)$ is not degenerate with respect to the symmetry, i.e., the representation of the group $G(Z_2^0)$ is irreducible in $W(\mathfrak{G}; Z_2^0)$.

We denote by $\mathfrak{G}_0' = (\alpha_0', \ell_0')$ the type of the representation of $G(Z_2^0)$ in $W(\mathfrak{G}; Z_2^0)$ and write

$$\lambda_n = \mu^6 - M(\bar{Z}_2^0)\cdot Q^2(\bar{Z}_2^0)\cdot 2^{-1}(n+L-\ell')^{-2};$$

$$E_n[c,\lambda] = \left[\lambda_n - \frac{c}{(n+L)^3 L}\ ,\ \lambda_n + \frac{c}{(n+L)^3 L}\right].$$

THEOREM 1. There are numbers $C>0$, L_0 such that for each $L > L_0$

i) every discrete eigenvalue of H^6 belongs to one of the intervals $E_n[c,L]$, $n = 1, 2, \ldots$.

ii) every interval $E_n[c,L]$ contains exactly $dim\,\mathfrak{G}\cdot t(n)$ of eigenvalues (with account of their multiplicity), where $t(n) = \min\{n+1,\ 2\ell'+1\}$.

The next proposition follows from Theorem 1.

THEOREM 2. The following statement is valid:

a) $\quad dim\,A_1^{\mathfrak{G}} = dim\,\mathfrak{G}$

b) $\quad \varkappa_1^6 = \mu^6 - \dfrac{M(\bar{Z}_2^0)\,Q^2(\bar{Z}_2^0)}{2(L-\ell'+1)^2} + O(L^{-4}).$

Theorem 2 demonstrates, for example, that the representation of the group G on $A_1^{\mathfrak{G}}$ is irreducible under the conditions 1,2. Let us apply the obtained results to Hamiltonian of the helium atom. In this case m and e are the mass and charge of an electron, M is the nucleus mass, $M_1 = mM(m+M)^{-1}$ and $M_2 = mM_1(m+M_1)^{-1}$. Let $\alpha_\pm$ be the types of symmetric and antisymmetric representations of the group of permutations for two electrons.

Denoting $\mathfrak{G} = (\alpha_\pm, L)$, we have $\mu^6 = -2M_1 e^4$, $\ell' = 0$, $M(\bar{Z}_2^0) = M_2$ and $Q(\bar{Z}_2^0) = e^2$.

THEOREM 3. Let $H^{(\alpha_\pm, L)}$ be the Hamiltonian of helium atom. Then one can find such $L_0 > 0$ that for each $L > L_0$

(i) the discrete spectrum H^6 consists of an infinite number of
   eigenvalues

$$\mathcal{æ}_n^6 = -2M_1 e^4 - \frac{M_2 e^4}{2(n+L+1)^2} + \frac{1}{(n+L)^3} \cdot O(L^{-1}) \, ,$$

(ii) the multiplicity of $\mathcal{æ}_n^6$ is equal to $(2L+1)$.

Similar results may be obtained also for the operator
H on the space of functions having the types of the symmetry
α , L and parity ω .

The proof of the theorems will be published in [1].

REFERENCES

1. Vugal'ter S.A. On the asymptotic behaviour of eigenvalues of
   many-body Hamiltonians on symmetry spaces. Teor. Mat. Fiz.
   (to appear, in Russian).

Institute of Applied Physics, Academy of Sciences of the USSR,
Uljanov Street 46, 603600 Gorky, USSR

Operator Theory:
Advances and Applications, Vol. 46
© 1990 Birkhäuser Verlag Basel

SPECTRAL ASYMPTOTICS WITH HIGHLY ACCURATE
REMAINDER ESTIMATES

V.Ivrii, A.Kachalkina

We present local semiclassical spectral asymptotics with highly accurate remainder
estimates in the interior of the domain and near the boundary and their applications
to asymptotics of an eigenvalue counting function.

Let X be a domain in $\mathbb{R}^d$, $d \geq 2$,

$$(1) \quad A = \sum_{|\alpha| \leq m} a_\alpha(x,h)(hD)^\alpha$$

an operator in $L^2(X)$ with $h \in (0,1)$; for a sake of simplicity we consider here only
scalar operators. We assume that

(2) A is a self-adjoint operator in $L^2(X)$,

$$(3) \quad | D^\beta_{x,h} a_\alpha(x,h) | \leq c,$$

$$(4) \quad a(x, \xi) = \sum_{|\alpha| \leq m} a_\alpha(x,0)\xi^\alpha \geq |\xi|^m/c$$

$\forall\, x \in B(0,1) \;\; \forall\, \xi: |\xi| \geq c \;\; \forall\alpha: |\alpha| \leq m \;\; \forall\beta: |\beta| \leq K$

where $B(0,r) = \{x, |x| \leq r \}$ and K is a large enough number. Let $e(x,y,\lambda)$ be a
Schwartz kernel of the spectral projector of A and $N(\lambda)$ its eigenvalue counting
function. We are interested in semiclassical asymptotics of $\int \psi(x)e(x,x,\lambda)dx$ where
$\psi \in C_0^K(B(0,1/2))$ is an arbitrary fixed function. Moreover let us assume that

(5) If $x \in B(0,1)$, $|a(x,\xi) - \lambda| + |d\, a(x,\xi)| \leq 1/c$ then $\mathrm{Hess}\, a(x,\xi)$ has two
eigenvalues f_1 and f_2 with $f_1 f_2 \geq 1/c$.

(6) $X \supset B(0,1)$

and let $1 < T_1 \leq T_2 \leq ... \leq T_n \leq \rho = h^{-\sigma}$ and $\Lambda_1,...,\Lambda_n$ be closed subsets of $\Sigma_\lambda = \{a = \lambda\}$
with the following properties:

(7) $\Lambda_j \subset \{ |d\, a| \geq \gamma = h^{1/2-\delta}\}$,

(8) Through every point (x,ξ) of the γ-neighbourhood of Λ_j there passes a
Hamiltonian trajectory $(x(t),\xi(t))$ of $a(x,\xi)$ with $(x(0),\xi(0)) = (x,\xi)$ and either
$t \in (0,T_j)$ or $t \in (-T_j,0)$ such that

(9) $|D^{\alpha}_{x,h}D^{\beta}_{\xi} A(x(t),\xi(t),h)| \leq c\rho$ $\forall \beta: |\beta|\leq m$ $\forall \alpha:|\alpha|\leq K,$

(10) $|D(x(t),\xi(t))/(x,\xi)| \leq c\rho,$

(11) $\text{dist }(x(t),\partial X) \geq \gamma/c,$

(12) $|x(t) - x| +|\xi(t) - \xi|\geq \gamma|t|/c$

where here and below $D(y,\eta)/D(x,\xi)$ means Jacobi matrix, $\delta > 0$ is arbitrary, $K = K(d,\delta)$ is large enough and $\sigma = \sigma(d,\delta)>0$, $\beta = \beta(d,\delta)>0$ small enough numbers.

Moreover, let us consider a Schrödinger operator

(13) $A = \sum_{j,k} g^{jk}(hD_j - V_j)(hD_k - V_k) + V$

with real-valued functions $g^{jk} = g^{kj}$, V_j and V such that

(14) $\sum_{j,k} g^{jk}\eta_j\eta_k \geq |\eta|^2/c$ $\forall x\in B(0,1) \forall \xi\in \mathbb{R}^d.$

Then one can replace (6) and (11) by

(6)' $X\cap B(0,1) =\{\varphi\geq 0\}\cap B(0,1)$ with $|d\varphi| \geq 1$, $|D^{\alpha}\varphi| \leq c$ $\forall \alpha: |\alpha|\leq K$

and on $\partial X\cap B(0,1)$ either Dirichlet or Neumann boundary condition is considered,

(11)$_1$' In the γ-neighbourhood of $x(t)$ X coincides with $\{\varphi\geq 0\}$ where $|d\varphi|\geq 1$,

$|D^{\alpha}\varphi| \leq c\rho$ $\forall \alpha: |\alpha|\leq K$, φ depends on $x(t)$ and

$\qquad \sum_{j,k} g^{j,k}\eta_j\eta_k \geq |\eta|^2/c\rho$ $\forall \eta\in \mathbb{R}^d,$

(11)$_2$' $\text{dist }(x(t),\partial X) \leq \gamma/c$ $\Rightarrow$ $|d\varphi/dt| \geq 1/c\rho$ along Hamiltonian trajectory,

(11)$_3$' In the γ-neighbourhood of $x(t)$ either Dirichlet or Neumann boundary condition is considered.

Here $(x(t),\xi(t))$ is a Hamiltonian trajectory with reflections at ∂X.
Our main result is the following

THEOREM. Let conditions (1)-(5),(7)-(10),(12) and either (6),(11) or (13),(14),(6)',(11)$_{1-3}$' be fulfilled. Then the following estimate holds

$$\left|\int \psi(x)(e(x,x,\lambda) - h^{-d}\kappa_0(x,\lambda) - h^{1-d}\kappa_1(x,\lambda))dx\right| \leq$$

$$\sum_{0\leq j\leq n}CT_j h^{1-d}\int_{\Lambda_j} dx\, d\xi{:}da + Ch^{1-d+\beta}$$

where $T_0 =1$ and $\Lambda_0 = \Sigma\lambda\backslash\cap\Lambda_1\cup...\cup\Lambda_n).$

APPLICATIONS. (i) Let $A = \Delta +$ lower order terms where Δ is a positive Laplacian in the domain X, X is either polyhedral (polygonal for d=2) bounded domain, or a ball, or a planar elliptic domain and on $Y_0\subset\partial X$ and on $Y = \partial X\backslash Y_0$ respectively the Dirichlet and the Neumann boundary conditions are given. Let us assume that

$$\text{mes}_{\partial X}\{x \in \partial X,\ \text{dist}\,(x, \Upsilon_0 \cap \Upsilon_1) \le \varepsilon\} = O(\varepsilon^\delta) \quad \text{as } \varepsilon \text{ tends to } 0.$$

where $\text{mes}_{\partial X}$ denotes (d-1)-dimensional measure at ∂X. Then

$$N(\lambda) = c_0 \lambda^{d/2} + (c_1 + O(\lambda^{-\beta}))\lambda^{(d-1)/2} \quad \text{as } \lambda \to +\infty .$$

where c_0 and c_1 are Weylian coefficients.

(ii) Let $A = \Delta + V(x)$ in $\mathbb{R}^d$ where $V \in C^K$ and $V = (\varepsilon + O(|x|^{-\delta}))|x|^{2m}$ as $|x| \to \infty$. Then for $\varepsilon = 1$, $m > 0$ and $m \ne 1$

$$N(\lambda) = \mathcal{N}(\lambda) + O(\lambda^{q(d-1)-\beta}) \quad \text{as } \lambda \to +\infty .$$

where $q = (m+1)/2m$ and

$$\mathcal{N}(\lambda) = \int\limits_{|\xi|^2 + V(x) \le \lambda} dx\, d\xi \ \cong\ \lambda^{qd}$$

and for $\varepsilon = -1$, $-1 < m < 0$, $m \ne -1/2$

$$N(\lambda) = \mathcal{N}(\lambda) + O(\lambda^{q(d-1)+\beta}) \quad \text{as } \lambda \to -0.$$

This investigation was inspired by the paper of A. Volovoi.

Department of Mathematics,
Institute of Mining and Metallurgy
Magnitogorsk 455000 USSR

Operator Theory:
Advances and Applications, Vol. 46
© 1990 Birkhäuser Verlag Basel

BOUND STATES AND RESONANCES IN QUANTUM WIRES

Pavel Exner

In this lecture, we discuss spectral properties of Hamiltonians describing the electron motion in a curved quantum wire of a pure semiconductor material. The bound states in a planar strip whose existence has been established recently are analyzed further, and analogous results for a curved tube in three dimensions are presented. When a finite-length quantum wire is attached to a pair of macroscopic electrodes, the natural conjecture is that the bound states turn into resonances ; we show that it is true in a simple model of this scattering system.

1. INTRODUCTION

The quantum wires represent one of the new objects of experimental investigation in solid-state physics whose appearance has been made possible by rapid development of fabrication techniques - cf., e.g., [1-3]. At the same time, they are attractive from the mathematical point of view as a new class of quantum mechanical systems exhibiting, in particular, interesting relations between their geometric and spectral properties.

A quantum wire is a strip of a pure semiconductor material prepared on a suitable substrate. Due to its crystallic structure, the electrons move within it as free particles of some effective mass. Since the wavefunction is known to be suppressed near the boundary, it is natural to associate with such a system a multiple of Dirichlet Laplacian as its Hamiltonian. In this lecture, we are going to discuss some relations between its spectral properties and the shape of the wire.

First of all, we shall treat the simplest case of a two-dimensional planar strip. It has been shown recently [4] that if the strip is curved and thin enough, the corresponding Hamiltonian has at least one bound state, i.e., an eigenvalue below the threshold of the essential spectrum. Here we shall discuss further properties of these bound states, in particular, their behaviour as the strip width $d \rightarrow 0$ and an upper bound on their number.

Then we shall use the same ideas to treat the case of a three-dimensional curved tube. Here again a non-straight tube can bind particles provided it is thin enough, however, the problem is more complicated because in addition to the curvature also the torsion of the tube plays role.

One is naturally interested how these bound states might be observed. The first thing to be realized is that actually it is a scattering problem because in reality each quantum wire is of a finite length. Typically it is attached to a pair of "macroscopic" wires which have a much larger transversal size, and therefore a lower threshold of the essential spectrum. In this way, the bound state appears to be embedded into continuum and one expects it to turn into a resonance exhibiting possibly a spectral concentration as the length of the quantum wire tends to infinity. In the last part of the lecture, we are going to confirm this expectation in a model setting.

2. A CURVED STRIP IN $\mathbb{R}^2$

2.1 Existence of the bound states

Let us recall briefly the main result of Ref.4 . We consider a curved strip Σ in $\mathbb{R}^2$ of a constant width d , and choose one of its boundaries as the reference curve Γ . Up to Euclidean transformations, Σ is uniquely characterized by d and the function $s \longmapsto \gamma(s)$ where s is the arc length of Γ and $\gamma(s)$ is the (signed) curvature. We assume that

(a1) $\gamma \in C^2(\mathbb{R})$,

(a2) γ, γ' and γ'' are bounded in $\mathbb{R}$,

(a3) $d\|\gamma\|_\infty < 1$,

(a4) Σ is not self-intersecting

We are interested in the Dirichlet Laplacian [5] $-\Delta_D^\Sigma$ on $L^2(\Sigma)$ *excluding always the trivial case of a straight* Σ . Using the natural curvilinear coordinates s,u in Σ , one can show that it is unitarily equivalent to the operator

$$H = - \frac{\partial}{\partial s} (1+u\gamma)^{-2} \frac{\partial}{\partial s} - \frac{\partial^2}{\partial u^2} + V(s,u) \tag{1a}$$

on $L^2(\mathbb{R} \times [0,d])$ with

$$V(s,u) = - \frac{\gamma^2}{4(1+u\gamma)^2} + \frac{u\gamma''}{2(1+u\gamma)^3} - \frac{5}{4} \frac{u^2\gamma''^2}{(1+u\gamma)^4} \tag{1b}$$

which is e.s.a. on $D(H) = \{ \psi : \psi \in C^\infty , \psi(s,0) = \psi(s,d) = 0 , H\psi \in L^2 \}$. Suppose now that Σ is curved substantially in a bounded region only, e.g.,

(b1) $\gamma^2, \gamma'' \in L(\mathbb{R}, (1+|s|)ds)$

or

(b2) $\gamma, \gamma' \in L^2(\mathbb{R}, (1+s^2)ds)$ and $\gamma'' \in L(\mathbb{R}, (1+s^2)ds)$.

Using minimax estimates combined with Birman-Schwinger principle, we proved in [4]

THEOREM 1 : *(a) Assume (a) and (b1). Then* $\sigma_{ess}(H) = [\pi^2/d^2, \infty)$ *and there is* $d_0 > 0$ *such that for each* $d < d_0$ *the operator* H *has at least one eigenvalue* $\varepsilon \in [0, \pi^2/d^2)$.
(b) If, in addition, (b2) is valid an isolated eigenvalue exists iff

$$\int_{\mathbb{R}\times[0,d]} \left\{ -\frac{\gamma^2}{(1+u\gamma)^2} + \frac{u^2\gamma'^2}{(1+u\gamma)^4} \right\} \sin^2\left(\pi\,\frac{u}{d}\right)\, ds\, du < 0 \ . \tag{2}$$

2.2 <u>Behaviour of the eigenvalues for small d</u>

An inspection of the relations (1) shows that for a thin strip, the operator H is nearly decoupled to the sum of the transversal and the longitudinal part, the latter being associated with the one-dimensional Schroedinger operator

$$h = -\frac{d^2}{ds^2} - \frac{1}{4}\,\gamma(s)^2 \tag{3}$$

on $L^2(\mathbb{R})$ with the domain $AC^2(\mathbb{R})$. Now we want to make this relationship more precise.

Let us denote by λ_j the eigenvalues of h and by $\varepsilon_j(d)$ the eigenvalues of the operator H for a given d, both in the ascending order. A tedious but straightforward minimax estimation yields

PROPOSITION 2 : *In addition to (a) and (b), assume that*

$$d < \frac{c_\gamma}{1 + c_\gamma\|\gamma_-\|_\infty} \ ,$$

where

$$c_\gamma := \min\left\{ \frac{2\pi}{\|\gamma\|_\infty} \ , \ \left(\frac{2\pi^2}{\|\gamma''\|_\infty}\right)^{1/3} , \ \left(\frac{2\pi}{\sqrt{5}\,\|\gamma'\|_\infty}\right)^{1/2} \right\}$$

Then the eigenvalues $\varepsilon_j(d)$ of H fulfill the inequalities

$$
- \frac{\lambda_j}{(1-d\|\gamma_-\|_\infty)^2} - \frac{d\|\gamma\|_\infty^3}{(1-d\|\gamma_-\|_\infty)^3} - \frac{d\|\gamma''\|_\infty}{2(1-d\|\gamma_-\|_\infty)^3} - \frac{d^2\|\gamma''\|_\infty\|\gamma_-\|_\infty}{2(1-d\|\gamma_-\|_\infty)^4} \le
$$

$$
\le \frac{\pi^2}{d^2} - \varepsilon_j(d) \le \tag{4a}
$$

$$
\le - \frac{\lambda_j}{(1+d\|\gamma_+\|_\infty)^2} + \frac{d(2\|\gamma\|_\infty^3+\|\gamma''\|_\infty)}{2(1-d\|\gamma_-\|_\infty)^3} + \frac{5}{2}\frac{d^2(\|\gamma''\|_\infty\|\gamma\|_\infty+\|\gamma'\|_\infty^2)}{2(1-d\|\gamma_-\|_\infty)^4} +
$$

$$
+ \frac{15}{2}\frac{d^3\|\gamma'\|_\infty^2\|\gamma\|_\infty}{(1-d\|\gamma_-\|_\infty)^5} .
$$

Here $\gamma_\pm$ are the positive and negative part of γ, respectively, $\gamma = \gamma_+ - \gamma_-$. In particular, one has

$$
\lim_{d\to 0} \left(\frac{\pi^2}{d^2} - \varepsilon_j(d) \right) = \lambda_j \tag{4b}
$$

for $j = 1,2,\ldots$.

This result gives a rigorous meaning to some limiting procedures [6-8] invented for the purpose of quantization of motion on a curve. At the same time, it is just the first step in developing the perturbation theory for this class of operators ; more about it can be found in [9].

2.3 Bounds on the number of eigenvalues

Proof of Theorem 1a uses a known property of one-dimensional Schroedinger operators, namely a criterion for existence of a bound state for such an operator [10,11]. In the same way, one can use other results concerning one-dimensional Schroedinger operators combined with minimax estimates to derive properties of the operator H .

As an example, consider the Birman-Schwinger bound. It is known to be inapplicable in one and two dimensions, however, since the the singularity of Birman-Schwinger kernel is

independent of the spectral parameter, one can separate it and
modify the method to yield a bound to the number of eigenvalues
[12,13]. In combination with a minimax estimate, we get the
following assertion [9] :

PROPOSITION 3 : *Let the assumptions of Proposition 2 be
fulfilled, then the number* $N(H)$ *of isolated eigenvalues
including multiplicity obeys*

$$N(H) \leq 1 + \frac{1}{8} \left(\frac{1+d\|\gamma_+\|_\infty}{1-d\|\gamma_-\|_\infty} \right)^2 \frac{\int_{\mathbb{R}^2} \gamma(s)^2 |s-t| \, \gamma(t)^2 \, ds \, dt}{\int_{\mathbb{R}} \gamma(s)^2 \, ds} \, . \quad (5)$$

Combining the mentioned methods, one can find
conditions under which just one bound state exists. Let us
concentrate for simplicity on the case of a simple "outward"
bend, i.e., $\gamma(s) \geq 0$ for all s . We know from [4] lower
estimates to the critical width d_0 , hence one has to find
conditions ensuring that the second term on the *rhs* of (5) is
less than one. There are several reasons that can cause existence
of more than one bound state :

(i) the overall bending angle $\beta = \int_{\mathbb{R}} \gamma(s) \, ds$ too large ; one
    can check that it plays no role as long as we avoid self-
    intersecting stripes,

(ii) the width d too big so that isolated eigenvalues
     corresponding to higher transversal modes may exist ; this
     is excluded by the assumptions of Proposition 2,

(iii) extended shape of the bend with several, possibly well
      separated maxima,

(iv) the fact that the boundary is not smooth enough, or more
     exactly, existence of regions where γ exhibits sheer
     variations accompanied by large values of γ', γ'' .

The last point is most probably only technical : while the method
does not work for a non-smooth boundary, and moreover, the bounds
on d_0 spoil if γ has sheer variations [4], one can prove,
e.g., directly existence of a bound state for an L-shaped strip

[14]. In order to avoid (iii), and at the same time, to single out some geometric factors, we adopt the following conventions

$$\gamma(s) = \frac{\alpha}{\rho}\, \phi(s/\rho)\ ,$$

where the function is assumed to be integrable with

$$\int_{\mathbb{R}} \phi(x)\ dx = \pi$$

and obeys

$$0 \le \phi(x) \le \phi(0) \le 1$$

Hence the minimum curvature radius of the inner boundary is ρ/α (at the point s=0) and the total bending of the strip is $\beta = \pi\alpha$ Then we have

PROPOSITION 4 : *In addition to (a) and (b),suppose that (a) there are $0 < b \le 1 \le a < 2$ such that*

$$1 - (x/b)^2 \le \phi(x) \le \frac{a^2}{a^2 + x^2}$$

holds for all $x \in \mathbb{R}$,
(b) the strip width satisfies

$$\frac{d}{\rho} < \min \left\{ \frac{1}{2}\left[\left(1 + \frac{4\sqrt{b}}{\|\phi'\|_2}\right)^{1/2} - 1\right]\ ,\ \frac{2-a}{a}\ ,\ \left(\frac{2\pi^2}{\|\phi''\|_\infty}\right)^{1/3}\ , \right.$$

$$\left. \left(\frac{2\pi^2}{\sqrt{5}\,\|\phi'\|_\infty}\right)^{1/2} \right\}$$

Then the corresponding operator H has just one simple eigenvalue below the threshold of $\sigma_{ess}(H)$.

Let us finally remark that the operator H for a sufficiently thin strip with several well separated bends is in the above described sense similar to a multi-well Schroedinger operator ; it suggests it could be treated by the method of tunneling estimates [15-17].

$$3.\ \text{A CURVED TUBE IN } \mathbb{R}^3$$

3.1 Preliminaries

Now we are going to apply the same ideas to the three-dimensional case. Let $\Gamma \in C^k(\mathbb{R},\mathbb{R}^3)$, $k \geq 4$, be a curve and (T,N,B) the corresponding Frenet triedre. For $a > 0$, we denote $B_a := \{\ x \in \mathbb{R}^2 : |x| < a\ \}$ and define the map $f : \mathbb{R} \times B_a \longrightarrow \Sigma$ by

$$f(s,\rho,\theta) := \Gamma(s) + \rho\ [N \cos(\theta-\alpha) + B \sin(\theta-\alpha)] \ , \tag{6}$$

where ρ,θ are polar coordinates in the transversal planes and the function α will be specified later. Using the standard technique, one can check that f is a local C^{k-2} diffeomorphism provided $\alpha \in C^{k-2}$ and $a\|\gamma\|_\infty < 1$; it becomes global if f is required to be injective.

Next we consider the operator $U : L^2(\Sigma) \longrightarrow L^2(\mathbb{R} \times B_a)$ defined by $Ug := |f'|^{1/2}\ g \circ f$, where $|f'| = \rho(1 - \rho\gamma \cos(\theta-\alpha))$ is the corresponding Jacobian and γ is the curvature of Γ . The operator U is unitary and transforms $-\Delta_D^\Sigma$ into an operator on $L^2(\mathbb{R} \times B_a)$ just as in the two-dimensional case.

There is an important difference, however. The transformed operator contains, in general, mixed derivatives which make its analysis difficult. There is only one way how to get rid of them, namely to choose the function α so that

$$\alpha'(s) = \tau(s) \ , \tag{7}$$

where τ is the torsion of Γ . In that case, the original Dirichlet Laplacian $-\Delta_D^\Sigma$ on $L^2(\Sigma)$ is transformed by U to

$$H = -\frac{\partial}{\partial s} [1 - \rho\gamma \cos(\theta-\alpha)]^{-2} \frac{\partial}{\partial s} - \Delta_{B_a} + V(s,\rho,\theta) \tag{8a}$$

where the second term is Laplacian on the circle,

$$-\Delta_{B_a} = - \frac{\partial^2}{\partial\rho^2} - \rho^{-2} \frac{\partial^2}{\partial\theta^2} - \frac{1}{4}\rho^{-2} \tag{8b}$$

with Dirichlet boundary conditions at $\rho = a$. In order to express the effective potential V appearing in (8a), we introduce the function $h := 1 - \rho\gamma \cos(\theta-\alpha)$. In view of (7), its derivatives with respect to s equal

$$h_s = - \rho\gamma\tau \sin(\theta-\alpha) - \rho\gamma'\cos(\theta-\alpha) ,$$

$$h_{ss} = \rho(\gamma\tau^2-\gamma'') \cos(\theta-\alpha) - \rho(2\gamma'\tau + \gamma\tau') \sin(\theta-\alpha)$$

With this notation, the effective potential is equal to

$$V(s,\rho,\theta) = - \frac{\gamma^2}{4h^2} + \frac{1}{2}\frac{h_{ss}}{h^3} - \frac{5}{4}\frac{h_s^2}{h^4} . \tag{8c}$$

In this way, spectral analysis of $-\Delta_D^\Sigma$ is again reduced to that of the operator (8). Notice the similarity between (8c) and the two-dimensional potential (1b). In particular, their expansions in powers of ρ and u , respectively, have the same structure which makes it possible to develop the perturbation theory simultaneously for both cases.

Let us finish these preliminaries with two remarks. First of all, it is easy to replace the polar coordinates used above by the Cartesian ones. In particular, for $\alpha'= 0$ one recovers the formula derived in [18] but this case is not interesting with the exception of the curves which are essentially planar. On the other hand, in addition to the tubes considered above one can treat in this way also some tubes of a non-circular section ; for this one has to replace B_a by the corresponding open connected set G in the transversal plane. If Σ is obtained by moving G along Γ rotating it simultaneously with respect to the Frenet triedre with the angular velocity $\alpha' = \tau$, then one has only to replace (8b) by the Dirichlet Laplacian $-\Delta_D^G$.

3.2 <u>Existence of bound states</u>

The basic assumptions in the three-dimensional case are the following :

(a1) $\tau \in C^1(\mathbb{R})$ and γ is modulus of a C^2 function,

(a2) $\gamma, \gamma', \gamma''$ and τ, τ' are bounded,

(a3) $a\|\gamma\|_\infty < 1$,

(a4) the map (6) is injective

In addition, we shall again consider tubes whose curvature is localized in some sense, e.g.,

(b1) $\gamma, \gamma', \gamma'' \in L(\mathbb{R}, (1+|s|)ds)$

or

(b2) $\gamma, \gamma', \gamma'' \in L(\mathbb{R}, (1+s^2)ds)$.

Comparing to the two-dimensional case, these assumptions require a faster decay of γ at infinity. If one wants to relax these additional restrictions, it is necessary to impose a suitable decay condition on the torsion, e.g.,

(c1) $\gamma^2, \gamma'' \in L(\mathbb{R}, (1+|s|)ds)$ and $\tau, \tau' \in L^2(\mathbb{R}, (1+|s|)ds)$

or

(c2) $\gamma, \gamma', \tau, \tau' \in L^2(\mathbb{R}, (1+s^2)ds)$ and $\gamma'' \in L(\mathbb{R}, (1+s^2)ds)$

Mimicking the proof of Theorem 1, we arrive at the following assertion :

THEOREM 5 : *(a) Assume (a) and (b1) or (c1), then*
$\sigma_{ess}(H) = [\kappa_0^2, \infty)$, *where* $\kappa_0 = 2.40483\ a^{-1}$ *(first zero of the corresponding Bessel function). Furthermore, there is a positive* a_0 *such that for each* $a < a_0$, *H has at least one eigenvalue in* $[0, \kappa_0^2)$.
(b) If, in addition, (b2) or (c2) is valid, then an isolated eigenvalue exists iff

$$\int_0^a \rho \, d\rho \, J_0(\kappa_0\rho)^2 \int_{\mathbb{R}} \left\{ - \frac{\gamma^2}{(1-\rho^2\gamma^2)^{3/2}} + \frac{\rho^2}{(1-\rho^2\gamma^2)^{5/2}} \right.$$

$$\tag{9}$$

$$\left. \times \left[\gamma^2\tau^2 + \gamma'^2 \frac{3+2\rho^2\gamma^2}{1-\rho^2\gamma^2} - \frac{32 \, \rho\gamma^2\gamma'\tau}{3\pi(1-\rho^2\gamma^2)^{1/2}} \right] \right\} ds < 0$$

The criterion (9) can be used to derive a lower bound to the critical radius a_0 – cf.[9]. One gets, e.g.,

COROLLARY 6 : *Under the assumptions of Theorem 5b, one has*

$$a_0 \geq \|\gamma\|_\infty^{-1} \left[1 - \left(\sqrt{\tfrac{1}{4} A^2 + A} - \frac{A}{2} \right)^{4/7} \right]^{1/2},$$

where

$$A = \frac{5}{3\|\gamma\|_2^2\|\gamma\|_\infty^2} \left[\|\gamma\tau\|_2^2 + 3\|\gamma'\|_2^2 + \frac{32\||\gamma|\gamma'\tau|^{1/2}\|_2^2}{3\pi\|\gamma\|_\infty} \right] .$$

3.3 Bounds on the number of eigenvalues

As in the two-dimensional case one can prove [9]

PROPOSITION 7 : *Assume* (a) *and* (b1) *or* (c1), *and furthermore, let*

$$a < \frac{c_{\gamma,\tau}}{1 + c_{\gamma,\tau}\|\gamma\|_\infty}$$

with

$$c_{\gamma,\tau} := \min \left\{ \frac{\sqrt{10}}{\|\gamma\|_\infty} \ , \ \left(\frac{5}{\|\gamma\tau'-\gamma''\|_\infty + \|2\gamma'\tau+\gamma\tau'\|_\infty} \right)^{1/3} , \right.$$

$$\left. \left(\frac{\sqrt{2}}{\|\gamma\tau\|_\infty + \|\gamma'\|_\infty} \right)^{1/2} \right\} ,$$

then $N(H)$ *fulfills the inequality*

$$N(H) \leq 1 + \frac{1}{8} \left(\frac{1+a\|\gamma_+\|_\infty}{1-a\|\gamma_-\|_\infty} \right)^2 \frac{\int_{\mathbb{R}^2} \gamma(s)^2 |s-t| \ \gamma(t)^2 \ ds \ dt}{\int_{\mathbb{R}} \gamma(s)^2 \ ds} . \quad (10)$$

Notice that the torsion enters this bound only through the range of a for which it is valid. This is connected with the fact that for a thin tube Σ the spectral properties of $-\Delta_D^\Sigma$ are given primarily by the curvature of Γ . One can see it from the relation (8c) in which the first term on the *rhs* that is decisive for small a does not contain τ ; for more details see [9].

4. RESONANCES IN FINITE-LENGTH WIRES

4.1 Preliminaries

The question naturally arises how the bound states discussed above can be observed. Now we want to show that it is actually a scattering problem, because in reality each quantum wire has a finite length (we denote it as 2D) and is attached to a pair of macroscopic electrodes of a width a . Since a » d, the bottom of the spectrum at the electrodes is much lower than $\mu_1 = \pi^2/d^2$ independently of the electrode material ; one may suppose that it is zero. The eigenvalues referring to an infinitely long wire thus appear to be embedded into continuum for a finite D and one expects them to turn into resonances.

It is not easy, of course, to describe the associated scattering process generally, and we are not going to attempt it here. Our aim is to treat the problem in a simplified setting that preserves nevertheles its essential physical featurs adopting a few model assumptions. First of all, we shall regard the whole situation as two-dimensional and the electrodes as infinitely thick ; they will be described as halfplanes with Neumann boundary condition on the borderline. They are joined by a line segment of length 2D representing the quantum wire. The fact that in reality it has a finite thickness is expressed by keeping the first-transversal-mode energy μ_1 as a fixed number. We shall also replace the Hamiltonian (1) of the wire by a one-dimensional Schroedinger operator. In view of Proposition 2, this operator can be chosen as

$$H_\infty = -\frac{d^2}{ds^2} + \mu_1 - \frac{1}{4}\, \gamma(s)^2 \, , \tag{11}$$

i.e., the potential supported by the segment is the sum of μ_1 and the curvature-induced attractive term $-\frac{1}{4}\,\gamma(s)^2$. For simplicity, we shall suppose that

(d) the function γ is infinitely smooth, non-zero, has a compact support and $|\gamma(s)| \leq 2\lambda_1^{1/2}$ for all s .

In fact, one may require γ to be c^2-smooth and to decay sufficiently fast as $|s| \to \infty$ as in the previous sections ; it would make only the argument technically more complicated. Let us remark also that the Neumann conditions are essential for the present model since no other boundary conditions allow a nontrivial coupling of the halfplanes to the line segment - cf.[19] for details. It may not be so if the macroscopic wires are assumed to be three-dimensional [20].

4.2 <u>Description of the model</u>

The first question is how to connect the segment to the halfplanes. It can be done in a close analogy to the problem concerning a plane and a halfline [21]. In the present case only the even partial waves are present in the decomposition with respect to the connection point (but it makes no difference because only the s-wave can be coupled non-trivially to the segment) and also the normalization is different. The coupling is described by boundary conditions which we shall write down in a while.

We can therefore limit our attention to the nontrivial part of the problem concerning the s-wave part of the wavefunction in the halfplanes and write the electron wavefunction as $\psi = (u_1, f_2, u_3)$ with $u_j \in L^2(\mathbb{R}_+, r\, dr)$ and $f_2 \in L^2(-D, D)$. The model Hamiltonian H_D is of the form

$$H_D \begin{pmatrix} u_1 \\ f_2 \\ u_3 \end{pmatrix} = \begin{pmatrix} -u_1'' - ru_1' - (4r^2)^{-1}u_1 \\ -f_2'' + (\lambda_1 - \gamma(s)^2/4)f_2 \\ -u_3'' - ru_3' - (4r^2)^{-1}u_3 \end{pmatrix} \tag{12}$$

with the boundary conditions

$$L_0(u_j) = af_2(\mp D) \pm bf_2'(\mp D) \ ,$$
$$L_1(u_j) = cf_2(\mp D) \pm df_2'(\mp D) \tag{13}$$

at the junctions, where the minus (plus) sign corresponds to $j = 1,3$, respectively and L_k are the regularized boundary values

$$L_0(u) = \lim_{r \to 0} \frac{u(r)}{\ln r} \ , \tag{14}$$

$$L_1(u) = \lim_{r \to 0} \left[u(r) - L_0(u) \ln r \right]$$

The requirement of probability current conservation at the junctions selects a four-parameter family of the boundary conditions (13), namely

$$a = \frac{i}{\pi} \, 2^{5/4} \, e^{i(\delta-\alpha)} \, \frac{\mathcal{J}}{\sin \beta} \; ,$$

$$b = \frac{i}{\pi} \, 2^{5/4} \, e^{i(\delta-\alpha)} \, \frac{\mathcal{K}}{\sin \beta} \; ,$$

$$c = \frac{i}{\pi} \, 2^{5/4} \, e^{i(\alpha-\delta)} \, \frac{(\gamma - \ln 2) - \pi\mathcal{L}/4}{\sin \beta} \; ,$$

$$d = \frac{i}{\pi} \, 2^{5/4} \, e^{i(\alpha-\delta)} \, \frac{(\gamma - \ln 2) - \pi\mathcal{M}/4}{\sin \beta}$$

(15a)

for real α, δ, ξ and $\beta \neq 0$, where $\gamma = 0.577\ldots$ is the Euler's constant and

$$\mathcal{J} = \sin(\alpha + \delta + \tfrac{\pi}{4}) \, \cos \beta - \sin(\xi + \tfrac{\pi}{4}) \; ,$$

$$\mathcal{K} = \sin(\alpha+\delta) \, \cos \beta - \sin \xi \; ,$$

$$\mathcal{L} = \cos(\alpha + \delta + \tfrac{\pi}{4}) \, \cos \beta + \cos(\xi + \tfrac{\pi}{4}) \; ,$$

$$\mathcal{M} = \cos(\alpha+\delta) \, \cos \beta + \cos \xi$$

(15b)

We have adopted in (13) the natural assumption that both junctions are the same. In fact, one should choose some particular values of a,b,c,d, motivating the choice by a low-energy limit of a more realistic solution to the problem of injection of electrons into the quantum wire. This is not an easy task and we avoid this problem by proving the existence of resonances for any boundary conditions of the described type ; the scattering on the junctions plays then role of a background.

4.3 The results

The infinite wire is in our model described by the Hamiltonian (11). Under the stated assumptions about the curvature γ , it has a finite number of simple eigenvalues,

$$0 < \lambda_1 < \lambda_2 < \ldots < \lambda_n < \mu_1 , \tag{16}$$

where $n \geq 1$. In order to find the resonances of H_D and their relations to the eigenvalues (7), we use the factorization technique - cf., e.g., [22-24]. We write the Hamiltonian as

$$H_D = T_D - \tfrac{1}{2}|\gamma_D| \left(\tfrac{1}{2}|\gamma_D| \right) , \tag{17}$$

where T_D is the "kinetic" part and $\gamma_D := \gamma \mid [-D,D]$ .The resonances are then poles of the function

$$Q_1(z,D) := - \tfrac{1}{2}|\gamma_D| \ (H_D - z)^{-1} \left(\tfrac{1}{2}|\gamma_D| \right) \tag{18}$$

continued analytically to the lower complex halfplane, which can be expressed by means of the perturbation and the "free" resolvent as

$$Q_1(z,D) = I - \left[I + Q(z,D) \right]^{-1} , \tag{19a}$$

$$Q(z,D) := - \tfrac{1}{2}|\gamma_D| \ (T_D - z)^{-1} \left(\tfrac{1}{2}|\gamma_D| \right) . \tag{19b}$$

Hence the resonances are also points where the analytically continued function (10b) is not invertible. Then one has

THEOREM 8 : *Assume (d) and choose*

$$\Omega = \mathbb{C} \setminus \left[\, \mathbb{R}_- \cup (\, z : \mathrm{Re}\ z \geq \lambda_1 - \eta^2 + (\mathrm{Im}\ z)^2 \,) \right] \qquad (20)$$

for some $\eta < (\mu_1 - \lambda_n)^{1/2}$. *Then* $Q(.,D)$ *can be continued from the upper complex halfplane to* Ω . *For all large enough* D, *the function* $Q_1(.,D)$ *has just* n *simple poles in the lower-halfplane part of* Ω *at* $z_j(D) = \xi_j(D) - \frac{i}{2}\,\Gamma_j(D)$ *and*

$$\lim_{D \to \infty} z_j(D) = \lambda_j \qquad (21)$$

for $j = 1,\ldots,n$. *Moreover, the resonances* $z_j(D)$ *exhibit spectral concentration : choose a one-parameter family* { $\delta_j(D)$: $D > 0$ } *of positive numbers such that*

$$\lim_{D \to \infty} \delta_j(D) = \lim_{D \to \infty} \frac{\Gamma_j(D)}{\delta_j(D)} = 0 \qquad (22a)$$

and denote $\Delta_j(D) = (\xi_j(D) - \delta_j(D), \xi_j(D) + d_j(D))$, *then the corresponding spectral measure converges strongly,*

$$\text{s-}\lim_{D \to \infty} E_{H_D}(\Delta_j(D)) = P_j \; , \qquad (22b)$$

to the projection on the eigenspace associated with the eigenvalue λ_j.

Proof of the theorem is based on a general factorization-technique theorem by Howland [23]. In order to express $Q_1(z,D)$ from (19), one has to know the "free" resolvent $(T_D - z)^{-1}$. We introduce the auxiliary operator $T_D^{(0)}$ which differs from T_D by the boundary conditions which are changed to the separated ones, $f_2(\mp D) = 0$ and $L_0(u_j) = \dot{0}$ for $j=1,3$. The resolvent $(T_D^{(0)}-z)^{-1}$ is a known integral operator ; in particular, its "inner" part has the kernel

$$K(x,y;z) = \frac{ch(\kappa(2D-|x-y|)) - ch(\kappa|x+y|)}{2\kappa \; sh(2\kappa D)} \tag{23}$$

for $-D < x,y < D$, where $\kappa = (\mu_1-z)^{1/2}$. The sought resolvent is then given by Krein formula,

$$(T_D-z)^{-1} = (T_D^{(0)}-z)^{-1} + \sum_{j,k=1}^{4} \mu_{jk}(z) \; |g_j(z)><g_k(\bar{z})| \tag{24}$$

where the coefficients $\mu_{jk}(z)$ and the vectors $g_j(z)$ can be found from the requirement that $(T_D-z)^{-1}$ maps $\mathcal{H}$ into the domain $D(T_D)$. This yields a set of linear equations which can be solved explicitly [19].

The most important step in application of the Howland's theorem is to check that $Q(z,D)$ tends to $Q(z,\infty)$ corresponding to the operator (11) in the operator norm uniformly in Ω . Since (23) gives

$$- \tfrac{1}{2} \, |\gamma_D(x)| \; K(x,y;z) \left(\tfrac{1}{2}|\gamma_D(y)| \right) \; \rightarrow \; - \tfrac{1}{4} \, |\gamma(x)| \; \frac{e^{-\kappa|x-y|}}{2\kappa} \, |\gamma(y)|$$

as $D \to \infty$, where the *rhs* is just the kernel of $Q(z,\infty)$, it remains to check the contribution from the finite-rank part in (24) is zero. It follows from the explicit form of $\mu_{jk}(z)$ and $g_j(z)$ mentioned above. The formulae (23) and (24) allow to verify also the remaining hypotheses of the Howland's theorem, in particular, compactness of the operator $- \tfrac{1}{2}|\gamma_D| \; (T_D-z)^{-1}\left(\tfrac{1}{2}|\gamma_D| \right)$

for z from the resolvent set $\rho(T_D)$, existence of analytic continuation of $Q(.,D)$ from the upper complex halfplane to Ω and the strong convergence $(T_D-z)^{-1} \rightarrow (T_\infty-z)^{-1}$ as $D \rightarrow \infty$ for $\mathrm{Im}\, z \neq 0$, where T_∞ is the "free" counterpart of the operator (11). Details of the proof can be found in the paper [19].

Let us finish with two remarks. First of all, we have not mentioned the rate of spectral concentration. A semiclassical estimate shows that it should be exponential, $\Gamma_j(D) \propto \exp(-\mathrm{const.}$ $(\mu_1-\lambda_j)^{1/2})$. This is a positive feature from the experimentalist point of view because it gives hope for observation of sharp resonances on not very long quantum wires.

The last remark concerns the ways in which the resonance effect can be manifested. It certainly leads to a sheer variation in the transmission coefficient between the halfplanes which is related to the conductance by Landauer formula [25]

$$G = \frac{2e^2}{h} \; \frac{T(\varepsilon)}{1-T(\varepsilon)} \tag{25}$$

Tuning the applied voltage, one can change the electron energy, and by (25) the conductance of the whole structure. Let us mention an indirect but clear indication for curvature-induced resonances in quantum wires in a recent experiment by Timp et al. [3] : they have found that the resistance of a many-probe junction depends on the number of right-angle turns the electron must pass on its way between a pair of electrodes.

ACKNOWLEDGMENT

A part of the results discussed here has been obtained in collaboration with Petr Šeba and Pierre Duclos to whom I am sincerely grateful. I am indebted also to Sergio Albeverio for the hospitality extended to me in the Mathematical Institute at Bochum where the model discussed in Section 4 was constructed.

REFERENCES

1   H.Sakaki, in *Proceedings of the International Symposium on Foundations of Quantum Mechanics in the Light of New Technology* (S.Kamefuchi et al., eds.), Physical Society of Japan, Tokyo 1984 ; pp.94-110.
2   H.Temkin et al., Appl.Phys.Lett.50 (1987), 413-415.
3   G.Timp et al., Phys.Rev.Lett.60 (1988), 2081-2084.
4   P.Exner, P.Šeba, J.Math.Phys.30 (1989), No.11.
5   M.Reed, B.Simon : *Methods of Modern Mathematical Physics IV, Analysis of Operators*, Academic Press, New York 1978 ; Sec. XIII.15.
6   J.Marcus, J.Chem.Phys.45 (1966), 4493-4499.
7   H.Koppe, H.Jensen, Sitzungberichte der Heidelberger Akademie der Wissenschaften, 1971, pp.127-140.
8   R.C.T.da Costa, Phys.Rev.A23 (1981), 1982-1987.
9   P.Duclos, P.Exner, *Curved quantum waveguides in two and three dimensions,* in preparation
10 B.Simon, Ann.Phys.97 (1976), 279-287.
11 R.Blanckenbecler, M.L.Goldberger, B.Simon, Ann.Phys.108 (1977), 69-77.
12 M.Klaus, Ann.Phys.108 (1977), 288-300.
13 R.G.Newton, J.Operator Theory 10 (1983), 119-125.
14 P.Exner, P.Šeba, P.Šťovíček, Czech.J.Phys.B39 (1989), 1181-1191.
15 J.-M.Combes, P.Duclos, R.Seiler, J.Funct.Anal.52 (1983), 257-301.
16 J.-M.Combes, P.Duclos, M.Klein, R.Seiler, Commun.Math.Phys. 110 (1987), 215-236.
17 P.Briet, J.-M.Combes, P.Duclos, in Proceedings of the Conference on Partial Differential Equations (Holzhau 1988), Teubner, Leipzig 1989
18 D.Picca, Lett.N.Cim.34 (1982), 449-452.
19 P.Exner, *A model of resonance scattering in curved quantum wires,* Annalen der Physik (1990), to appear
20 P.Exner, P.Šeba, Czech.J.Phys.B38 (1988), 1095-1110.
21 P.Exner, P.Šeba, J.Math.Phys.28 (1987), 386-391, 2304.
22 T.Kato, Math.Ann.162 (1966), 258-279.
23 J.S.Howland, Trans.Amer.Math.Soc.162 (1971), 141-156.
24 H.Baumgärtel, M.Demuth, J.Func.Anal.22 (1976), 187-203.
25 R.Landauer, Phys.Lett.A85 (1981), 91-93.

Laboratory of Theoretical Physics
Joint Institute for Nuclear Research
141980 Dubna, USSR

Operator Theory:
Advances and Applications, Vol. 46
© 1990 Birkhäuser Verlag Basel

SPECTRAL PROPERTIES OF THE OPERATORS
$H\psi = -\psi_{xx} + p(x)\psi + v(\varepsilon x)\psi$, p IS PERIODIC

V.S.Buslaev

We describe the spectrum in $L_2(\mathbb{R})$ and the asymptotic behavior of the eigenfunctions of the operator indicated in the title as $\varepsilon \longrightarrow 0$.

1. INTRODUCTION

This report will be devoted to the discussion of spectral properties in $L_2(\mathbb{R})$ of the differential operators

$$H\psi = -\psi_{xx} + p(x)\psi + v(\varepsilon x)\psi, \quad x \in \mathbb{R}.$$

The coefficients p and v will be supposed to be real smooth functions, the function p, in addition, will be supposed periodic, $p(x + a) = p(x)$, where $a > 0$ is the period. The parameter ε, $\varepsilon > 0$, will be a small number, i.e. $\varepsilon \ll 1$.

We will use methods described in [1-3], see also [4]. These methods allow to investigate the asymptotic behavior of the solutions of the equation

$$(1) \qquad -\psi_{xx} + p(x)\psi + v(\varepsilon x)\psi = E\psi,$$

as $\varepsilon \longrightarrow 0$. The character of the spectrum of the operator H depends on the behavior of the function $v(\xi)$ when $\xi \longrightarrow \infty$ and, of course, on the properties of the spectrum of the operator

$$H_o\psi = -\psi_{xx} + p(x)\psi.$$

The particular case $v(\xi) = -\xi$, i.e.,

$$H\psi = -\psi_{xx} + p(x)\psi - \varepsilon x\psi,$$

is of interest in applications to problems of the quantum solid-state physics. This last case was described in the report presented at the previous conference [5], see also [6].

Here we will consider two essentially different and in some sense more general cases:

A. $v(\xi) \longrightarrow +\infty$ as $\xi \longrightarrow \infty$,
B. $v(\xi) \longrightarrow 0$ as $\xi \longrightarrow \infty$.

that have been discussed briefly earlier in [7,8].

We are interested in 1) the character of the spectrum of the operator H, 2) the asymptotic behavior of the spectral characteristics of the operator H and 3) the asymptotic behavior of the eigenfunctions as $\varepsilon \longrightarrow 0$.

The outline of the paper is as follows. In Section 2 we will shortly describe some properties of the solutions of Eq.(1), in Sections 3 and 4 we will consider the operator H the cases A and B, respectively.

Since we cannot present in this lecture complete proofs, we restrict ourselves to description of the results in general terms, their relations and motivations.

The author is grateful to M.Sh.Birman and L.A.Dmitrieva for helpful discussions.

2. ON THE ASYMPTOTIC BEHAVIOR OF THE SOLUTIONS OF EQ.(1)

2.1. The equations with periodic potentials

We will reproduce here in short some results of the papers [1-3] restricting ourselves to those of them which will be used in the following.

We need some general facts about the solutions of the equation

$$(2) \qquad -\chi_{xx} + p(x)\chi = E\chi$$

with a pure periodic potential p. This equation has solutions of the following form:

$$\chi(x,k) = e^{ikx} \varphi(x,k), \quad \varphi(x + a,k) = \varphi(x,a), \quad E = \mathcal{E}(k),$$

where χ are called Bloch solutions, $\mathcal{E}$ is called dispersion function and the parameter k is called quasimomentum. The latter is a complex number, more precisely, it is a point on some Riemann surface $\mathbb{K}$, see Fig. 1 [9,10]. Positions of the branching points b_1, b_1^*, $b_1 = k_1 + ih_1$, $k_1 = l\pi/a$, $h_1 \geq 0$, $l \in \mathbb{Z}$, on the figure depend on the potential p.

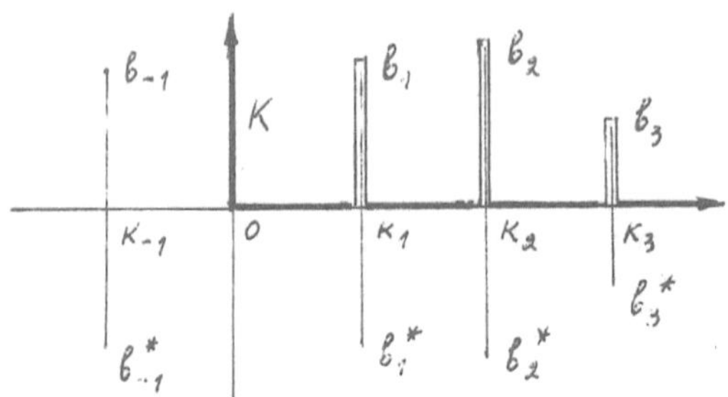

Fig. 1

Let us suppose for simplicity that all $h_1 > 0$. We recall that on the surface $\mathbb{K}$ the edges of the cuts, which are related by the transformation $k \longrightarrow -k^*$, are identified. Moreover, we note that the function $\mathcal{E}$ has the following properties

$$\mathcal{E}^*(k^*) = \mathcal{E}(k), \quad \mathcal{E}(-k) = \mathcal{E}(k).$$

Furthermore, let us introduce the points $\sigma_o = 0$, $\sigma_1 = k_1-0$, $\sigma_2 = k_1+0$, $\sigma_3 = k_2-0$, $\sigma_4 = k_2+0,\ldots$, on the surface $\mathbb{K}$. The dispersion function $\mathcal{E}$ restricted to the broken line K marked in Fig. 1 maps this line into $\mathbb{R}$ monotonically. If we introduce the points $E_m = \mathcal{E}(\sigma_m)$, then the intervals $\Delta_1 = [E_{m-1},E_m]$, $m = 2l-1$, $l = 1,2,\ldots$, constitute the spectrum of the operator H_o. The intervals are divided by the gaps $\bar{\Delta}_1 = (E_m,E_{m+1})$. The spectrum of the operator H_o is continuous of multiplicity two and we can choose $\chi(x,k)$ and $\chi^*(x,k)$, $k \in \mathbb{R}_+$, as the eigenfunctions.

The dispersion function $\mathcal{E}$ restricted to the interval Δ_1 can be analytically continued to the whole real axis $\mathbb{R}$. This corresponds to another method of choosing the cuts between the branching points. After this continuation we obtain an even smooth function $\mathcal{E}_1$ on the k-axis, see Fig. 2.

In Fig. 3 we can see the character of the relation between $E = \mathcal{E}(k)$ and $\mathrm{Im}(k)$ on the imaginary axis and on the loops enveloping the edges of the cuts indicated in Fig. 1.

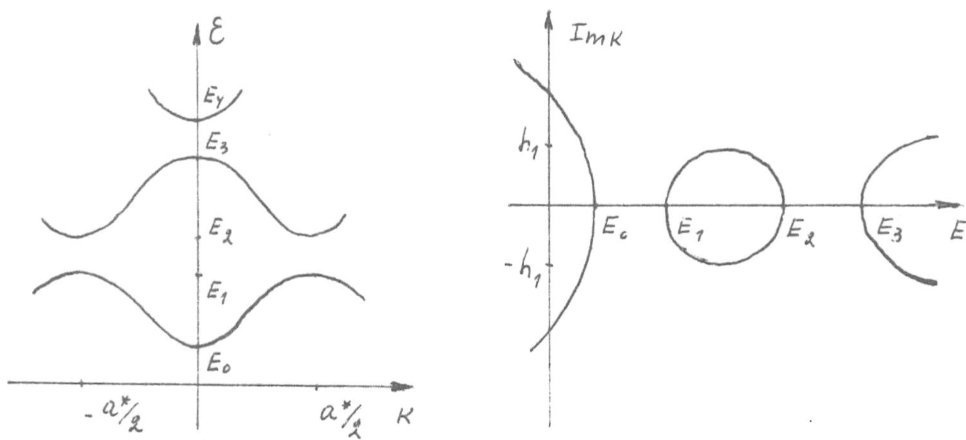

Fig. 2   Fig. 3

2.2 The asymptotic behavior of the
solutions of Eq.(1)

The main element of asymptotic formulas for the solutions of Eq.(1) is the corresponding isoenergy curve characterized by the equation

$$\mathcal{E}(k) + v(\xi) = E,$$

where $E \in \mathbb{R}$ is a fixed parameter.

An interval of the isoenergy curve

$$\gamma = \{k, \xi : k = \kappa(\xi), \ \xi \in \Delta = (a,b)\}$$

is called a regular branch of the isoenergy curve if $\kappa \in C^{\infty}(\Delta)$. With any regular branch γ it is possible to associate a formal solution f of the Eq.(1). The main term of this solution is given by the formula

$$f \sim |\mathcal{E}_k(k)|^{-1/2}\, \varphi(x,k)\, \exp\left\{\, i\int_{\gamma(\xi)} \Omega \,\right\}\, , \quad \Omega = \frac{1}{\varepsilon}\, k\, d\xi - \omega,$$

$$\omega = -\,\frac{\langle d\varphi(.,k),\varphi(.,-k)\rangle}{\langle\varphi(.,k),\varphi(.,-k)\rangle}, \quad d\varphi(.,k) = \varphi_k(.,k)\,dk,$$

$$\langle f,g\rangle = \int_0^a f(x)g(x)\,dx, \quad k = \kappa(\xi),$$

see Fig. 4.

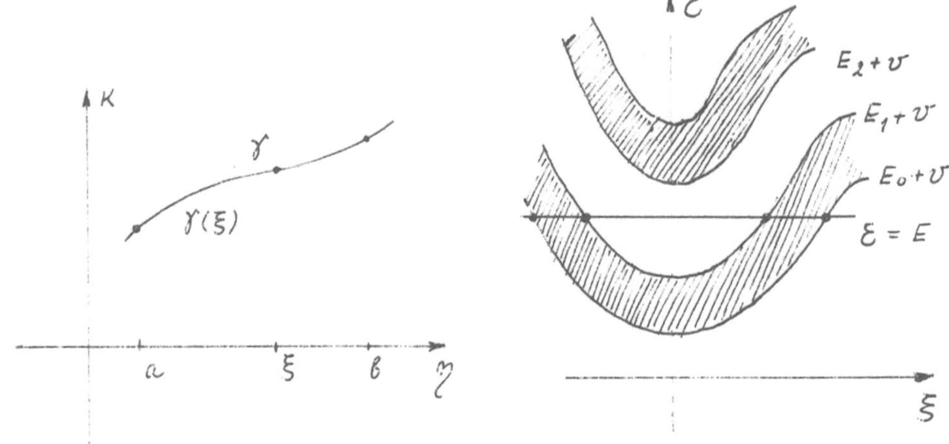

Fig. 4   Fig. 5

Obvious representation of the isoenergy curve can be obtained if
the function v consists of a finite number of strictly monotonous
branches divided by nondegenerated critical points. Let us
consider the graphs of the functions

$$\mathcal{E} = E_m + v(\xi),$$

see Fig. 5. The strips corresponding to the spectral intervals Δ_1
are shaded. Consider the line $\mathcal{E} = E$. The ξ-coordinates of its

intersection points with indicated graphs are called the turning points. They are determined by the equations

$$E_m + v(\xi) = E \ , \ m = 0,1,\ldots \ .$$

The turning points separate the projections of the real and complex branches of the isoenergy curve to the ξ-axis.

There are four typical cases, see Fig. 6.

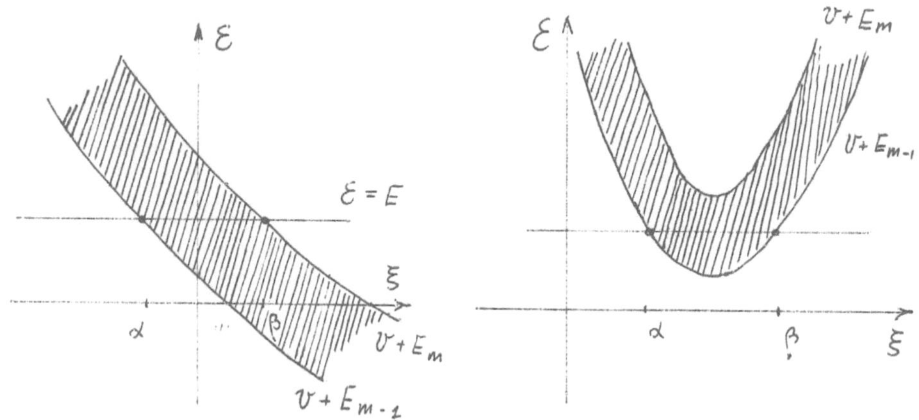

Fig. 6a   Fig. 6b

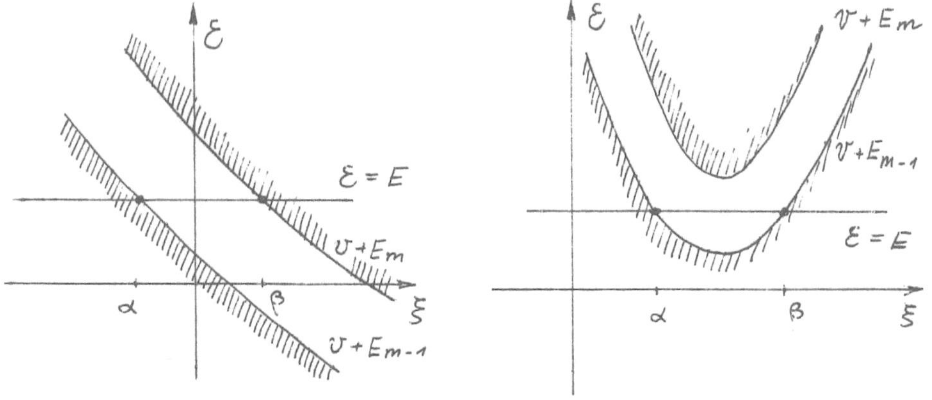

Fig. 6c   Fig. 6d

In the case 6a it is quite reasonable to include the real branches of the isoenergy curve on the phase plane into the periodic (with respect to k) curve, in other words to consider the curve given by the equation

$$\mathcal{E}_1(k) + v(\xi) = E , \quad m = 2l - 1,$$

see Fig 7a.

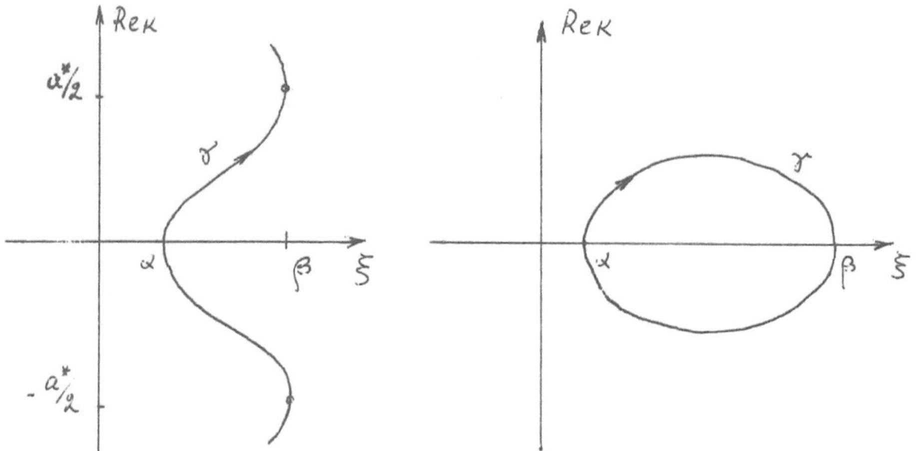

Fig. 7a   Fig. 7b

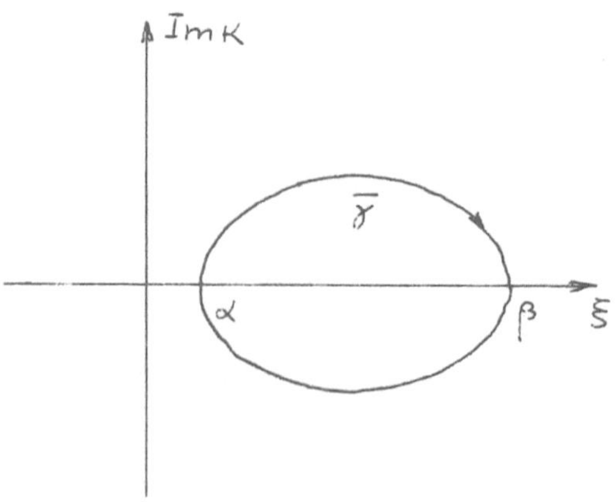

Fig. 7c

The curve presented in Fig.7b corresponds to the case 6b. Above the intervals (α,β) in Fig.6c, 6d there lie complex branches of the isoenergy curve. Their graphs in the plane $(\xi,\mathrm{Im}(k))$ are presented in Fig.7c.

2.3 Quantization conditions

The formal solutions f corresponding to real branches do not change their order as $\varepsilon \rightarrow 0$ inside of the interval (α,β). For complex branches this order differs exponentially in different points of (α,β). In asymptotically small vicinities of the turning points the formal solutions f indicated earlier must be changed to more complicated formal solutions g, see [1,2]. In a vicinity of the turning point ξ_0 the main term has the following form

$$g \sim \varepsilon^{-1/6} A(\zeta)\, \chi(x,\sigma_m).$$

Here σ_m corresponds to $E_m \sim \xi_0$, χ is the periodic solution of Eq.(2), $\zeta = \varepsilon^{-2/3}(\xi - \xi_0)$, $A(\zeta)$ is a solution of the equation

$$-\frac{1}{2\kappa_m} A'' + \lambda\zeta A = 0,$$

where $\lambda = v'(\xi_0)$, κ_m is the effective mass of the point E_m, i.e.

$$\mathcal{E}(k) = E_m + \frac{1}{2\kappa_m}(k - \sigma_m)^2 + \ldots .$$

Let us consider some interval (α,β) bounded by turning points and covered by a real branch of the isoenergy curve. It is possible to construct a linear combination f_+ (f_-) of the solutions f and f^* which after continuation through the vicinity of the points α (β) becomes exponentially small as $\varepsilon \rightarrow \infty$ and $\xi > \beta$ $(\xi < \alpha)$.

The solutions f_+ and f_- coincide (more exactly, they are linearly dependent) if certain conditions, which might be called quantization conditions, are fulfilled. These conditions are of the the form

$$(3) \qquad \frac{1}{\varepsilon} \int_\gamma k\, d\xi - \int_\gamma \omega + \ldots = 2\pi n + \mathrm{ind}(\gamma), \quad n \in \mathbf{Z}.$$

Here γ denotes the closed branch of the isoenergy curve situated above the interval (α,β), see Fig. 7a, 7b. In the case 7a the curve γ becomes a closed curve after the identification of the points k and $k+na$, $n \in \mathbf{Z}$. In the last formula $\mathrm{ind}(\gamma)$ denotes the Maslov index of the curve γ. In the cases 7a and 7b $\mathrm{ind}(\gamma)$ equals 0 and 2, respectively.

The quantization conditions can be regarded as a restriction imposed to the spectral parameter E. The values of the spectral parameter obeying the quantization conditions can be understood in another way. They can provide an asymptotic description of eigenvalues or resonances. For such E the corresponding formal solutions $f_+ = f_-$ give the asymptotic descriptions either of the eigenfunctions or of the resonance states as $\varepsilon \to \infty$. In a natural sense their supports lie essentially in the intervals (α,β).

Turning again to the linear combination f_+ we continue this solution through the interval (β,α_1), covered by the complex branch $\bar{\gamma}$ of the isoenergy curve, to the next interval (α_1,β_1) covered by the real branch. The relative order of the solution in the interval (α,β) is $\exp\{-\frac{1}{2}\mu_{\bar{\gamma}}\}$,

$$\mu_{\bar{\gamma}} = \frac{1}{\varepsilon} \int_\gamma \mathrm{Im}(k)\, d\xi.$$

Such a type of continuation of the solution can be called a tunneling.

3. THE SPECTRUM OF $-\partial_x^2 + p + v$ IF $v \longrightarrow \infty$

If $v(\xi) \longrightarrow +\infty$ as $\xi \longrightarrow \infty$ the spectrum of the operator H is simple and purely discrete. In this case the periodic potential $p(x)$ plays role of a perturbation to the operator

$$H_1 = -\frac{d^2}{dx^2} + v(\varepsilon x).$$

This perturbation has a substantial influence on the structure of the spectrum.

Let us suppose that $v(0) = 0$ and that $v(\xi)$ is strictly decreasing when $\xi < 0$ and strictly increasing when $\xi > 0$. In addition, we suppose that $\xi = 0$ is a nondegenerated critical point.

We point out that the operator H_1 has the standard semiclassical form, i.e., the solutions of the equation

$$-\psi_{xx} + v(\varepsilon x)\psi = E\psi$$

can be described asymptotically in terms of standard semi-classical constructions. Fig. 5 in this case looks as it is shown on Fig. 8.

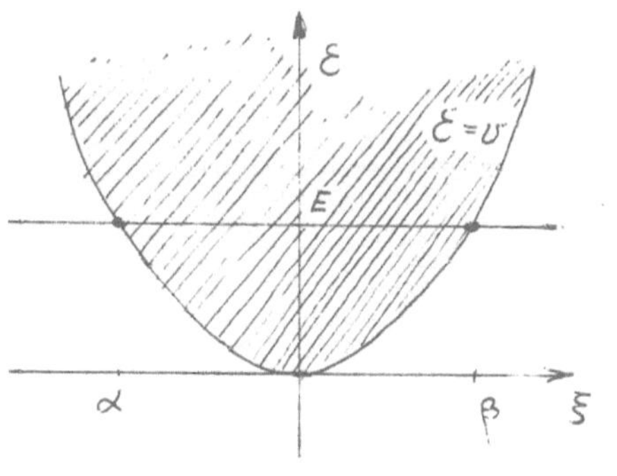

Fig.8

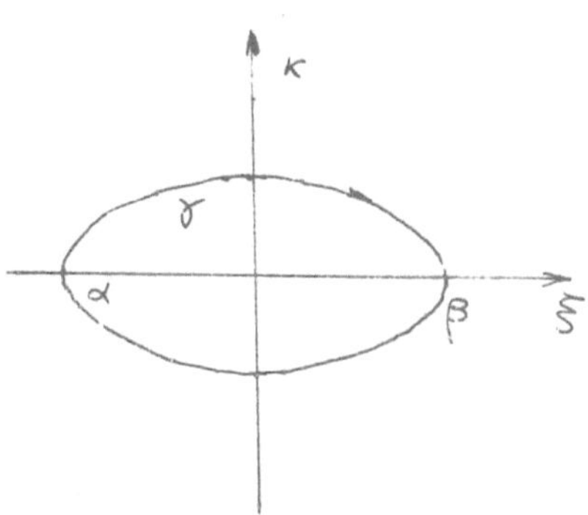

Fig.9

In other words the usual isoenergy curve

$$k^2 + v(\xi) = E$$

has a nontrivial real branch if $E > 0$ which is oval, see Fig.9. We have to select the ovals obeying the quantization conditions

$$\frac{1}{\varepsilon} \int_\gamma k \; d\xi + O(\varepsilon) = 2\pi n + \pi.$$

These conditions determine asymptotically the eigenvalues of H_1. The supports of the corresponding eigenfunctions coincide asymptotically with the projections of the ovals to the ξ-axis.

Switching the perturbation p in, Fig.8 transforms into Fig 10.

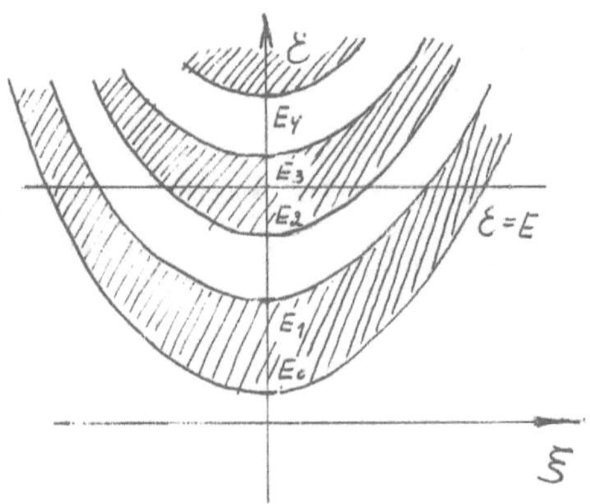

Fig.10

It is clear that for all E smaller than some E_o, i.e. $E < E_o$, real branches of the isoenergy curve $\mathcal{E}(k) + v(\xi) = E$ are absent, and therefore these points cannot belong to the spectrum of H. For $E_o < E < E_1$ the real branch of the isoenergy curve is an oval as before, see Fig. 9, but the equation describing the oval is naturally quite different,

(4) $$\mathcal{E}_1(k) + v(\xi) = E$$

from the previous one and the quantization conditions depend, of course, on the potential p, see (3). We denote the corresponding eigenvalues by E_n^1, $n = 0,\ldots,N_1(\varepsilon)$. The quantization conditions give their asymptotic description. If $E = E_o$ the real part of isoenergy curve reduces to a point. At the opposite end $E = E_1$ of the interval $\Delta_1 = [E_o,E_1]$ the isoenergy curve possesses a singularity, see Fig. 11. If E is close to E_o or E_1 (n is close to 0 or $N_1(\varepsilon)$), we have to use more complicated asymptotic formulas for the eigenvalues and the eigenfunctions, see [1,2].

If $E \in \bar{\Delta}_1 = (E_1,E_2)$ the real part of the isoenergy curve splits into two periodic branches γ_1^- and γ_1^+ described by the equation (4), see Fig. 11. They generate two series of eigenvalues $E_n^{1,-}$, $N_1(\varepsilon) < n < N_2^-(\varepsilon)$, and $E_n^{1,+}$, $N_1(\varepsilon) < n < N_2^+(\varepsilon)$ the asymptotes of which are described by the quantization conditions.

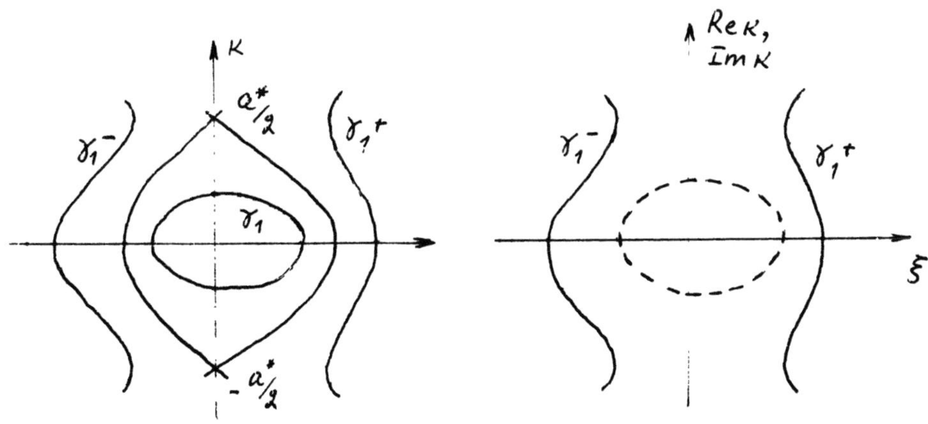

Fig.11   Fig.12

In general, the two indicated series differ already in the main order, but if v is an even function, the asymptotic series for them are absolutely identical. To calculate the splitting of the eigenvalues it is necessary to take into account the tunneling between the branches γ_1^- and γ_1^+. Using standard arguments it can be shown that the order of the splitting $\exp\{-\mu_1\}$ is determined by the integral

$$\mu_1 = \frac{1}{\varepsilon} \int_{\gamma_1} \mathrm{Im}(k)\ d\xi$$

taken along the complex branch $\bar{\gamma}_1$ of the isoenergy curve which covers the interval of the ξ-axis between the real branches, see Fig.12.

Further change of the isoenergy curve with increasing E can be understood from Fig. 10. If $E \in \Delta_2$, then there are three real branches of the isoenergy curve described by the equations

$$\mathcal{E}_1(k) + v(\xi) = E \quad \text{and} \quad \mathcal{E}_2(k) + v(\xi) = E,$$

see Fig. 13. In Fig. 13 we have two ovals. One of them differs from the other by a translation with the period a^* along the k-axis. However, we have to take into consideration only one of them.

It is clear that in this case there are three sets of eigenvalues: the two previous sets $E_n^{1,-}$ and $E_n^{1,+}$ determined by the periodic branches γ_1^- and γ_1^+ and the new set E_n^2 determined by the ovals. The value of the splitting between $E_n^{1,-}$ and $E_n^{1,+}$ can be estimated in terms of the tunneling through the two intervals covered by the complex branches $\bar{\gamma}_1^-$ and $\bar{\gamma}_1^+$ of the isoenergy curve. This splitting has the order $\exp\{-\mu_1^- - \mu_1^+\}$.

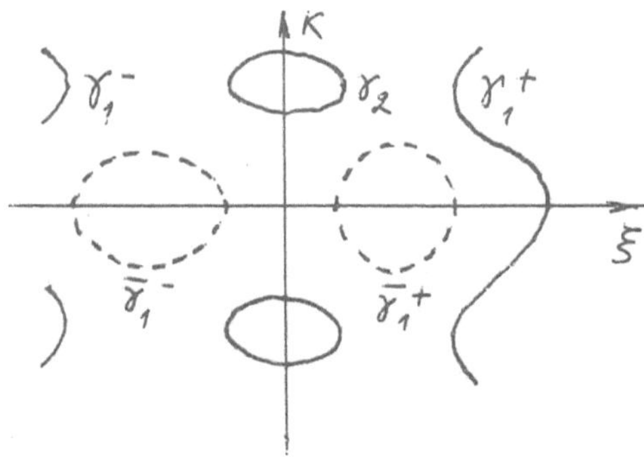

Fig. 13

Summarizing the previous consideration we can say that with each spectral interval $\Delta_1 = [E_{21-2}, E_{21-1}]$ there is connected

a sequence of eigenvalues $E_n^1 > E_{21-2}$, $n > N_{21-2}(\varepsilon)$. For $E \in \Delta_1$ they correspond to the ovals while for $E > E_{21-1}$ they are divided into two groups $E_n^{1,-}$ and $E_n^{1,+}$, $n > N_{21-1}(\varepsilon)$, which correspond to two periodic branches.

The interaction between two intervals of eigenvalues corresponding to different real branches of the isoenergy curve can be estimated in terms of tunneling through all the intervals separating the asymptotic supports of the eigenfunctions and covered by the complex branches of the isoenergy curve. This interaction leads to exponentially small displacement of the eigenvalues. The displacement must be neglected if the asymptotic expansions for the eigenvalues, which are given by the quantization conditions, are different. More complicated formulas has to be used if E belongs to a vicinity of E_1, $1 = 0,1,\ldots$.

The total picture is presented in Fig 14.

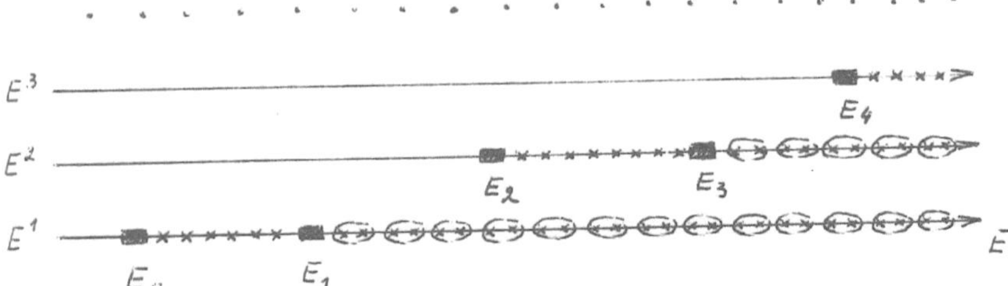

Fig. 14

The supports of the corresponding eigenfunctions have a more complicated structure than in the case $p = 0$, see Fig. 10.

Concerning the asymptotic behavior of the eigenfunctions we remark the following. Denote by $P^1(\Delta)$ the spectral projection of H with respect to the interval Δ and to the 1-branch E^1 of the eigenvalues, and let g and g be two smooth finite functions depending on x, then

$$P^1(\Delta)f,g) \longrightarrow (P_O(\Delta \cap \Delta_1)f,g) \quad \text{as } \varepsilon \longrightarrow 0$$

where P_O is the spectral projection of H_O.

Finally we should say something about large values of the parameter E. For such E the tunneling between all real branches of the isoenergy curve becomes essential, and therefore these branches cannot be considered separately. However, we are not going to discuss it in more detail.

4. THE SPECTRUM IF $v \longrightarrow 0$

4.1 Contribution of a separate branch

Let $v(\xi)$ be a decreasing function as $\xi \longrightarrow \infty$ satisfying the condition

$$|v^{(r)}(\xi)| \leq C|\xi|^{-\sigma-r}, \quad \sigma > 0, \quad r = 0,1,\ldots .$$

It appears that the roles of the potentials $p(x)$ and $v(\varepsilon x)$ are reversed in a sense in comparison with the previous case: the operator H_0 has to be regarded now as the unperturbed one while the potential $v(\varepsilon x)$ must be interpreted as a perturbation. The operators H and H_0 have identical essential spectra which are continuous and twice degenerated in the intervals Δ_1, Δ_2,.... In addition, the operator H can have eigenvalues in the intervals $\bar{\Delta}_0 = (-\infty, E_0)$, $\bar{\Delta}_1$, $\bar{\Delta}_2$,..., see Fig 15.

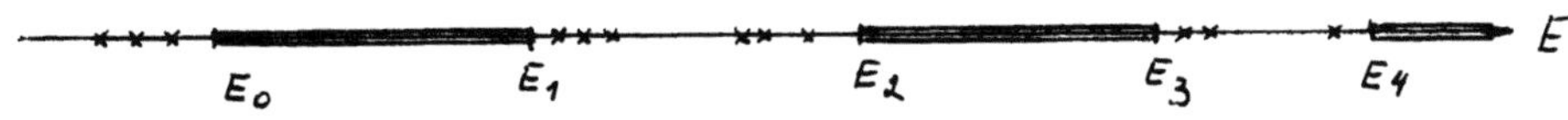

Fig. 15

Let us specify the assumptions concerning v. We assume that it has only one nongenerating critical point at $\xi = 0$ such that $v(0) = -v_0 < 0$ and, moreover, that it is strictly de(in)creasing if $\xi < 0$ (> 0). Sometimes we will suppose additionally that the potentials p and v are analytical functions in a strip surrounding the real axis.

Under these assumptions Fig. 15 acquires the form presented in Fig. 16.

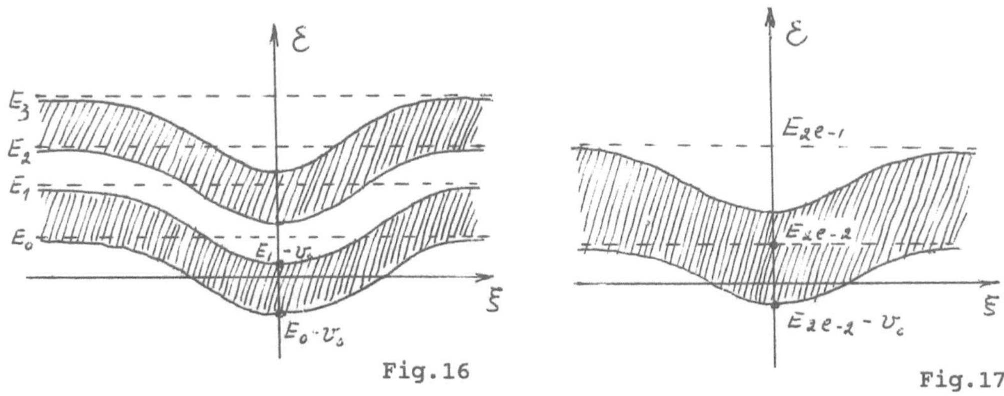

Again each spectral interval $\Delta_1 = [E_{2l-2}, E_{2l-1}]$ of the operator H_o yields a branch of the spectrum of H. This spectral branch lies in the interval $[E_{2l-2}-v_o, E_{2l-1}]$. Its structure depends on the correlation of v_o and $|\Delta_1| = E_{2l-1}-E_{2l-2}$. If $v_o < |\Delta_1|$, then the relative position of the graphs of the functions $v(\xi) + E_{2l-2}$ and $v(\xi) + E_{2l-1}$ looks as presented in Fig. 17.

If $E \in (E_{2l-2}-v_o, E_{2l-2})$, then the real part of the isoenergy curve generated by the spectral interval Δ_1 is an oval, see 1,2 in Fig.18. The corresponding branch of the spectrum consists of discrete eigenvalues E_n^1 asymptotical formulas of which can be obtained from the quantization conditions. If $E \to E_{2l-2}$, then isoenergy curve becomes infinite, see Fig.18. The question about the eigenvalues E_n^1 as $E_n^1 \to E_{2l-2}$ will be discussed later. The isoenergy curve is unbounded if $E \in [E_{2l-2}, E_{2l-1}]$ and disappears tending to infinity if $E \to E_{2l-1}$, see 7 in Fig. 18.

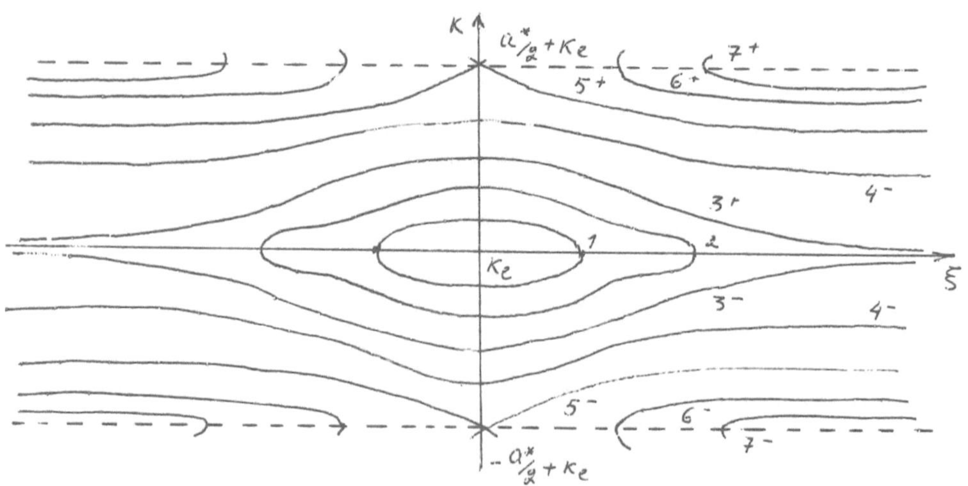

Fig. 18

Infiniteness of the isoenergy curve means that the corresponding part of the spectrum is continuous. The structure of the infinite isoenergy curve is different in the cases $E < E_{2l-1}$ and $E >$

E_{21-1}, see 4 and 6 of Fig. 18.

With infinite branches we can connect a scattering picture. It can be described by the unitary scattering matrix

$$\begin{pmatrix} s^1(E) & r^1_{12}(E) \\ r^1_{21}(E) & s^1(E) \end{pmatrix}.$$

It is possible to give explicit asymptotic formulas for the eigenfunctions of the continuous spectrum and for the scattering matrix. In particular, if $E < E_{21-1}$, then r_{12}, $r_{21} = O(\varepsilon^\infty)$, if $E > E_{21-1}$, then $s = O(\varepsilon^\infty)$. A symbolic picture of the spectrum is presented in Fig. 19.

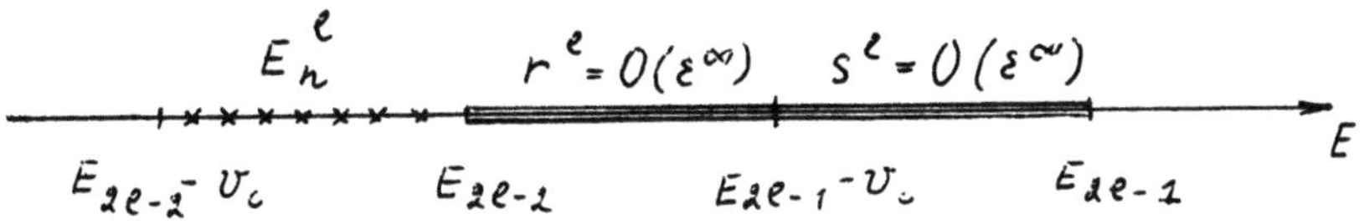

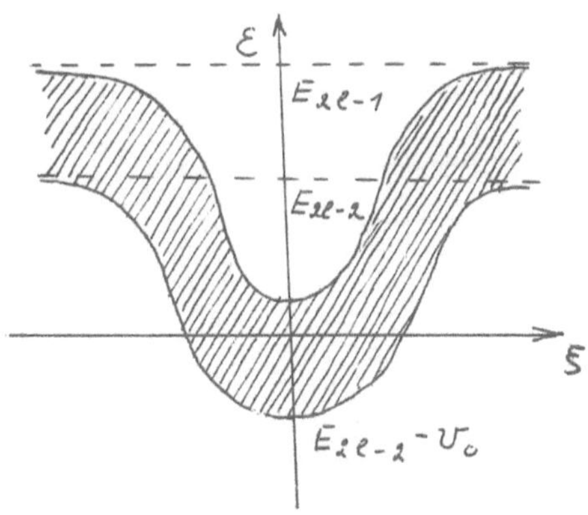

Fig. 19

The analogous picture for the case $v_o > |\Delta_1|$ is quite different. In this case Fig. 17 transforms to Fig. 20.

Fig. 20

The character of the isoenergy curve   in this case is given in
Fig. 21.

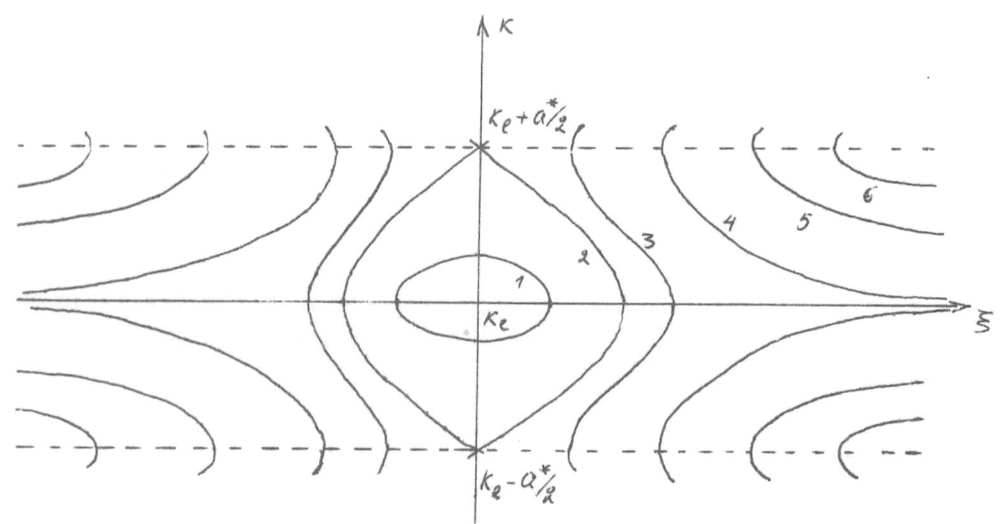

Fig. 21

The symbolic picture of Fig. 19 transforms to Fig. 22

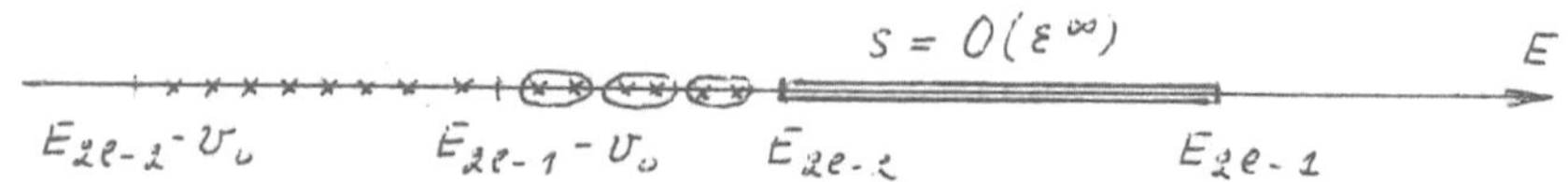

Fig. 22

4.2 General picture

Summarizing the different pictures we obtain total
description of the spectrum.

Before that, however, let us say some words about the

interaction of the different branches. In general it is true what we have said concerning this interaction in the previous part of the paper but with one essential correction. The tunneling between two real bounded branches of the isoenergy curve results into exponentially small displacements of the eigenvalues along the real axis. In contrast to that the tunneling between a real bounded branch and a real unbounded branch yields exponentially small displacements of the eigenvalues from the real axis to the complex spectral plane so the eigenvalues become resonances of the operator H. In other words, they are poles of the resolvent kernel $R(x,y;E)$ of the operator H,

$$R(E) = (H - E)^{-1}.$$

These poles lie on the second sheet of the spectral plane and can be obtained by analytical continuation from the main spectral sheet through the continuous spectrum described asymptotically by the unbounded branch of the isoenergy curve. We can write down the asymptotic representation of these poles-resonances and the corresponding resonance states.

If the number 1 is large, then the length $|\bar{\Delta}_1|$ of the interval $\bar{\Delta}_1$ is sufficiently small. Consequently, the above constructions are not uniform with respect to E because for large E we have to take effects of the strong tunneling between different real branches of the isoenergy curve into account.

Let us now return to the question about the asymptotic behavior of the eigenvalues of a given branch E^1 of the continuous spectrum. We restrict our discussion to the case $v_o < |\Delta_1|$. If $\sigma > 2$, then the whole area bounded by the branches 3^+ and 3^- of the isoenergy curve for $E = E_{21-2}$, see Fig. 18, is finite and the total number of eigenvalues E_n^1 in the interval $[E_{21-2}-v_o, \ E_{21-2}]$ is finite. In the case $\sigma < 2$ this area is infinite and the number of eigenvalues E_n^1 is infinite. The asymptotic behavior of the eigenvalues E is the same as for the operator

$$E_{21-2} - \frac{1}{2\kappa_{21-2}} \frac{d^2}{dx^2} + v(\varepsilon x).$$

Our last remark is concerned with the asymptotic behavior of the eigenfunctions. The main result says that

$$(P(\Delta)f,g) \longrightarrow (P_0(\Delta \cap [E_{21-2} - v_0, E_{21-1} - v_0])f,g)$$

as $\varepsilon \longrightarrow 0$. Here P^1 is the spectral projection of the operator H corresponding to the part of the spectrum of H represented in Fig. 19 or 22. Owing to a possible interaction between this part and an infinite branch of the isoenergy curve, the contribution of resonances arising from eigenvalues must be included in $(P^1(\Delta)f,g)$ but only from one half-plane, say, from the lower one.

REFERENCES

[1] Buslaev, V.S., *Teor. Mat. Fiz.* 58(1984), 233-243 (in Russian).

[2] Buslaev, V.S., *Sov. Mat. Uspekhi* 42(1987), 77-98 (in Russian).

[3] Buslaev, V.S.; Dmitrieva, L.A., *Teor. Mat. Fiz.* 73(1987), 430-442 (in Russian).

[4] Guillot, J.C.; Ralston, J.; Trubowitz, E., *Comm. Math. Phys.* 116(1988), 401-415.

[5] Buslaev, V.S.; Dmitrieva, L.A., *Bloch electrons in the external electric field*, in Schroedinger operators, standard and non-standard (eds P.Exner and P.Seba), World Scientific, Singapore, 1989, pp. 103-129.

[6] Buslaev, V.S.; Dmitrieva, L.A., *Algebra and Analysis* 1 (1989), N.2, 1-29 (in Russian).

[7]   Buslaev, V.S., *Semi classical approach to the equations with periodic coefficients*, in Proceedings of the XI school on diffraction and wave propagation (ed. B.Kinber), Kazan, 1988 (in Russian).

[8]   Buslaev, V.S.; Dmitrieva, L.A., *Spectral properties of the Bloch electrons in external fields*, Preprint, Berkely, 1989.

[9]   Firsova, N.E., *Zapiski Nauchn. Sem. LOMI* 51(1975), 183-196 (in Russian).

[10] Marchenko, V.A.; Ostrovskii, I.V., *Mat. Sbornik* 97(1975), 540-606 (in Russian).

Institute of Physics
Leningrad State University
198904 Leningrad - St.Peterhof
USSR

Operator Theory:
Advances and Applications, Vol. 46
© 1990 Birkhäuser Verlag Basel

DISCRETE SPECTRUM FOR A PERIODIC SCHROEDINGER OPERATOR PERTURBED BY A DECREASING POTENTIAL

S.V.Khryashchev

One studies a periodic Schroedinger operator in $L^2(\mathbb{R})$ perturbed by a potential decreasing at infinity in the mean.Criteria of infinitude and finiteness of the discrete spectrum in gaps are given. Similar results for a semiinfinite gap of the perturbed periodic Schroedinger operator in $L^2(\mathbb{R}^n)$, $n \geq 3$, are obtained.

1. We consider the selfadjoint Schroedinger operator $A(q_1,q_2) = -d^2/dx^2 + p + q_1 - q_2$ in $L^2(\mathbb{R})$, where $p : \mathbb{R} \longrightarrow \mathbb{R}$ is a periodic bounded function, $p(x+1) = p(x)$; $q_{1,2} \in L^1_{loc}(\mathbb{R})$, $q_{1,2} \geq 0$, $\int_a^{a+1} q_{1,2}(x)dx \longrightarrow 0$ as $|a| \longrightarrow \infty$. The operator $A(q_1,q_2)$ is defined in the form sense. Its continuous spectrum coincides (see [1]) with the spectrum of the operator $-d^2/dx^2 + p$, consisting of intervals (bands) separated by gaps. The operator $A(q_1,q_2)$ has a discrete spectrum in the gaps (see [1] and also the reviews [2], [3]). The purpose of the present paper is to find out if the discrete spectrum near a given edge of a gap is finite or infinite.

Let $\lambda_1 < \lambda_2$ and the interval (λ_1,λ_2) be a gap, $\delta = (\lambda_2 - \lambda_1)/2$. Let ω_k, $k=1,2$, be a bounded (normalized in $L^2(0,1)$) solution to the equation $-\omega'' + p\omega = \lambda_k\omega$; m_k be an effective mass (see [3]) for the operator $-d^2/dx^2 + p$ at the point $\lambda = \lambda_k$. Let $N_k(q_1,q_2)$ be the total spectral multiplicity for the operator $A(q_1,q_2)$ in the interval $(\lambda_1,\lambda_1+\delta)$ if $k=1$ and $(\lambda_2-\delta,\lambda_2)$ if $k=2$. Let $\hat{N}_k(q_1,q_2)$, $k=1,2$, be the total multiplicity of the negative spectrum for the "model" operator $-(1/2|m_k|)d^2/dx^2 + |\omega_k|^2(q_1-q_2)$.

For the left edge λ_1 of the gap the following theorem holds.

THEOREM 1. *For any* $\varepsilon, 0 < \varepsilon < 1$, *the implications*

$$(a)\quad \left\{\ \hat{N}_1(q_2,(1+\varepsilon)q_1)<\infty\ \right\} \Rightarrow \left\{\ N_1(q_1,q_2)<\infty\ \right\},$$

$$(b)\quad \left\{\ \hat{N}_1(q_2,(1-\varepsilon)q_1)=\infty\ \right\} \Rightarrow \left\{\ N_1(q_1,q_2)=\infty\ \right\}$$

are valid.

For the right edge λ_2 of the gap the analogous theorem holds.

THEOREM 2. *For any* $\varepsilon, 0 < \varepsilon < 1$, *the implications*

$$(a)\quad \left\{\ \hat{N}_2(q_1,(1+\varepsilon)q_2)<\infty\ \right\} \Rightarrow \left\{\ N_2(q_1,q_2)<\infty\ \right\},$$

$$(b)\quad \left\{\ \hat{N}_2(q_1,(1-\varepsilon)q_2)=\infty\ \right\} \Rightarrow \left\{\ N_2(q_1,q_2)=\infty\ \right\}$$

are valid.

The assertions presented below are consequences of Theorem 2. Similar consequences of Theorem 1 are not cited here. The subscript 2 for $N,\hat{N},\omega$ and m is omitted.

COROLLARY 1. *The relations*

$$(a)\quad \left\{\ \forall\alpha>0\ :\ N(q_1,\alpha q_2)<\infty\ \right\} \Leftrightarrow \left\{\ \forall\alpha>0\ :\ \hat{N}(q_1,\alpha q_2)<\infty\ \right\},$$

$$(b) \quad \left\{ \ \forall \alpha > 0 \ : \ N(q_1, \alpha q_2) = \infty \ \right\} \ \leftrightarrow \ \left\{ \ \forall \alpha > 0 \ : \ \hat{N}(q_1, \alpha q_2) = \infty \ \right\}$$

are valid.

$$\text{Set} \quad J_{\pm}(\eta) = \limsup_{x \to \pm \infty} \ x \int_{x}^{+\infty} \eta(t) \, dt \quad \text{for } \eta \geq 0.$$

COROLLARY 2. *The implications*

$$(a) \quad \left\{ \ J_{+}(q_2) + J_{-}(q_2) = 0 \ \right\} \ \Rightarrow \ \left\{ \ \forall \alpha > 0 \ : \ N(0, \alpha q_2) < \infty \ \right\},$$

$$(b) \quad \left\{ \ \forall \alpha > 0 \ : \ N(0, \alpha q_2) = \infty \ \right\} \ \Rightarrow \ \left\{ \ J_{+}(q_2) + J_{-}(q_2) = \infty \ \right\}$$

are valid.

(c) If the function q_2 is monotonous near infinity or the gap is semiinfinite, then one may replace $\Rightarrow$ by $\leftrightarrow$ in (a) and (b).

For a semiinfinite gap the assertion of Corollary 2(c) has been established in [1]. If the gap is finite then the monotonicity of q_2 in Corollary 2(c) is essential: there exists a function q_2 non-monotonous near infinity such that $q_2(x) \longrightarrow 0$ as $|x| \longrightarrow \infty$, $J_{\pm}(q_2) = \infty$, but $N(0, \alpha q_2) < \infty$ for $\forall \alpha > 0$.

COROLLARY 3. *Let the function q_2 be monotonous near infinity. Then*

$$(a) \quad \left\{ \ 8mJ_{\pm}(q_2) < 1 \ \right\} \ \Rightarrow \ \left\{ \ N(0, q_2) < \infty \ \right\},$$

$$(b) \quad \left\{ \ 2mJ_{-}(q_2) > 1 \ \text{or} \ 2mJ_{+}(q_2) > 1 \ \right\} \ \Rightarrow \ \left\{ \ N(0, q_2) = \infty \ \right\}.$$

REMARK. *The monotonicity of q_2 in Corollary 3 is also essential:*

(a) if $\max\limits_{0\leq x\leq 1}|\omega(x)|^2 > 4$, then there exists a function q_2 non-monotonous near infinity such that $q_2(x) \longrightarrow 0$ as $|x| \longrightarrow \infty$, $J_\pm(q_2)<1/8m$, but $N(0,q_2) = \infty$;

(b) if $\min\limits_{0\leq x\leq 1}|\omega(x)|^2 < 1/4$ then there exists a function q_2 non-monotonous near infinity such that $q_2(x) \longrightarrow 0$ as $|x| \longrightarrow \infty$, $J_\pm(q_2)>1/2m$, but $N(0,q_2) < \infty$.

COROLLARY 4. *Denote* $q = q_1-q_2$. *Then*

(a) $\left\{\, q(x) \geq -(1-\varepsilon)/(8mx^2), \forall x, |x|>R \text{ for some } \varepsilon,R>0 \,\right\} \Rightarrow$ $\left\{N(q_1,q_2)<\infty\right\},$

(b) $\left\{\, q(x) \leq -(1+\varepsilon)/(8mx^2), \forall x>R \text{ (or } \forall x<-R \text{) for}$ $\text{some } \varepsilon,R>0 \,\right\} \Rightarrow \left\{\, N(q_1,q_2) = \infty \,\right\}.$

Corollary 4 and a similar assertion for the left edge of the gap have been obtained earlier in [3].

2. Let us consider now the selfadjoint Schroedinger operator $B(\alpha) = -\mathrm{div}(a\nabla) + p + q_1 - \alpha q_2$ in $L^2(\mathbb{R}^n)$, $n\geq 3$, where a is a positive linear operator in $\mathbb{R}^n$, $\alpha>0$, p and $q_{1,2}$ are bounded real-valued functions, p being periodic in $\mathbb{R}^n$ with a basic period cell Q ; $q_k(x)\geq 0$, $q_k(x) \longrightarrow 0$ as $|x| \longrightarrow \infty$, $k=1,2$. The essential spectrum of $B(\alpha)$ coincides with the spectrum of the operator $-\mathrm{div}(a\nabla) + p$, whose lower bound is assumed without loss of generality to be zero. The operator $B(\alpha)$ may have discrete eigenvalues below zero. Let ω be a periodic (with the period cell Q) solution to the equation $-\mathrm{div}(a\nabla\omega) + p\omega = 0$, normalized in $L^2(Q)$ by $(\mathrm{mes}Q)^{1/2}$, $(2b)^{-1}$ be the effective-mass tensor (see [4]) for the operator $-\mathrm{div}(a\nabla) + p$ at the point $\lambda=0$. Let $M(\alpha)$ be the total multiplicity of the negative spectrum of $B(\alpha)$, $\hat{M}(\alpha)$ be that of the operator $-\mathrm{div}(b\nabla) + |\omega|^2(q_1-\alpha q_2)$.

THEOREM 3. *For any $\varepsilon, 0 < \varepsilon < 1$, the implications*

(a) $\left\{ \hat{M}(1+\varepsilon) < \infty \right\} \Rightarrow \left\{ M(1) < \infty \right\}$,

(b) $\left\{ \hat{M}(1-\varepsilon) = \infty \right\} \Rightarrow \left\{ M(1) = \infty \right\}$

hold.

COROLLARY 5. *The relations*

(a) $\left\{ \forall \alpha > 0 : M(\alpha) < \infty \right\} \Leftrightarrow \left\{ \forall \alpha > 0 : \hat{M}(\alpha) < \infty \right\}$,

(b) $\left\{ \forall \alpha > 0 : M(\alpha) = \infty \right\} \Leftrightarrow \left\{ \forall \alpha > 0 : \hat{M}(\alpha) = \infty \right\}$

hold.

COROLLARY 6. *Let* $c = b^{-1/2}$, $q = q_1 - q_2$. *Then*

(a) $\left\{ \liminf_{r \to \infty} r^2 \inf_{|cx|=r} q(x) > -(n-2)^2/4 \right\} \Rightarrow \left\{ M(1) < \infty \right\}$,

(b) $\left\{ \limsup_{r \to \infty} r^2 \sup_{|cx|=r} q(x) < -(n-2)^2/4 \right\} \Rightarrow \left\{ M(1) = \infty \right\}$.

ACKNOWLEDGMENT

The author would like to thank his supervisor Professor M.Sh.Birman for his guidance and kind attention to this work.

REFERENCES

1   Birman M.Sh. *On the spectrum of singular boundary-value problems*, Sov.Math.Sbornik 55(97),2,125-174 (1961) (in Russian).

2   Rofe-Beketov F.S. *Spectral analysis of the Hill operator and its perturbations*, Funct.Anal.9,144-155 (1977) (in Russian).
3   Rofe-Beketov F.S. *Spectrum perturbations, the Kneser-type constants and the effective masses of the zone-type potentials*, in *Constructive theory of functions*, Sofia 1984, pp.757-766.
4   Anselm A.I. *Introduction to the theory of semiconductors*, Nauka, Moscow 1978 (in Russian).

Department of Mathematical Physics
Institute of Physics
Leningrad State University
198904 Leningrad - St.Peterhof
USSR

Operator Theory:
Advances and Applications, Vol. 46
© 1990 Birkhäuser Verlag Basel

ON A COMPLETE DESCRIPTION OF THE PRINCIPAL DISCRETE SERIES OF SPECTRAL INVARIANTS OF THE HILL OPERATOR.

Michail Novitskii

The spectral invariants of Hill's operator which are generated by the generalized Jacobi formula are studied. The connection with the main polynomial series of first integrals of the Korteweg-de Vries equation is established.

Let $u(x)$ be an infinitely differentiable function with the period one, i.e. $u(x+1)=u(x)$. Let $\{\lambda_i(H)\}_{i=0}$ be the spectrum of the periodic boundary value problem in $L^2[0,1]$ for the Hill operator. *The functional $F(u)$ is called the spectral invariant of H if $F(u)$* has the following property: if the spectra of operators $H_i = -\dfrac{d^2}{dx^2} + u_i(x)$, $i=1,2$, coincide, then $F(u_1)=F(u_2)$. The most familiar discrete series of spectral invariants of the Hill operator is the polynomial series $\{I_k\}_{k=0}$ of first integrals of the Korteweg-de Vries equation $u_t = 6uu_x - u_{xxx}$, where

$$I_n = \frac{1}{2n+1} \int_0^1 P_{n+1}(u,u',\dots)\ dx\ .$$

Here $P_n(u,u',\dots)$ are universal homogeneous polynomials of degrees equal exactly to $n+1$ which satisfy the following recursion formula

$$\left(-\frac{d^3}{dx^3} + 4u\frac{d}{dx} + 2\frac{d}{dx}u\right)P_n = 4\frac{d}{dx}P_{n+1}$$

with the condition $P_0 \equiv 1$. It is important that the series $\{I_k\}_{k=0}^{\infty}$ is the basis in the linear space of local conserved functionals [1]. The main discrete series of spectral invariants have the natural parametrization $n=0,1,2,\ldots$ in which the series $\{I_k\}_{k=0}^{\infty}$ corresponds to the parameter $n=0$. To determine the n-th series for the operator H, we set $\theta(t)=\sum_{l=1}^{\infty} \exp(-\lambda_l t)$, $t>0$. The generalized Jacobi formula for the Hill operator may be written as

$$\theta(t)= \sum_{n=-\infty}^{\infty} \theta_n(t), \quad \text{where } \theta_n(t)= (4\pi t)^{-1/2}\exp(-\frac{n^2}{4t})F_n(t).$$

The function $F_n(t)$ has a representation of the form $\int_0^1 M\,[\exp(-t\int_0^1 u(x+n\tau+t^{1/2}w(\tau))d\tau)]\,dx$. Here M is the mathematical expectation, $w(\tau)$ is the "Wiener bridge", i.e. one-dimensional conditional Wiener process determined by the requirements:

i) $w(0)=w(1)=0$;

ii) $w(\tau)$ is the Gaussian process with zero mean and correlation function $B(s,t)=(s\wedge t)(1-s\vee t)$.

DEFINITION. The n-th series of spectral invariants of the Hill operator are the quantities $\{b_k^n\}_{k=1}^{\infty}$ determined by the asymptotic expansion

$$F_n(t) \sim 1 + \sum_{k=1}^{\infty} b_k^n t^k, \quad t \searrow 0$$

It is known [2] that $\{b_k^0\}_{k=1}^{\infty}$ differ from $\{I_k\}_{k=0}^{\infty}$ by some universal multipliers,

$$b_{k+1}^0 = \frac{(-1)^{k+1}\,2^k}{(2k-1)!!}\,I_k, \quad k=0,1,2,\ldots.$$

THEOREM 1. *The coefficients of the n-th series* $\{b_k^n\}_{k=1}^{\infty}$ *are related to* $\{I_k\}_{k=0}^{\infty}$ *by the following equality*

$$b_k^n = \sum_{l=1}^{k} \frac{n^{2(k-l)}\,(-1)^l\,2^{l-1}}{(2l-3)!!}\,I_{l-1}$$

CORROLARY 1. *Every series* $\{b_k^n\}_{k=1}^{\infty}$ *is a basis in the linear space of local conserved functionals.*

It follows from the results of [3] and theorem 1 that

CORROLARY 2. *The spectrum* $\{\lambda_k\}_{k=0}^{\infty}$ *of the Hill operator* $H = -\dfrac{d^2}{dx^2} + u(x)$ *is recovered from the n-th series* $\{b_k^n\}_{k=0}^{\infty}$ *if and only if the function u(x) is quasianalytic.*

The eigenvalues $\{\lambda_k\}_{k=0}^{\infty}$ of the periodic boundary value problem generated by the operator H satisfy the well-known asymptotic formulas

$$\lambda_{2n-1}^{1/2} \sim 2\pi n + \sum_{k=0}^{\infty} \frac{a_k(u)}{n^{2k+1}}, \quad n \longrightarrow \infty .$$

The same asymptotic formulas are valid for $\lambda_{2n}^{1/2}$, $n \longrightarrow \infty$.

THEOREM 2. *The following equality is valid,*

$$a_k(u) = (-1)^k 2\pi \ (4\pi^2)^k \ I_k(u)$$

Proofs of Theorems 1 and 2 are based on analysis of algebraic properties of polynomials, which generate the spectral invariants $\{b_k^n\}_{k=1}^{\infty}$ and $\{a_k\}_{k=0}^{\infty}$.

REFERENCES

1. Serre D., C.R.Acad. Sci., Paris, <u>239</u>, Ser.I., 505-508 (1981).
2. McKean H.P., Moerbeke van P., Inventions Math., <u>30</u>, 217-241 (1975).
3. Novitskii M.V., Funkcional. Anal. i Prilozen., <u>23</u>, №3 (1980), 78-79 (in Russian).

Mathematical Department,
Institute of Low Temperature Physics and Engeneering,
Ukrainian Academy of Sciences,
Lenin Ave. 47, Kharkov 310164, USSR.

Operator Theory:
Advances and Applications, Vol. 46
© 1990 Birkhäuser Verlag Basel

WKB-APPROXIMATIONS FROM THE PERTURBATION THEORY VIEWPOINT

N.A.Chernyavskaya and L.A.Shuster

In this work analogs of the classical WKB-approximation theorems covering nonsmooth, rapidly growing and oscillating potentials are obtained.

Consider the equation

$$y''(x) = q(x)y(x), \qquad 0 < q(x) \in L_1^{loc}(R), \ x \in R \tag{1}$$

THEOREM 1. *Assume there exists a function* $q_1(x)$, $0 < q_1(x) \in C^2(R_+)$, *satisfying the following conditions*

$$q_1^{1/2}(x) \notin L_1(R_+).$$

$$K(x;q_1) = q_1^{-1/4}(x)\left[q_1^{-1/4}(x)\right]'' + \left[q_1(x) - q(x)\right]q_1^{-1/2}(x) \in L_1(R_+)$$

Denote

$$\varphi(x) = \int_x^\infty |K(t;q_1)|\,dt. \qquad \psi(x) = \varphi(x) + |q_1'(x)|q_1^{-3/2}(x)$$

Then

(A) There exists a fundamental system of solutions (FSS) of the equation (1), which for $x \geq x_0$ *(* x_0 *being a sufficiently large number) has the form*

$$y_i(x) = q_1^{-1/4}(x) \, exp\left\{(-1)^{i+1}\int_{x_0}^{x} q_1^{1/2}(t)dt\right\}\left[1+\varepsilon_i(x)\right], \quad i=1,2;$$

furthermore, $\lim\limits_{x\to\infty}\varepsilon_i(x)=0$, $i=1,2$, *and* $|\varepsilon_2(x)|\leqslant c_1\varphi(x)$.
In addition, if the condition

$$\lim_{x\to\infty} q_1'(x)q_1^{-3/2}(x)=0 \qquad\qquad (2)$$

is fulfilled, then

    (B) The derivatives $y_{1,2}'(x)$ *for* $x\geqslant x_0$ *have the form*

$$y_i'(x)=q_1^{1/4}(x) \, exp\left\{(-1)^{i+1}\int_{x_0}^{x} q_1^{1/2}(t)dt\right\}\left[(-1)^{i+1}+\varepsilon_{i+2}(x)\right], \quad i=1,2$$

and moreover, $\lim\limits_{x\to\infty}\varepsilon_{i+2}(x)=0$, $i=1,2$, $|\varepsilon_4(x)|\leqslant c_2\psi(x)$. *Here and
later on,* $c_1,c_2,\ldots$ *are positive constants.*

    Notice that if $q(x)\in C^2(R_+)$ then assuming $q_1(x)=q(x)$ in Theorem 1 we get the classical WKB-approximation theorem [1].

    For sufficiently wide classes of potentials the function $q_1(x)$ can be found explicitly.

    Let us introduce the averages $q_*(x),q_{**}(x)$ of the potential $q(x)$ by the formulas [2]

$$q_*(x) = \inf_{d>0}\left\{d^{-2}: 1 \geqslant \int_0^d\int_{x-t}^{x+t}q(\xi)d\xi dt\right\}, \quad q_{**}(x)=\left[q_*(\ )\right]_*(x)$$

    THEOREM 2. Let $q(x)=\tilde{q}(x)+h(x)$, where $\tilde{q}(x)\geqslant c_3>1$ and assume that there exist numbers $\varepsilon>1/4$, $\gamma>1/2$ such that

$$|\tilde{q}(t)-\tilde{q}(x)|\leqslant c_4\tilde{q}^{1-\varepsilon}(x), \qquad |x-t|\leqslant c_5\tilde{q}^{-1/2}(x) \qquad (3)$$

$$|\tilde{q}(x+t)-2\tilde{q}(x)+\tilde{q}(x-t)|\leqslant c_6\tilde{q}^{1-\gamma}(x), \qquad |t|\leqslant c_5\tilde{q}^{-1/2}(x) \qquad (4)$$

$$\tilde{q}^{-\delta}(x) \in L_1(R_+), \quad \delta = \min\left\{\gamma, 2\varepsilon\right\} - 1/2 \qquad (5)$$

$$h(x)\tilde{q}^{-1/2}(x) \in L_1(R_+) \qquad (6)$$

Assume $q_1(x) = \tilde{q}_{**}(x)$ *and denote*

$$\varphi(x) = \int_x^\infty \left[\tilde{q}^{-\delta}(t) + |h(t)|\tilde{q}^{-1/2}(t)\right] dt, \quad \psi(x) = \varphi(x) + \tilde{q}^{-\varepsilon}(x)$$

Then the statements (A),(B) of Theorem 1 are true.

EXAMPLE 1. Let q(x) be one of the functions:

$$\exp(x) + \exp(\alpha x)|\sin[\exp(x^k)]|, \quad \exp(x) + \exp(\alpha x)\sin[\exp(x^k)],$$

$\alpha < 1/2$, $k \geqslant 0$. If we put $q_1(x) = \exp(x)$, then in both cases Theorem 1 can be applied.

EXAMPLE 2. Let $q(x) = \exp(x)\{1 + \sin[\exp(x^k)]\}$, $k \geqslant 1$. In this case Theorem 1 is inapplicable; such potentials satisfy the hypotheses of Theorem 3.

THEOREM 3. *Assume there exists a function* $q_1(x) > 0$, $q_1(x) \in C^2(R_+)$, *satisfying the following conditions*

$$q_1^{1/2}(x) \notin L_1(R_+), \quad q_1^{-1/2}(x) \in L_1(R_+)$$

$$A(x) = \sup_{t \geqslant x} \left|\int_t^\infty K(\xi; q_1)d\xi\right| \leqslant c_7 q_1^{-1/2}(x)$$

and denote

$$\varphi(x) = A(x) + \int_x^\infty q_1^{-1/2}(t)dt, \quad \psi(x) = \varphi(x) + |q_1'(x)|q_1^{-3/2}(x)$$

Then the statement (A) of Theorem 1 is valid. In addition, if the condition (2) is fulfilled, then the statement (B) of the theorem is true.

THEOREM 4. *Let* $q(x)=\tilde{q}(x)+h(x)$, *and moreover, let the function* $\tilde{q}(x)$ *satisfy the conditions* (3)-(5), $\tilde{q}(x)\geqslant c_3>1$ *and the relations*

$$B(x) = \sup_{t\geqslant x} \left|\int_t h(\xi)\,\tilde{q}_{**}^{-1/2}(\xi)d\xi\right| \leqslant c_8 \tilde{q}_{**}^{-1/2}(x) \in L_1(R_+)$$

are fulfilled. Assume $q_1(x)=\tilde{q}_{**}(x)$ *and denote*

$$\varphi(x)=B(x)+\int_x^\infty \tilde{q}^{-\delta}(t)dt, \quad \psi(x) = \varphi(x)+\tilde{q}^{-\varepsilon}(x)$$

Then the statements A),(B) of Theorem 1 are valid.

It is interesting to compare Theorems 1-4 with the apriori estimates of the FSS of the equation (1). One can show [2] that there exists FSS $\{u(x),v(x)\}$ of the equation (1) with the properties $u(x),v(x)>0$, $u'(x)<0$ and $v'(x)>0$ for $x\in R$.

Let us denote

$$d_+(x) = \sup\left\{ d:\, d>0,\ 1 \geqslant \int_0^d \int_x^{x+t} q(\xi)d\xi dt \right\}$$

$$d_-(x) = \sup\left\{ d:\, d>0,\ 1 \geqslant \int^d \int^x q(\xi)d\xi dt \right\}$$

THEOREM 5. *For* $x\in R$, *the inequalities*

$$d_+^{-1}(x) \leqslant |u'(x)|u^{-1}(x) \leqslant 2d_+^{-1}(x)$$

$$d_-^{-1}(x) \leqslant v'(x)v^{-1}(x) \leqslant 2d_-^{-1}(x)$$

are fulfilled.

REFERENCES

1. M.V.Fedoryuk : *Asymptotic Methods for Linear Ordinary
   Differential Equations*, Nauka, Moscow 1983 (in Russian).
2. N.A.Chernyavskaya and L.A.Shuster, On the WKB-method,
   Sov.J.Diff.Eq.25 (1989),1826-1821.

Institute of Mathematics and Mechanics
of the Kazakh Academy of Sciences,
ul.Pushkina, 125, 480021 Alma-Ata, USSR.

Operator Theory:
Advances and Applications, Vol. 46
© 1990 Birkhäuser Verlag Basel

MOVING POTENTIALS AND THE COMPLETENESS OF WAVE OPERATORS. PART II: PROPAGATING OBSERVABLES ON SCATTERING STATES

Hagen Neidhardt

The paper is devoted to the time evolution of the position and impulse operators as well as certain functions of them, for example the dilation and classical kinetic energy operator, in quantum mechanical scattering theory with time-dependent and travelling along fixed trajectories long range potentials.

1. INTRODUCTION

In this paper, we study the time evolution of certain observables in quantum mechanical potential scattering theory for time-dependent and moving potentials which are long range in space direction. Thus, we intend to generalize the results of V.Enss [2] to a very general case of time-dependent potentials. In a subsequent paper [7], we will apply these results to the description of the ranges of wave operators.

Mathematically spoken, in the Hilbert space $\mathfrak{h} = L^2(\mathbb{R}^n)$, $n \geq 1$, we consider the Schrödinger equation

$$(1.1) \qquad i\frac{\partial}{\partial t}\, u = H(t)\, u \equiv (H_o + V(t))u, \qquad u\big|_{t=s} = u_o,$$

where H_o is the free Hamiltonian defined as usual, i.e. $H_o = -\frac{1}{2}\Delta$, and $\{V(t)\}_{t\in\mathbb{R}^1}$ is a time-dependent perturbation of the structure

$$(1.2) \qquad V(t) = \sum_{j=1}^{N} V_j(t)$$

where the time-dependent perturbations $\{V_j(t)\}_{t\in\mathbb{R}^1}$, $j = 1,\dots,N$, arise from time-dependent potentials q_j as follows:

$$(1.3) \qquad (V_j(t)f)(x) = q_j(t, x - x_j(t))f(x),$$

$t \in \mathbb{R}^1$, $x \in \mathbb{R}^n$, $x_j(.): \mathbb{R}^1 \longrightarrow \mathbb{R}^m$, $j = 1,2,\dots,N$.

In order to describe the conditions satisfying the potentials we introduce the sets $C^1_{loc}(\mathbb{R}^m)$ and $C^1_{loc}(\mathbb{R}^m;\mathbb{R}^k)$, $m \geq 1$, of all functions defined on $\mathbb{R}^m$ with values in $\mathbb{R}$ and $\mathbb{R}^k$, $k \geq 1$, respectively, whose first partial derivatives everywhere exist and are continuous on every bounded region of $\mathbb{R}^m$.

ASSUMPTION P. The potentials q_j, $j = 1,2,\dots,N$, belong to $C^1_{loc}(\mathbb{R}^{n+1})$ and satisfy the properties

$$(1.4) \qquad q_j \in L^\infty(\mathbb{R}^{n+1}), \quad \lim_{|x|\to\infty} q_j(t,x) = 0, \quad t \in \mathbb{R}^1, \quad j=1,2,\dots,N,$$

$$(1.5) \qquad |x|\,|\nabla q_j| \in L^\infty(\mathbb{R}^{n+1}), \quad \lim_{|x|\to\infty} x\nabla q_j(t,x) = 0, \quad t \in \mathbb{R}^1, \quad j=1,2,\dots,N,$$

$$(1.6) \qquad \dot{q}_j \in L^\infty(\mathbb{R}^{n+1}), \quad \sup_{x\in\mathbb{R}^n}|\dot{q}_j(t,x)| \in L^1(\mathbb{R}^1), \quad j = 1,2,\dots,N,$$

where we have used the notation $\dot{q}_j = \frac{\partial}{\partial t}\, q_j$.

It is not our aim maximally to sharp the potential conditions. Thus, it is quite possible, using suitable methods, to weaken Assumption P.

The notion "moving potentials" in the title arises from the special structure (1.3) of the time dependent perturbation which allows one to regard the function $x_j(.)$: $\mathbb{R}^1 \to \mathbb{R}^n$ as a trajectory along which the potentials q_j move. Concerning the trajectories, we assume the following.

ASSUMPTION T. The trajectories $x_j(.)$, $j = 1,2,\ldots,N$, belong to $C^1_{loc}(\mathbb{R}^1;\mathbb{R}^n)$ and obey

$$(1.7) \qquad \sup_{t\in\mathbb{R}^1} |x_j(t) - t\dot{x}_j(t)| < +\infty, \quad j = 1,2,\ldots,N,$$

where we have used the notation $\dot{x}_j(t) = \frac{d}{dt} x_j(t)$.

Taking into account Assumptions P and T it is not hard to see that by (1.3) we define strongly continuous families $\{V_j(t)\}_{t\in\mathbb{R}^1}$, $j = 1,2,\ldots,N$, of bounded operators which are uniformly bounded in $t \in \mathbb{R}^1$. Hence, the family $\{H(t)\}_{t\in\mathbb{R}^1}$ of Schrödinger operators given by $H(t) = H_o + V(t)$, $\mathcal{D}(H(t)) = \mathcal{D}(H_o)$, is well-defined; and by [12], there exists a strongly continuous family $\{U(t,s)\}_{(t,s)\in\mathbb{R}^2}$ of unitary operators on $\mathfrak{h}$ such that in a suitable sense $u(t) = U(t,s)u_o$ gives a solution of the Schrödinger equation (1.1). A family $\{U(t,s)\}_{(t,s)\in\mathbb{R}^2}$ like that is usually called the propagator of Eq. (1.1).

In terms of the propagator we characterize the subspace $\mathfrak{h}^{sc}_\pm(s)$, $s \in \mathbb{R}^1$, of the so-called scattering or asymptotically free states.

DEFINITION 1.1. The state $f \in \mathfrak{h}$ belongs to the subspace of scattering states $\mathfrak{h}^{sc}_\pm(s)$, $s \in \mathbb{R}^1$, if for every $R > 0$ we have

$$(1.8) \qquad \lim_{T\to\pm\infty} \frac{1}{T} \int_0^T dt \; \|F(x \in \mathcal{B}_R) \, e^{ix_j(t)P} \, U(t,s)f\|^2 = 0,$$

$j = 1,2,\ldots,N$, where $F(x \in \mathcal{B}_R)$ indicates the characteristic function of the ball $\mathcal{B}_R = \{x \in \mathbb{R}^n: |x| \leq R\}$.

Throughout the paper by P we denote the impulse operator $P \equiv -i\nabla \equiv \{-i\frac{\partial}{\partial x_1}, \ldots, -i\frac{\partial}{\partial x_n}\}$.

Physically, Definition 1.1 excludes a possibility that in far past and far future the scattering state is localized in any neighbourhood of any moving power center q_j, $j = 1,2,\ldots,N$.

Thus, for large times we can believe that the particle is not influenced by the power centers and, consequently, must be regarded as an asymptotically free particle. On the other hand, a state, which does not satisfy the conditions of Definition 1.1, has to be localized in a neighbourhood of at last one of the moving power centers during the evolution. Obviously, such a situation can be characterized only as a bounded state of the particle together with this moving power center.

The orthogonal projection from $\mathfrak{h}$ onto $\mathfrak{h}_\pm^{sc}(s)$ is denoted by $P_\pm^{sc}(s)$, $s \in \mathbb{R}^1$. Introducing the family $U(s) = U(s,0)$, $s \in \mathbb{R}^1$, of unitary operators simple computations show that

$$(1.9) \qquad P_\pm^{sc}(s) = U(s) \, P_\pm^{sc} \, U(s)^*, \qquad s \in \mathbb{R}^1,$$

where for simplicity we have set $P_\pm^{sc} \equiv P_\pm^{sc}(0)$. Relation (1.9) allows one to restrict the consideration in Section 3 and 4 to the special case $s = 0$.

REMARK 1.2. (i) If the potentials q_j, $j = 1,2,\ldots,N$, are nonmoving, i.e., $x_j(t) \equiv 0$, $j = 1,2,\ldots,N$, and time-independent, i.e., $q_j(t,x) \equiv q_j(x)$, $j = 1,2,\ldots,N$, then our definition of the scattering subspace coincides with those by Ruelle [9] and Amrein-Georgescu [1].

(ii) If the potentials are nonmoving but time-dependent, the definition coincides with that by Yafaev [10] and covers the definition introduced by H.Kitada and K.Yajima [4,5].

(iii) Moving but time-independent potentials are considered by K.Yajima [11]. It can be shown that our definition of the scattering subspace gives a correct description of the time behavior of states of the so-called asymptotically free channel in [11], i.e., the subspace of all states which allow an asymptotically free evolution of all involved particles.

The aim of the paper will be to show that for every scattering state $f \in \mathfrak{h}_\pm^{sc} \equiv \mathfrak{h}_\pm^{sc}(0)$ we have

(1.10) $(\frac{X}{t} - P)U(t)f \longrightarrow 0$ as $t \longrightarrow \pm\infty$,

in some generalized sense where X denotes the so-called position operator $X \equiv \{X_1, X_2, \ldots, X_n\}$. Thus, we considerably extend Theorem 2.4 of [2] to a new large class of physical systems. Using (1.9) the assertion (1.10) can easily be reformulated for $f \in \mathfrak{h}_{\pm}^{sc}(s)$, $s \in \mathbb{R}^1$.

The outline of the paper is as follows. In Section 2 we establish the existence of the propagator of Eq. (1.1) and, moreover, indicate certain properties of it which are necessary in the following. Section 3 is devoted to the propagating properties of the position, dilation and energy operators. Thus we generalize Theorem 2.1 of [2]. In Section 4 we give the correct statement of (1.10) and its proof.

2. THE PROPAGATOR

In this section we collect some properties of the propagator of Eq. (1.1) which are necessary in the following. First of all, we precisely formulate the conditions under which the propagator exists.

PROPOSITION 2.1. *If* $q_j \in C^1_{loc}(\mathbb{R}^{n+1})$ *and* $x_j(.) \in C^1_{loc}(\mathbb{R}^1;\mathbb{R}^n)$ *and* q_j, $|\nabla q_j|$, $\dot{q}_j \in L^\infty(\mathbb{R}^{n+1})$, $j = 1,2,\ldots,N$, *then there is a family* $\{U(t,s)\}_{(t,s)\in\mathbb{R}^2}$ *of unitary operators on* $\mathfrak{h}$ *obeying the following properties.*

(i) $U(t,r)U(r,s) = U(t,s)$, $t,r,s \in \mathbb{R}^1$, $U(t,t) = I$, $t \in \mathbb{R}^1$.

(ii) $U(t,s)$ *is strongly continuous in* $\mathfrak{h}$ *jointly in t and s.*

(iii) $U(t,s)\mathcal{D}(H_o) \subseteq \mathcal{D}(H_o)$, $t,s \in \mathbb{R}^1$, *and* $U(t,s)$ *is strongly continuous in the graph norm of* P^2 *jointly in t and s.*

(iv) *For every* $f \in \mathcal{D}(H_o)$ $U(t,s)f$ *is everywhere continuously differentiable in* $t,s \in \mathbb{R}^1$ *and*

(2.1) $i\frac{\partial}{\partial t} U(t,s)f = H(t)U(t,s)f$, $(t,s) \in \mathbb{R}^2$,

and

$$(2.2) \qquad -i\frac{\partial}{\partial t} U(t,s)f = U(t,s)H(s)f, \qquad (t,s) \in \mathbb{R}^2.$$

The family $\{U(t,s)\}_{(t,s)\in\mathbb{R}^2}$ *is uniquely determined by* (i) - (iv).

PROOF. It is easy to see that the Assumption (A.2) of [12] is fulfilled. Therefore, applying Theorem 1.3 of [12] we establish (i) - (iv) and, moreover, the uniqueness.∎

As usual, we call the family $\{U(t,s)\}_{(t,s)\in\mathbb{R}^2}$ of unitary operators obeying (i) - (iv) the propagator of Eq. (1.1). However, in the following we need some additional properties of the propagator of Eq. (1.1) which we only indicate. For the proofs of this facts the reader is referred to [6]. In order to formulate these properties, let us introduce the linear subsets $\mathcal{D}_1 = \mathcal{D}(|P|) \cap \mathcal{D}(|X|)$ and $\mathcal{D}_2 = \mathcal{D}(H_o) \cap \mathcal{D}(X^2)$ which are dense in $\mathfrak{h}$. Equipping the subsets $\mathcal{D}_1$ and $\mathcal{D}_2$ with the norms $\|f\|_1^2 = \|f\|^2 + \||P|f\|^2 + \||X|f\|^2$, $f \in \mathcal{D}_1$ and $\|f\|_2^2 = \|f\|^2 + \|P^2 f\|^2 + \|X^2 f\|^2$, $f \in \mathcal{D}_2$, respectively, the pairs $\mathfrak{h}_1 = \{\mathcal{D}_1, \|.\|_1\}$ and $\mathfrak{h}_2 = \{\mathcal{D}_2, \|.\|_2\}$ define two Hilbert spaces such that $\mathfrak{h}_1$ and $\mathfrak{h}_2$ as well as $\mathfrak{h}_2$ are densely and continuously embedded in $\mathfrak{h}$ and $\mathfrak{h}_1$, respectively, i.e., $\mathfrak{h}_2 \subset \mathfrak{h}_1 \subset \mathfrak{h}$.

PROPOSITION 2.2. *If* $q_j \in C^1_{loc}(\mathbb{R}^{n+1})$, $x_j(.) \in C^1_{loc}(\mathbb{R}^1;\mathbb{R}^n)$ *and* $q_j, |\nabla q_j|, \dot{q}_j \in L^\infty(\mathbb{R}^{n+1})$, $j = 1,2,\ldots,N$, *then the propagator of Eq.* (1.1) *satisfies the following additional properties.*

(i) $U(t,s)\mathcal{D}(|P|)=\mathcal{D}(|P|)$, $(t,s)\in\mathbb{R}^2$, *and* $\{U(t,s)|\mathcal{D}(|P|)\}_{(t,s)\in\mathbb{R}^2}$ *is strongly continuous in the graph norm of* $|P|$ *jointly in t and s.*

(ii) $U(t,s)\mathcal{D}_1 = \mathcal{D}_1$, $(t,s) \in \mathbb{R}^2$, *and* $\{U(t,s)|\mathcal{D}_1\}_{(t,s)\in\mathbb{R}^2}$ *is strongly continuous in* $\|.\|_1$*-norm jointly in t and s.*

(iii) $U(t,s)\mathcal{D}_2 = \mathcal{D}_2$, $(t,s) \in \mathbb{R}^2$, *and* $\{U(t,s)|\mathcal{D}_2\}_{(t,s)\in\mathbb{R}^2}$ *is strongly continuous in* $\|.\|_2$*-norm jointly in t and s.*

Introducing the so-called dilation operator D defined as the closure of the symmetric operator $\tilde{D}$,

$$(2.3) \qquad \tilde{D}f = \frac{1}{2}(XP + PX)f, \qquad f \in \mathcal{D}(\tilde{D}) = \mathscr{S}(\mathbb{R}^n),$$

i. e. $D = (\tilde{D})^-$, it can be shown that D^- is a self-adjoint operator. Now Proposition 2.2 admits the following

COROLLARY 2.3. *If* $f \in \mathcal{D}_2$, *then*

(i) $U(t,s)f \in \mathcal{D}(D)$ *for every* $(t,s) \in \mathbb{R}^2$ *and*

(ii) *the function* $DU(.,.)f: \mathbb{R}^2 \to \mathfrak{h}$ *is continuous.*

PROOF. First of all we show that the operator XPf, $f \in \mathcal{D}_2$, is well defined. We introduce $X_\lambda = X(I + \lambda X^2)^{-1/2}$, $\lambda > 0$. A straightforward calculation gives

$$(2.4) \qquad \|X_\lambda Pf\|^2 \leq \frac{n+1}{2} \|f\|_1^2 + \frac{1}{2} \|f\|_2^2, \quad f \in \mathcal{D}_2.$$

Tending $\lambda \to +\infty$ (2.4) yields $\mathcal{D}(XP) = \{f \in \mathcal{D}(|P|): P_i f \in \mathcal{D}(X_i), i = 1,2,\ldots,n\} \supseteq \mathcal{D}_2$. Moreover, we get

$$(2.5) \qquad \|XPf\|^2 \leq \frac{n+1}{2} \|f\|_1^2 + \frac{1}{2} \|f\|_2^2, \quad f \in \mathcal{D}_2.$$

Taking into account the Fourier transform we obtain $\mathcal{D}(PX) = \{f \in \mathcal{D}(|X|): X_i f \in \mathcal{D}(P_i), i = 1,2,\ldots,n\} \supseteq \mathcal{D}_2$ and

$$(2.6) \qquad \|PXf\|^2 \leq \frac{n+1}{2} \|f\|_1^2 + \frac{1}{2} \|f\|_2^2, \quad f \in \mathcal{D}_2.$$

Thus we can introduce the symmetric operator D_0,

$$(2.7) \qquad D_0 f = \frac{1}{2}(XP + PX)f, \quad f \in \mathcal{D}(D_0) = \mathcal{D}_2.$$

Since $D_0 \supseteq \tilde{D}$ we have $D = (\tilde{D})^- = (D_0)^-$. Therefore, $\mathcal{D}(D) \supseteq \mathcal{D}(D_0) = \mathcal{D}_2$. But $\mathcal{D}_2$ is invariant by Proposition 2.2(iii). Hence (i) holds. Since $\|f\|_1^2 \leq 2 \|f\|_2^2$, $f \in \mathcal{D}_2$, we obtain

$$(2.8) \qquad \|Df\|^2 \leq \frac{1}{2}\{\|XPf\|^2 + \|PXf\|^2\} \leq (n + \frac{3}{2}) \|f\|_2^2, \quad f \in \mathcal{D}_2,$$

where we have used (2.5) and (2.6). But (2.8) immediately yields that D can be regarded as a bounded operator acting from $\mathfrak{h}_2$ into $\mathfrak{h}$. On the other hand, by Proposition 2.2(iii) $\{U(t,s)|\mathcal{D}_2\}_{(t,s)\in\mathbb{R}^2}$ is strongly continuous in the $\|.\|_2$-norm. Both results immediately imply (ii). ∎

Furthermore, in the following we need a modified definition of the scattering subspace.

LEMMA 2.4. *If* $q_j \in C^1_{loc}(\mathbb{R}^{n+1})$, $j = 1,2,\ldots,N$, *and if the Assumption T is satisfied, then* $f \in \mathfrak{h}^{sc}_{\pm}(s)$, $s \in \mathbb{R}^1$, *if and only if*

$$(2.9) \qquad \lim_{T \to \pm\infty} \frac{1}{T} \int_0^T dt \, \|F(x \in \mathcal{B}_R) e^{it\dot{x}_j(t)P} U(t,s) f\|^2 = 0,$$

$j = 1,2,\ldots,N$, *for every* $R > 0$.

PROOF. $f \in \mathfrak{h}^{sc}_{\pm}(s) \Rightarrow (2.9)$. Obviously, we have

$$\|F(x \in \mathcal{B}_R) e^{i(t\dot{x}_j(t) - x_j(t))P} e^{ix_j(t)P} U(t,s) f\|^2 =$$

$$(2.10)$$

$$\|F(x - (t\dot{x}_j(t) - x_j(t)) \in \mathcal{B}_R) e^{itx_j(t)P} U(t,s) f\|^2,$$

$j = 1,2,\ldots,N$, $R > 0$. But using Assumption T we get $\sup_{t\in\mathbb{R}^1} |x_j(t) - t\dot{x}_j(t)| = c_j < +\infty$, $j = 1,2,\ldots,N$. Thus, one has $\{x \in \mathbb{R}^n : x - (t\dot{x}_j(t) - x_j(t)) \in \mathcal{B}_R\} \subseteq \mathcal{B}_{R+c_j}$, $j = 1,2,\ldots,N$. Hence we find

$$\|F(x \in \mathcal{B}_R) e^{itx_j(t)P} U(t,s) f\|^2 \leq$$

$$(2.11)$$

$$\|F(x \in \mathcal{B}_{R+c_j}) e^{ix_j(t)P} U(t,s) f\|^2,$$

$j = 1,2,\ldots,N$, which immediately proves (2.9).

$(2.9) \Rightarrow f \in \mathfrak{h}^{sc}_{\pm}(s)$. The proof of this direction is analogous to the previous one. Thus, we omit this proof. $\blacksquare$

3. PROPAGATING PROPERTIES

Now we are going to generalize Theorem 2.1 of Enss [2] to moving and time-dependent potentials. To this end we need some precise understanding of the well-known commutation relations $H_o D - D H_o = -2iH_o$ and $H_o X^2 - X^2 H_o = -2iD$ which will be given in the following lemma. We left the proof of the lemma to the reader.

LEMMA 3.1. (i) *If* $f,g \in \mathcal{D}(H_o) \cap \mathcal{D}(D)$, *then*

$$(3.1) \qquad (Df, H_o g) - (H_o f, Dg) = -2i(H_o f, g).$$

(ii) *If* $f, g \in \mathcal{D}(H_o) \cap \mathcal{D}(X^2)$, *then*

$$(3.2) \qquad (X^2 f, H_o g) - (H_o f, X^2 g) = -2i(Df, g).$$

Notice that by Corollary 2.3(i) we have $\mathcal{D}_2 = \mathcal{D}(H_o) \cap \mathcal{D}(X^2) \subseteq \mathcal{D}(D)$.

Now we intend to generalize Theorem 2.1 of [2]. To his end let us introduce the family $\{\tilde{A}(t)\}_{t \in \mathbb{R}^1}$ of linear operators on $\mathfrak{h}$ given by

$$
\tilde{A}(t)f = U(t)^* D U(t) f - Df - 2\int_0^t ds\, U(s)^* H_o U(s) f +
$$

$$
(3.3) \qquad \sum_{j=1}^N t \int_0^t dr\, U(r)^* x_j(r) \nabla V_j(r) U(r) f -
$$

$$
\sum_{j=1}^N \int_0^t ds \int_0^s dr\, U(r)^* x_j(r) \nabla V_j(r) U(r) f,
$$

$f \in \mathcal{D}(\tilde{A}(t)) = \mathcal{D}_2$, $t \in \mathbb{R}^1$. We note that by Assumptions P and T $\{\nabla V_j(t)\}_{t \in \mathbb{R}^1}$, $j = 1, 2, \ldots, N$, are families of bounded operators which are strongly continuous and uniformly bounded in $t \in \mathbb{R}^1$. Thus, the last integrals of (3.3) are well-defined operators for every $t \in \mathbb{R}^1$. The integral $\int_0^t ds\, U(s)^* H_o U(s) f$, $f \in \mathcal{D}_2$, makes sense by Proposition 2.1(iii). The expression $U(t)^* D U(t) f$, $f \in \mathcal{D}_2$, is well-defined by Proposition 2.2(iii). Hence, the family $\{\tilde{A}(t)\}_{t \in \mathbb{R}^1}$ is meaningful.

PROPOSITION 3.2. *If the Assumptions P and T are satisfied, then the family* $\{\tilde{A}(t)\}_{t\in\mathbb{R}^1}$ *can uniquely be extended to a strongly continuous family* $\{A(t)\}_{t\in\mathbb{R}^1}$ *of bounded operators such that*

(i) $\{\frac{1}{t}A(t)\}_{t\in\mathbb{R}^1}$ *is uniformly bounded in* $t\in\mathbb{R}^1$ *and*

(ii) *for every* $f\in\mathfrak{h}_{\pm}^{sc}$ *we have*

$$\lim_{t\to\pm\infty}\frac{1}{t}A(t)f = 0.$$

PROOF. By Proposition 2.1(iii) and Proposition 2.2(iii) the function $(DU(.)f,U(.)g):\ \mathbb{R}^1\ \to\ \mathbb{C}^1,\ \ f,g\ \in\ \mathcal{D}_2,$ is differentiable at every point $t\in\mathbb{R}^1$. We get

$$\frac{d}{dt}(DU(t)f,U(t)g) = i\{(DU(t)f,H_0U(t)g) -$$

(3.4)

$$(H_0U(t)f,DU(t)g)) + (i[V(t),D]U(t)f,U(t)g),$$

$f,g\in\mathcal{D}_2$. A straightforward calculation yields

$$(3.5)\qquad i[V(t),D] = -\sum_{j=1}^{N}(x - t\dot{x}_j(t))\nabla V_j(t) - \sum_{j=1}^{N}t\dot{x}_j(t)\nabla V_j(t),$$

$t\in\mathbb{R}^1$. Thus, applying Lemma 3.1 we have

$$\frac{d}{dt}(DU(t)f,U(t)g) = 2(H_0U(t)f,U(t)g) -$$

$$(3.6)\qquad \sum_{j=1}^{N}((x - t\dot{x}_j(t))\nabla V_j(t)U(t)f,U(t)g) -$$

$$\sum_{j=1}^{N}t(\dot{x}_j(t)\nabla V_j(t)U(t)f,U(t)g),$$

$f,g\in\mathcal{D}_2$. Taking into account Lemma III.1.35 of [3] it is not hard to see that (3.6) yields the representation

$$(3.7)\qquad U(t)^*DU(t)f - Df - 2\int_0^t ds\, U(s)^*H_0U(s)f +$$

$$\sum_{=1}^{N} \int_0^t ds \; sU(s)^* \dot{x}_j(s) \nabla V_j(s) U(s) f =$$

$$- \sum_{j=1}^{N} \int_0^t ds \; U(s)^* (x - s\dot{x}_j(s)) \nabla V_j(s) U(s) f,$$

$f \in \mathcal{D}_2$. Integrating by parts $\int_0^t ds \; sU(s)^* \dot{x}_j(s) \nabla V_j(s) U(s) f$ we get

$$(3.8) \qquad \tilde{A}(t) f = - \sum_{j=1}^{N} \int_0^t ds \; U(s)^* (x - s\dot{x}(s)) \nabla V_j(s) U(s) f,$$

$f \in \mathcal{D}_2$. Since the right hand side of (3.8) defines a bounded operator for every $t \in \mathbb{R}^1$, the family $\{\tilde{A}(t)\}_{t \in \mathbb{R}^1}$ can uniquely be extended to a family $\{A(t)\}_{t \in \mathbb{R}^1}$ of bounded operators. Since $\{\int_0^t ds U(s)^* (x - s\dot{x}_j(s)) \nabla V_j(s) U(s)\}_{t \in \mathbb{R}^1}$, $j = 1,2,\ldots,N$, are strongly continuous families in $t \in \mathbb{R}^1$, the family $\{A(t)\}_{t \in \mathbb{R}^1}$ has this property, too. Moreover, denoting by $\{\hat{V}_j(t)\}_{t \in \mathbb{R}^1}$ the family of multiplication operators induced by $q_j(t, x + t\dot{x}_j(t) - x_j(t))$, $j = 1,2,\ldots,N$, it is not hard to see that the representation

$$U(s)^* (x - s\dot{x}_j(s)) \nabla V_j(s) U(s) =$$

$$(3.9)$$

$$U(s)^* e^{-is\dot{x}_j(s)P} x\nabla\hat{V}_j(s) e^{is\dot{x}_j(s)P} U(s),$$

$j = 1,2,\ldots,N$, takes place. Thus, by Assumptions P and T $\{\frac{1}{t} \int_0^t ds \; U(s)^* (x - s\dot{x}_j(s)) \nabla V_j(s) U(s)\}_{t \in \mathbb{R}^1}$ are uniformly bounded families in $t \in \mathbb{R}^1$. Consequently, $\{\frac{1}{t} A(t)\}_{t \in \mathbb{R}^1}$ is also a uniformly bounded family in $t \in \mathbb{R}^1$ which proves (i).

In order to establish (ii) we note that by (1.5) and Lemma 2.4 we have

$$(3.10) \qquad \lim_{t \to \pm\infty} \frac{1}{t} \int_0^t ds \; \|x\nabla\hat{V}_j(s) e^{is\dot{x}_j(s)P} U(s) f\|^2 = 0,$$

$f \in \mathfrak{h}_{\pm}^{sc}$, $j = 1,2,\ldots,N$. Applying (3.9) we get

$$(3.11) \qquad \lim_{t\to\pm\infty} \frac{1}{t} \int_0^t ds \, \|(x - s\dot{x}_j(s))\nabla V_j(s)U(s)f\|^2 = 0,$$

$f \in \mathfrak{h}_{\pm}^{sc}$, $j = 1,2,\ldots,N$. But (3.8) and (3.11) immediately yield (ii). $\blacksquare$

Further, let us introduce the family $\{\tilde{B}(t)\}_{t\in\mathbb{R}^1}$ of linear operators on $\mathfrak{h}$ given by

$$\tilde{B}(t)f = U(t)^* \frac{x^2}{2} U(t)f - \frac{x^2}{2} f - tDf -$$

$$(3.12) \qquad 2 \int_0^t ds \int_0^s dr \, U(r)^* H_o U(r)f +$$

$$\sum_{j=1}^N t \int_0^t ds \int_0^s dr \, U(r)^* \dot{x}_j(r)\nabla V_j(r)U(r)f -$$

$$2 \sum_{j=1}^N \int_0^t ds \int_0^s dr \int_0^r dw \, U(w)^* \dot{x}_j(w)\nabla V_j(w)U(w)f,$$

$f \in \mathcal{D}(\tilde{B}(t)) = \mathcal{D}_2$. Taking into account Proposition 2.1(iii) and Proposition 2.2(iii) as well as the Assumptions P and T it can be shown that the family $\{\tilde{B}(t)\}_{t\in\mathbb{R}^1}$ is well defined.

PROPOSITION 3.3. *If the Assumptions P and T are satisfied, then the family* $\{\tilde{B}(t)\}_{t\in\mathbb{R}^1}$ *can uniquely be extended to a strongly continuous family* $\{B(t)\}_{t\in\mathbb{R}^1}$ *of bounded operators such that*

(i) $\{\frac{1}{t^2} B(t)\}_{t\in\mathbb{R}^1}$ *is uniformly bounded in* $t \in \mathbb{R}^1$ *and*

(ii) *for every* $f \in \mathfrak{h}_{\pm}^{sc}$ *we have*

$$\lim_{t\to\pm\infty} \frac{1}{t^2} B(t)f = 0.$$

PROOF. Analogously to the previous proposition, we find that the function $(U(.)^* \frac{x^2}{2} U(.)f,g): \mathbb{R}^1 \to \mathbb{C}$, $f,g, \in \mathcal{D}_2$, is differentiable at every point $t \in \mathbb{R}^1$. Therefore, we have

$$\frac{d}{dt}\,(U(t)^{*}\,\frac{x^{2}}{2}\,U(t)f,g) =$$

(3.13)

$$i\{(\frac{x^{2}}{2}\,U(t)f,H_{o}U(t)g) - (H_{o}U(t)f,\frac{x^{2}}{2}\,U(t)g)\},$$

$f,g \in \mathcal{D}_{2}$. By Proposition 2.2(iii), Corollary 2.3 and Lemma 3.1 we obtain

(3.14) $$\frac{d}{dt}\,(U(t)^{*}\,\frac{x^{2}}{2}\,U(t)f,g) = (DU(t)f,U(t)g),$$

$f,g \in \mathcal{D}_{2}$. Finally, using (3.7), where we have integrated by parts, one gets

$$\frac{d}{dt}(U(t)^{*}\,\frac{x^{2}}{2}\,U(t)f,g) = (Df,g) + (2\int_{0}^{t}ds\,U(s)^{*}H_{o}U(s)f,g) -$$

$$\sum_{j=1}^{N}t(\int_{0}^{t}ds\,U(s)^{*}\dot{x}_{j}(s)\nabla V_{j}(s)U(s)f,g) +$$

(3.15)

$$\sum_{j=1}^{N}(\int_{0}^{t}ds\int_{0}^{r}dr\,U(r)^{*}\dot{x}_{j}(r)\nabla V_{j}(r)U(r)f,g) -$$

$$\sum_{j=1}^{N}(\int_{0}^{t}ds\,U(s)^{*}(x - s\dot{x}_{j}(s))\nabla V_{j}(s)U(s)f,g),$$

$f,g \in \mathcal{D}_{2}$. Using previous considerations we show that (3.15) yields the representation

(3.16) $$\tilde{B}(t)f = -\sum_{j=1}^{N}\int_{0}^{t}ds\int_{0}^{s}dr\,U(r)^{*}(x - r\dot{x}_{j}(r))\nabla V_{j}(r)U(r)f,$$

$f \in \mathcal{D}_{2}$. Hence, by (3.8) we find the relation

(3.17) $$\tilde{B}(t)f = \int_{0}^{t}ds\,A(s)f, \quad f \in \mathcal{D}_{2},$$

which easily yields the proof of all assertions. ∎

Furthermore, we need the family $\{\tilde{C}(t)\}_{t\in\mathbb{R}^{1}}$ of linear operators on $\mathfrak{h}$ given by

$$\tilde{C}(t)f = U(t)^* H_o U(t)f - H(0)f +$$

$$(3.18)$$
$$\sum_{j=1}^{N} \int_0^t ds\ U(s)^* \dot{x}_j(s) \nabla V_j(s) U(s)f, \quad f \in \mathcal{D}(\tilde{C}(t)) = \mathcal{D}_2.$$

On account of Proposition 2.1(iii) and Assumptions P and T the family $\{\tilde{C}(t)\}_{t\in\mathbb{R}^1}$ is well defined.

Let $\{\hat{V}_j(t)\}_{t\in\mathbb{R}^1}$, $j = 1,2,\ldots,N$, be the family of multiplication operators on $\mathfrak{h}$ induced by $\dot{q}_j(t,x - x_j(t))$. By Assumptions P and T the family consists of bounded operators and is strongly continuous and uniformly bounded in $t \in \mathbb{R}^1$. Obviously, we have

$$(3.19) \qquad \|\hat{V}_j(t)\| = \sup_{x\in\mathbb{R}^n} |\dot{q}_j(t,x)|.$$

But (3.19) and (1.6) yield the property $\|\hat{V}_j(.)\| \in L^1(\mathbb{R}^1)$, $j = 1,2,\ldots,N$. Thus, we can introduce the bounded operators

$$(3.20) \qquad R_j^{\pm} = \text{s-}\lim_{t\to\pm\infty} \int_0^t ds\ U(s)^* \hat{V}_j(s) U(s), \quad j = 1,2,\ldots,N.$$

Furthermore, it is useful for us to introduce the so-called strong convergence in the Cesaro mean. Let $\{Y(t)\}_{t\in\mathbb{R}^1}$ be a strongly measurable family of bounded operators on $\mathfrak{h}$ which is uniformly bounded in $t \in \mathbb{R}^1$. We say $\{Y(t)\}_{t\in\mathbb{R}^1}$ strongly converges in the Cesaro mean to Y_+, in short $|A|\text{-}\lim_{t\to+\infty} Y(t) = Y_+$, if for every $f \in \mathfrak{h}$ we have

$$(3.21) \qquad \lim_{T\to+\infty} \frac{1}{T} \int_0^T dt\ \|(Y(t) - Y_+)f\|^2 = 0.$$

Similarly we define $|A|\text{-}\lim_{t\to-\infty} Y(t) = Y_-$. We note that $\text{s-}\lim_{t\to\pm\infty} Y(t) = Y_\pm$ implies $|A|\text{-}\lim_{t\to\pm\infty} Y(t) = Y_\pm$ while the converse is not true in general. Of course, this yields

$$(3.22) \qquad R_j^{\pm} = |A|\text{-}\lim_{t\to\pm\infty} \int_0^t ds\ U(s)^* \hat{V}_j(s)U(s), \qquad j = 1,2,\ldots,N.$$

PROPOSITION 3.4. *If the Assumptions P and T are satisfied, then the family* $\{\tilde{C}(t)\}_{t\in\mathbb{R}^1}$ *can uniquely be extended to a strongly continuous family* $\{C(t)\}_{t\in\mathbb{R}^1}$ *of bounded operators such that* $\{C(t)\}_{t\in\mathbb{R}^1}$ *is uniformly bounded in* $t \in \mathbb{R}^1$ *and*

$$(3.23) \qquad |A|\text{-}\lim_{t\to\pm\infty} C(t)P_\pm^{sc} = \sum_{j=1}^N R_j^{\pm}\ P_\pm^{sc}.$$

PROOF. By Proposition 2.1(iii) the function $(H(.)U(.)f,U(.)g)\colon \mathbb{R}^1 \to \mathbb{C}$, $f,g \in \mathcal{D}_2$, is differentiable at every point $t \in \mathbb{R}^1$. We obtain

$$\frac{d}{dt}(H(t)U(t)f,U(t)g) = -\sum_{j=1}^N (\dot{x}_j(t)\nabla V_j(t)U(t)f,U(t)g) +$$

$$(3.24)$$

$$\sum_{=1}^N (\hat{V}_j(t)U(t)f,U(t)g),$$

$f,g \in \mathcal{D}_2$. Hence we get

$$(3.25) \qquad \tilde{C}(t)f = -\sum_{j=1}^N U(t)^* V_j(t)U(t)f + \sum_{j=1}^N \int_0^t ds\ U(s)^* \hat{V}_j(s)U(s)f,$$

$f \in \mathcal{D}_2$. By Assumptions P and T the right hand side defines a family of bounded operators which is strongly continuous and uniformly bounded in $t \in \mathbb{R}^1$. Hence, $\{\tilde{C}(t)\}_{t\in\mathbb{R}^1}$ can be uniquely extended to a family $\{C(t)\}_{t\in\mathbb{R}^1}$ of bounded operators obeying the same properties. Since

$$(3.26) \qquad U(s)^* V_j(s)U(s) = U(s)^* e^{-is\hat{x}_j(s)P}\ \hat{V}_j(s)e^{is\hat{x}_j(s)P}\ U(s),$$

$s \in \mathbb{R}^1$, $j = 1,2,\ldots,N$, assumption (1.4) and Lemma 2.4 imply

$$(3.27) \qquad \lim_{t\to\pm\infty} \frac{1}{t} \int_0^t ds\ \|\hat{V}_j(s)\ e^{is\hat{x}_j(s)P}\ U(s)f\|^2 = 0,$$

$f \in \mathfrak{h}_\pm^{sc}$, $j = 1,2,\ldots,N$. Taking into account (3.26) we get

$$(3.28) \qquad \lim_{t \to \pm\infty} \frac{1}{t} \int_0^t ds \, \|V_j(s)U(s)f\|^2 = 0,$$

$f \in \mathfrak{h}_\pm^{sc}$, $j = 1,2,\ldots,N$. But (3.22), (3.25) and (3.28) imply (3.23). ∎

4. MAIN THEOREM

Now we generalize Theorem 2.4 of [2] to our case of moving and time-dependent potentials. To this end, we introduce the operators K_i,

$$(4.1) \qquad \tilde{K}_i f = P_i f - \frac{X_i}{t} f, \quad f \in \mathcal{D}(\tilde{K}_i) = \mathscr{S}(\mathbb{R}^n), \quad i = 1,2,\ldots,n.$$

Since $e^{i\frac{X_i^2}{2t}} \mathscr{S}(\mathbb{R}^n) = \mathscr{S}(\mathbb{R}^n)$, $t \in \mathbb{R}^1$, we find

$$(4.2) \qquad e^{-i\frac{X_i^2}{2t}} \tilde{K}_i \, e^{i\frac{X_i^2}{2t}} f = P_i f, \quad f \in \mathscr{S}(\mathbb{R}^n).$$

But $\mathscr{S}(\mathbb{R}^n)$ is a core of P_i, $i = 1,2,\ldots,n$. Hence, the closure K_i of $\tilde{K}_i$ is a self-adjoint operator which is unitarily equivalent to P_i, $i = 1,2,\ldots,n$. Moreover, we find $\mathcal{D}(K_i^2) \supseteq \mathscr{S}(\mathbb{R}^n)$, $K_i^2 f = \tilde{K}_i^2 f$ and

$$(4.3) \qquad e^{-i\frac{X_i^2}{2t}} K_i^2 \, e^{i\frac{X_i^2}{2t}} f = P_i^2 f, \quad t \in \mathbb{R}^1,$$

$f \in \mathscr{S}(\mathbb{R}^n)$, $i = 1,2,\ldots,n$. Setting $\tilde{L}f = \frac{1}{2} \sum_{i=1}^{N} K_i^2 f$, $f \in \mathscr{S}(\mathbb{R}^n)$, and taking into account $e^{i\frac{X_i^2}{2t}} \mathscr{S}(\mathbb{R}^n) = \mathscr{S}(\mathbb{R}^n)$, $t \in \mathbb{R}^1$, we obtain

$$(4.4) \qquad e^{-i\frac{X_i^2}{2t}} \tilde{L} \, e^{i\frac{X_i^2}{2t}} f = H_o f, \quad t \in \mathbb{R}^1, \quad f \in \mathscr{S}(\mathbb{R}^n).$$

Since $\mathscr{S}(\mathbb{R}^n)$ is a core of H_o the closure L of $\tilde{L}$ is a self-adjoint operator.

LEMMA 4.1. *The operator L obeys the following properties.*

(i) $\mathcal{D}(L) \supseteq \mathcal{D}_2$ *and* $\mathcal{D}_2$ *is a core of L.*

(ii) *For* $f \in \mathcal{D}_2$ *we have*

$$(4.5) \qquad Lf = \frac{X^2}{2t^2} f - \frac{D}{t} f + H_o f.$$

PROOF. If $f \in \mathcal{S}(\mathbb{R}^n)$, then we have

$$(4.6) \qquad \tilde{L} f = \frac{X^2}{2t^2} f - \frac{D}{t} f + H_o f.$$

Since by Corollary 2.3 the relation $\mathcal{D}_2 \subseteq \mathcal{D}(H_o^{\frac{1}{2}}) \cap \mathcal{D}(D) \cap \mathcal{D}(X^2)$ holds, we can introduce the symmetric operator L_o,

$$(4.7) \qquad \tilde{L}_o f = \frac{X^2}{2t} f - \frac{D}{t} f + H_o f, \quad f \in \mathcal{D}(L_o) = \mathcal{D}_2.$$

Therefore $L_o \supseteq \tilde{L}$ and we obtain $(L_o)^- = L$. But $L = (L_o)^-$ yields (i) and (ii).∎

LEMMA 4.2. *If the Assumptions P and T are satisfied, then*

$$(4.8) \qquad \begin{aligned} U(t)^* L U(t) &= \frac{X^2}{2t^2} + \frac{1}{t^2} B(t) - \frac{1}{t} A(t) + C(t) - \\ &\quad \frac{2}{t} \int_0^t ds\, C(s) + \frac{2}{t^2} \int_0^t ds \int_0^s dr\, C(r), \end{aligned}$$

$t \in \mathbb{R}^1$.

PROOF. Let $f \in \mathcal{D}_2$. By Proposition 2.2(iii) and Lemma 4.1(ii) we have

$$(4.9) \qquad \begin{aligned} U(t)^* LU(t)f &= U(t)^* \frac{X^2}{2t^2} U(t)f - U(t)^* \frac{D}{t} U(t)f + \\ &\quad U(t)^* H_o U(t)f. \end{aligned}$$

Taking into account Proposition 3.2, Proposition 3.3 and Proposition 3.4 a straightforward computation shows

$$U(t)^* \ L \ U(t)f = \frac{X^2}{2t^2} \ f + \frac{1}{t^2} \ B(t)f - \frac{1}{t} \ A(t)f + C(t)f -$$

(4.10)
$$\frac{2}{t} \int_0^t ds \ C(s)f + \frac{2}{t^2} \int_0^t ds \int_0^s dr \ C(r)f,$$

$f \in \mathbb{R}^1$. By Proposition 2.2(iii) and Lemma 4.1(i) we obtain that

(4.11) $(U(t)^* \ L \ U(t)|\mathcal{D}_2)^- = U(t)^* \ L \ U(t), \quad t \in \mathbb{R}^1.$

Moreover, since $\mathcal{D}_2$ is a core of X^2 the relation (4.10) immediately implies (4.8).∎

BY $C(\mathbb{R}^n)$ we denote the set of all bounded continuous functions defined on $\mathbb{R}^n$ while $C_\infty(\mathbb{R}^n)$ indicates the subset given by $C_\infty(\mathbb{R}^n) = \{\varphi \in C(\mathbb{R}^n): \lim_{|x| \to \infty} \varphi(x) = 0\}$.

LEMMA 4.3. *If the Assumptions P and T are satisfied, then for every* $\psi \in C(\mathbb{R}^1)$ *we have*

(4.12) $|A|-\lim_{t \to \pm\infty} U(t)^* \ \psi(L) \ U(t) \ P_\pm^{sc} = \psi(0)P_\pm^{sc}.$

PROOF. Let us introduce the family $\{\Gamma(t)\}_{t \in \mathbb{R}^1}$ of bounded self-adjoint operators defined by

$$\Gamma(t) = \frac{1}{t^2} \ B(t) - \frac{1}{t} \ A(t) + C(t) -$$

(4.13)
$$\frac{2}{t} \int_0^t ds \ C(s) + \frac{2}{t^2} \int_0^t ds \int_0^s dr \ C(r).$$

On account of Propositions 3.2 – 3.4 we find that $\{\Gamma(t)\}_{t \in \mathbb{R}^1}$ is a strongly continuous and uniformly bounded family in $t \in \mathbb{R}^1$ and obeys

(4.14) $|A|-\lim_{t \to \pm\infty} \Gamma(t)P_\pm^{sc} = 0.$

Since

$$(4.15) \qquad \sup_{t \in \mathbb{R}^1} \| (\frac{X^2}{2t^2} - z)^{-1} \, \Gamma(t) \| \leq \frac{1}{\text{Im}(z)} \, \sup_{t \in \mathbb{R}^1} \| \Gamma(t) \|$$

there is a z_o such that

$$(4.16) \qquad \sup_{t \in \mathbb{R}^1} \| (\frac{X^2}{2t^2} - z_o)^{-1} \, \Gamma(t) \| < 1.$$

On account of Lemma 4.2 one gets

$$U(t)^* \, (L - z_o)^{-1} \, U(t) = (\frac{X^2}{2t^2} + \Gamma(t) - z_o)^{-1} =$$

$$(4.17)$$

$$(I + (\frac{X^2}{2t^2} - z_o)^{-1} \Gamma(t))^{-1} \, (\frac{X^2}{2t^2} - z_o)^{-1},$$

$t \in \mathbb{R}^1$. Obviously, we have

$$(4.18) \qquad \text{s-}\lim_{t \to \pm\infty} (\frac{X^2}{2t^2} - z_o)^{-1} = - \frac{1}{z_o} \, I.$$

By (4.16) $\{ (I + (\frac{X^2}{2t^2} - z_o)^{-1} \Gamma(t))^{-1} \}_{t \in \mathbb{R}^1}$ is strongly continuous
and uniformly bounded in $t \in \mathbb{R}^1$. A straightforward calculation
shows that

$$(4.19) \qquad |A|\text{-}\lim_{t \to \pm\infty} (I + (\frac{X^2}{2t^2} - z_o)^{-1} \Gamma(t))^{-1} \, P_{\pm}^{sc} = P_{\pm}^{sc}.$$

Thus we find

$$(4.20) \qquad |A|\text{-}\lim_{t \to \pm\infty} U(t)^* \, (L - z_o)^{-1} \, U(t) \, P_{\pm}^{sc} = - \frac{1}{z_o} \, P_{\pm}^{sc}.$$

Similarly we prove

$$(4.21) \qquad |A|\text{-}\lim_{t \to \pm\infty} U(t)^* \, (L - \bar{z}_o)^{-1} \, U(t) \, P_{\pm}^{sc} = - \frac{1}{\bar{z}_o} \, P_{\pm}^{sc}.$$

Taking into account (4.20) and (4.21), it is not hard to see that
for every polynomial $P(x,y)$ the relation

$$(4.22) \qquad |A|\text{-}\lim_{t \to \pm\infty} U(t)^* \, P((L - z_o)^{-1}, (L - \bar{z}_o)^{-1}) \, U(t) \, P_{\pm}^{sc} =$$

$$P(-\frac{1}{z_o}, -\frac{1}{\overline{z}_o}) P_{\pm}^{sc}$$

holds. But by the Stone-Weierstrass Theorem [8] the set of polynomials $P(x,y)$ is dense in $C_{\infty}(\mathbb{R}^1)$. Hence we obtain

$$(4.23) \qquad |A|-\lim_{t\to\pm\infty} U(t)^* \varphi(L) P_{\pm}^{sc} = \varphi(0) P_{\pm}^{sc}$$

for every $\varphi \in C_{\infty}(\mathbb{R}^1)$.

Now let $\psi \in C(\mathbb{R}^1)$. Obviously, we have $\varphi(\lambda) = \psi(\lambda)(\lambda-z_o)^{-1} \in C_{\infty}(\mathbb{R}^1)$. Therefore, we get

$$|A|-\lim_{t\to\pm\infty} U(t)^* \psi(L) (-\frac{1}{z_o}) U(t) P_{\pm}^{sc} =$$

$$(4.24) \qquad |A|-\lim_{t\to\pm\infty} U(t)^* \psi(L) \{-\frac{1}{z_o} - (L - z_o)^{-1}\} U(t) P_{\pm}^{sc} +$$

$$|A|-\lim_{t\to\pm\infty} U(t)^* \varphi(L) U(t) P_{\pm}^{sc}.$$

Since the first summand of the right hand side tends to zero and the second summand can be calculated by (4.23) we find

$$(4.25) \qquad (-\frac{1}{z_o}) |A|-\lim_{t\to\pm\infty} U(t)^* \psi(L) U(t) P_{\pm}^{sc} = (-\frac{1}{z_o}) \psi(0) P_{\pm}^{sc}$$

which yields (4.17).∎

Now we prove our main

THEOREM 4.4. *Let φ be the Fourier transform of an integrable function $\hat{\varphi}$ defined on $\mathbb{R}^n$, i.e. $\hat{\varphi} \in L^1(\mathbb{R}^n)$. If the Assumptions P and T are satisfied, then*

$$(4.26) \qquad \lim_{T\to\pm\infty} \frac{1}{T} \int_0^T dt \; \|\{\varphi(\frac{X}{t}) - \varphi(P)\}U(t) f\|^2 = 0$$

for every $f \in \mathfrak{h}_{\pm}^{sc}$. The assertion remains true if $1 - \varphi$ possesses a summable Fourier transform.

PROOF. We have

$$\varphi\left(\frac{X}{t}\right) - \varphi(P) = \frac{1}{\sqrt{2\pi}} \int_{\mathbb{R}^n} dy\ \hat{\varphi}(y) \left\{ e^{iy\frac{X}{t}} - e^{iyP} \right\} =$$

(4.27)
$$\frac{1}{\sqrt{2\pi}} \int_{\mathbb{R}^n} dy\ \hat{\varphi}(y)\ e^{iyP} \left\{ e^{-iyP}\ e^{iy\frac{X}{t}} - 1 \right\} =$$

$$\frac{1}{\sqrt{2\pi}} \int_{\mathbb{R}^n} dy\ \hat{\varphi}(y)\ e^{iyP} \left\{ e^{iyK}\ e^{-i\frac{y^2}{2t}} - 1 \right\}$$

where $K = \{K_1, K_2, \ldots, K_n\}$ and where we have applied the Baker-Campbell-Hausdorff formula. By the dominated convergence theorem it is sufficient to show that

(4.28)
$$\lim_{T\to\pm\infty} \frac{1}{T} \int_0^T dt\ \|(e^{iyK} - I)U(t)f\|^2 = 0$$

for every $f \in \mathfrak{h}_\pm^{sc}$ and $y \in \mathbb{R}^n$. We note that

(4.29)
$$\|(e^{iyK} - I)U(t)f\|^2 = 4\|\,|\sin(\tfrac{1}{2}|yK|)|\,U(t)f\|^2,$$

$f \in \mathfrak{h}$, $t \in \mathbb{R}^1$. Let us introduce the function ψ_o,

(4.30)
$$\psi_o(\lambda) = \begin{cases} 0 & \lambda \leq 0 \\ \sin(\lambda) & 0 < \lambda \leq \pi/2 \\ 1 & \lambda > \pi/2 \end{cases}$$

which belongs to $C(\mathbb{R}^1)$. Since $|\sin(\lambda)| \leq \psi_o(\lambda)$, $\lambda \geq 0$, we have $|\sin(\tfrac{1}{2}|yK|)| \leq \psi_o(\tfrac{1}{2}|yK|)$. Moreover, on account of $|yK| \leq |y||K|$, $y \in \mathbb{R}^n$, we find

(4.31)
$$|\sin(\tfrac{1}{2}|yK|)| \leq \psi_o(\tfrac{1}{2}\,|y||K|), \quad y \in \mathbb{R}^n.$$

But (4.31) yields

(4.32)
$$\|(e^{iyK} - I)U(t)f\|^2 \leq 4\|\psi_o(\tfrac{1}{2}|y||K|)U(t)f\|^2,$$

$y \in \mathbb{R}^n$, $f \in \mathfrak{h}$. Setting $\psi(\lambda) = \psi_0(\sqrt{\lambda/2})$, $\lambda \geq 0$, we have

$$(4.33) \qquad \|(e^{iyK} - I)U(t)f\|^2 \leq 4\|\psi(|y|^2 L)U(t)f\|^2, \qquad y \in \mathbb{R}^n, \ f \in \mathfrak{h}.$$

Applying (4.29) and (4.33) one gets

$$(4.34) \qquad \begin{aligned} 0 &\leq \lim_{T \to \pm\infty} \frac{1}{T} \int_0^T dt \ \|(e^{iyK} - I)U(t)f\|^2 \leq \\ &4 \lim_{T \to \pm\infty} \frac{1}{T} \int_0^T dt \ \|\psi(|y|^2 L)U(t)f\|^2 = 0 \end{aligned}$$

for every $f \in \mathfrak{h}_{\pm}^{sc}$ and every $y \in \mathbb{R}^n$. ∎

COROLLARY 4.5. *Let* φ *be the Fourier transform of an integrable function* $\hat{\varphi}$ *defined on* $\mathbb{R}^1$, *i.e.* $\hat{\varphi} \in L^1(\mathbb{R}^1)$. *If the Assumptions P and T are satisfied, then*

$$(4.35) \qquad \lim_{T \to \pm\infty} \frac{1}{T} \int_0^T dt \ \|\{\varphi(\tfrac{X_i}{t}) - \varphi(P_i)\}U(t)f^2\| = 0$$

for every $f \in \mathfrak{h}_{\pm}^{sc}$ *and* $i = 1,2,\ldots,n$.

PROOF. Acting as in the previous case we have to show that

$$(4.36) \qquad \lim_{T \to \pm\infty} \frac{1}{T} \int_0^T dt \ \|(e^{iy_i K_i} - I)U(t)f\|^2 = 0$$

for every $f \in \mathfrak{h}_{\pm}^{sc}$, $y_i \in \mathbb{R}^1$ and K_i, $i = 1,2,\ldots,n$. But (4.34) holds for every $y \in \mathbb{R}^n$. Consequently, choosing $y = \{0,\ldots,0,y_i,0,\ldots 0\}$ it is immediately to see that (4.34) yields (4.36). ∎

REFERENCES

[1] Amrein,W.O.; Georgescu,V.: On the characterization of bounded states and scattering states in quantum mechanics, *Helv. Phys. Acta* 46(1973), 653-657.

[2] Enss,V.: Asymptotic observables on scattering states,
      Commun. Math. Phys. 89(1983), 245-268.268.

[3] Kato, T.: *Perturbation theory for linear operators*, Springer
      -Verlag, Berlin-Heidelberg-New York, 1966.

[4] Kitada,H,; Yajima,K.: A scattering theory for time-dependent
      long-range potentials, *Duke Math. J.* 49(1982), 341-376.

[5] Kitada,H.; Yajima,K.: Remarks on our paper "A scattering
      theory for time-dependent long-range potentials", *Duke Math.
      J.* 50(1983), 1005-1016.

[6] Neidhardt,H.: Moving potentials and the completeness of wave
      operators. Part I: The propagator, in preparation.

[7] Neidhardt,H.: Moving potentials and the completeness of wave
      operators. Part III: Existence and completeness, in
      preparation.

[8] Reed,M.; Simon,B.: *Methods of modern mathematical physics.
      I. Functional analysis*, Academic Press, New York-London,
      1972.

[9] Ruelle,D.: A remark on bounded states in potential scatter-
      ing theory, *Nuovo Cimento* 59A(1969), 655-662.

[10] Yafaev,D.R.: Asymptotic completeness for the multidimensio-
      nal time-dependent Schrödinger equation, *Dokl. Akad. Nauk
      SSSR* 21(1980), 545-549.

[11] Yajima,K.: A multiple-channel scattering theory for some
      time-dependent Hamiltonians, charge transfer problems,
      Commun. Math. Phys. 75(1980), 153-178.

[12] Yajima,K.: Existence of solutions for Schrödinger evolution
      equations, *Commun. Math. Phys.* 110(1987), 415-426.

Laboratory of Theoretical Physics
Joint Institute for Nuclear Research
141 980 Dubna, U.S.S.R.

Operator Theory:
Advances and Applications, Vol. 46
© 1990 Birkhäuser Verlag Basel

REPRESENTATION OF THE THREE-BODY S-MATRIX IN TERMS
OF EFFECTIVE AMPLITUDES

Yu.A.Kuperin, Yu.B.Melnikov

In the triangle representation of three-body scattering problem in $\mathbb{R}^3$, a relation between effective amplitudes and the three-body S-matrix for $2 \to (2,3)$ processes is obtained.

The main result of this paper are new formulae relating the amplitudes in the asymptotics of solutions of the effective equation generated by the triangle representation [1-3] to the total three-body S-matrix for $2 \to (2, 3)$ processes. Here we outline the results only; the proofs will be given elsewhere.

Let us recall the main notions of the triangle representation method [1-3]. For an eigenfunction Ψ of the discrete or continuous spectrum of the self-adjoint Hamiltonian H $= - \Delta_{x_\alpha} \otimes I_{y_\alpha} + I_{x_\alpha} \otimes (- \Delta_{y_\alpha}) + \sum_\gamma v_\gamma(x_\gamma)$, written in Jacobi coordinates $\{x_\alpha, y_\alpha\}$ [4], we use the so-called *triangle representation*

$$\Psi = \sum_n \chi_n(y_\alpha)\varphi_n(x_\alpha, y_\alpha) + \int \chi(y_\alpha, q)\varphi(x_\alpha, y_\alpha, q)\, d^3q$$

over the *moving frame* formed by the discrete-spectrum eigenfunctions $\varphi_n(x_\alpha, y_\alpha)$ and continuous-spectrum eigenfunctions $\varphi(x_\alpha, y_\alpha, q)$ of the self-adjoint *frame operator* $\mathscr{L}(y_\alpha) = - \Delta_{x_\alpha} + \sum_\gamma$

$v_\gamma(x_\gamma)$. In this representation the Schroedinger equation $H\Psi = z\Psi$ generates (by projecting onto the moving frame) for the set of coefficients $\chi(y_\alpha) = \{\chi_n(y_\alpha), \chi(y_\alpha, q)\}$, $\chi_n(y_\alpha) =$

$$\int \Psi(x_\alpha, y_\alpha)\overline{\varphi_n(x_\alpha, y_\alpha)}\, d^3x_\alpha \ , \ \chi(y_\alpha, q) = \int \Psi(x_\alpha, y_\alpha)\overline{\varphi(x_\alpha, y_\alpha, q)}\, d^3x_\alpha$$

the following *effective dynamical equation* [1-3]:

$$(-[\, \nabla_{y_\alpha}\otimes \mathbb{1} + A(y_\alpha)]^2 + \Lambda(y_\alpha) - z\mathbb{1}\,)\, \chi(y_\alpha) = 0 \ , \qquad (1)$$

where $A_{mn}(y_\alpha) = \int \nabla_{y_\alpha}\varphi_n(x_\alpha, y_\alpha)\overline{\varphi_m(x_\alpha, y_\alpha)}\, d^3x_\alpha$ are the matrix elements of the *connection operator* $A(y_\alpha)$, $\mathbb{1}$ is the identity operator, and $\Lambda(y_\alpha) = \mathrm{diag}\, \{\varepsilon_n(y_\alpha), |q|^2\}$ is the diagonal operator (with $\varepsilon_n(y_\alpha)$ being the eigenvalues of the frame operator $\mathscr{L}(y_\alpha)$ and $q \in \mathbb{R}^3$ parametrizing the continuous spectrum of $\mathscr{L}(y_\alpha)$).

In order to express the three-body S-matrix in terms of the amplitudes in asymptotics of the solutions $\chi(y_\alpha)$ of the effective equation (1) at $|y_\alpha| \to \infty$, one should project the asymptotics of the total wave function Ψ (which can be found, e.g., in [4]) on the asymptotics of the moving frame $\{\varphi_n(x_\alpha, y_\alpha), \varphi(x_\alpha, y_\alpha, q)\}$ at $|y_\alpha| \to \infty$ (see [3]). We introduce the two-body operators $h_{\alpha\alpha} = -\Delta_{x_\alpha} + v_\alpha(x_\alpha)$, $h_{\alpha\beta} = -\Delta_{x_\beta} + c_{\beta\alpha}^{-2}v_\beta(x_\beta)$, $\beta \neq \alpha$, where $c_{\beta\alpha}$ are the coefficients in the Jacobi coordinate transformation [4]: $x_\beta = c_{\beta\alpha}x_\alpha + s_{\beta\alpha}y_\alpha$, $y_\beta = -s_{\beta\alpha}x_\alpha + c_{\beta\alpha}y_\alpha$ parametrized by the masses of the particles. We denote as $\sigma_d(h_{\alpha\gamma})$ the discrete spectrum of the operator $h_{\alpha\gamma}$.

DEFINITION   We say that there is *no accidental degeneracy* in the system if $c_{\beta\alpha}^{-2}\, \sigma_d(h_{\alpha\beta}) \cap c_{\gamma\alpha}^{-2}\, \sigma_d(h_{\alpha\gamma}) = \varnothing$ for every $\beta, \gamma = 1,2,3$ ($c_{\alpha\alpha} = 1$).

THEOREM 1   *If for every* $\gamma = 1,2,3$:

1) $v_\gamma(x_\gamma) \in L_2(\mathbb{R}^3)$;
2) $v_\gamma(x_\gamma)$ *is continuous for sufficiently large* $|x_\gamma|$;
3) $v_\gamma(x_\gamma)$ $|x_\gamma|^{-3-\varepsilon(\gamma)}$, $\varepsilon(\gamma) > 0$, *as* $|x_\gamma| \to \infty$;

and if there is no accidental degeneracy in the system, then for the solutions $\chi(y_\alpha)$ of the effective equation (1) the following asymptotics at $|y_\alpha| \to \infty$ are valid:

$$\chi_n(y_\alpha, p_\alpha) \simeq \delta_{nj_0} \exp\{i(p_\alpha, y_\alpha)\} + \mathcal{A}_{\alpha,n}^{(n)}(\hat{y}_\alpha, p_\alpha) \exp\{i\sqrt{z+\kappa_A^2}\,|y_\alpha|\}|y_\alpha|^{-1}$$

$$+ \sum_{B \neq A} \mathcal{A}_B^{(n)}(\hat{y}_\alpha, p_\alpha) \exp\{i\sqrt{z+\kappa_A^2}\,|y_\alpha|/|c_{\beta\alpha}|\}\,|y_\alpha|^{-1} + O(|y_\alpha|^{-2})$$

$$\chi(y_\alpha, q, p_\alpha) \simeq \mathcal{A}_0(\hat{y}_\alpha, q, p_\alpha) \exp\{i\sqrt{z-|q|^2}\,|y_\alpha|\}\,|y_\alpha|^{-1} +$$

$$+ \sum_{B \neq A}(\mathcal{A}_B(\hat{y}_\alpha, q, p_\alpha) + \mathcal{A}_B'(\hat{y}_\alpha, q, p_\alpha) \exp\{i(q, y_\alpha)s_{\beta\alpha}/c_{\beta\alpha}\})$$

$$\times \exp\{i\sqrt{z+\kappa_A^2}\,|y_\alpha|/|c_{\beta\alpha}|\}\,|y_\alpha|^{-1} + O(|y_\alpha|^{-2})\,,$$

where $\hat{y}_\alpha = y_\alpha/|y_\alpha|$, $B = \{\beta, j\}$ is the multiindex of j-th bound state in the pair β with energy $-\kappa_B^2$ and eigenfunction $\psi_B(x_\beta)$. The initial state of the system is determined by the multiindex $A_0 = \{\alpha, j_0\}$ and the momentum p_α, i.e., the bounded pair α in the j_0-th state and the third particle free with the relative momentum p_α conjugated to y_α. The effective amplitudes $\mathcal{A}_{\alpha,n}^{(n)}$, $\mathcal{A}_B^{(n)}$, $\mathcal{A}_0$, $\mathcal{A}_B$, $\mathcal{A}_B'$ can be expressed in terms of the total amplitudes in the asymptotics of Ψ.

More interesting, however, is an inverse expression of the three-body S-matrix in terms of the effective amplitudes.

THEOREM 2 *Under the assumptions of Theorem 1 the components of the three-body S-matrix for $2 \to (2, 3)$ processes are expressed as*

$$\hat{S}_{BA_0}(\hat{y}_\beta, p_\alpha, z) = 2i(2\pi)^{-5/2} \rho_{A_0}(z)\rho_B(z) \, |c_{\beta\alpha}|^{-1} \int d^3x_\beta \, \overline{\psi_B(x_\beta)}$$

$$\times \exp\{i(q,y_\alpha)s_{\beta\alpha}/c_{\beta\alpha}\} \, \{ \sum_{B \neq A} \mathcal{A}_B^{(n)}(\hat{y}_\beta, p_\alpha) \, \psi_n^{(\alpha\beta)}(x_\beta) +$$

$$+ \int d^3q \, [\mathcal{A}_B'(\hat{y}_\beta, q, p_\alpha) \, \exp\{i(q,x_\beta) \, /|c_{\beta\alpha}|\} + \mathcal{A}_B(\hat{y}_\beta, q, p_\alpha)$$

$$\times J_\beta^{as}(x_\beta, \hat{y}_\beta, q)] \, \} \, , \quad B \neq A, \text{ for the rearrangement channel;}$$

$$\hat{S}_{AA_0}(\hat{y}_\alpha, p_\alpha, z) = \delta_{jj_0}\delta(\hat{y}_\alpha - \hat{q}) + 2i(2\pi)^{-5/2} \rho_{A_0}(z)\rho_B(z) \, \mathcal{A}_{\alpha,j}^{(j)}(\hat{y}_\alpha, p_\alpha) \, ,$$

$$A = \{\alpha,j\}, \text{ for the elastic and inelastic channels;}$$

$$\hat{S}_{0A_0}(\hat{X}, p_\alpha, z) = 2(2\pi)^{-2} \rho_{A_0}(z)\rho_0(z) \, e^{-i\pi/4} \, \{\mathcal{A}_0(\hat{y}_\alpha, \zeta_\alpha\hat{x}_\alpha, p_\alpha) +$$

$$+ 2\pi\zeta_\alpha \int_{S^2} \mathcal{A}_0(\hat{y}_\alpha, \zeta_\alpha\hat{q}, p_\alpha) \, f_\alpha^{as}(\hat{x}_\alpha, \hat{y}_\alpha, \zeta_\alpha\hat{q}) \, d^2\hat{q} \, \}$$

$$\text{for the break-up channel } 2 \to 3.$$

Here $\zeta_\alpha = \tau_\alpha(1+\tau_\alpha^2)^{-1/2}z^{1/2}$, $\tau_\alpha = |x_\alpha|/|y_\alpha|$, $X = \{x_\alpha, y_\alpha\} \in \mathbb{R}^6$, $\hat{X} = X/|X|$, $\hat{q} = q/|q|$, $\hat{x}_\beta = x_\beta/|x_\beta|$; furthermore,

$$J_\beta^{as} = \lim_{|y_\alpha| \to \infty} \, [(-\Delta_{x_\alpha} - |q|^2 - i0)^{-1} \, (v_\beta(x_\beta) \, \varphi(x_\alpha, y_\alpha, q))] \, ,$$

$$f_\alpha^{as} = \lim_{|y_\alpha| \to \infty} [|y_\alpha|e^{-i|q||y_\alpha|} \, (-\Delta_{x_\alpha} - |q|^2 - i0)^{-1}(v_\beta(x_\beta) \, \varphi(x_\alpha, y_\alpha, q))]$$

$$\rho_B = [(z+\kappa_B^2)/2]^{1/2}, \quad \rho_0 = (z/2)^{1/2}$$

and $\psi_n^{(\alpha\beta)}$ are the discrete-spectrum eigenfunctions of the operator $h_{\alpha\beta}$.

References

1. Yu.A.Kuperin, Yu.B.Melnikov, B.S.Pavlov. In: Schrodinger Operators, Standard and Non-Standard. (Eds. P.Exner and P.Seba). World Scientific, Singapore, 1989, pp.295-319.

2. Yu.A.Kuperin, P.B.Kurasov, Yu.B.Melnikov, S.P.Merkuriev. Connexions and Effective S-Matrix in Triangle Representation for Quantum Scattering. Preprint INFN-ISS 89/6, Roma, 1989, 52p.

3.Yu.A.Kuperin, Yu.B.Melnikov, S.L.Yakovlev. In: Proc. of the Workshop "Microscopic Methods in the Theory of Few-Body Systems", Kalinin, USSR, August 15-21, 1989, vol.1, pp.164-170.

4.S.P.Merkuriev, L.D.Faddeev Quantum Scattering Theory for the Few-Body Systems. Moscow, Nauka Publishing, 1985 (in Russian).

Department of Mathematical and Computational Physics,
Institute for Physics, Leningrad University,
198904 Leningrad, USSR .

Operator Theory:
Advances and Applications, Vol. 46
© 1990 Birkhäuser Verlag Basel

A THREE-BODY ONE-DIMENSIONAL SYSTEM WITH INCREASING INTERACTIONS: WAVE FUNCTION ASYMPTOTICS

E.A.Yarevsky

Asymptotics for the discrete-spectrum eigenfunctions in the system of three one-dimensional particles with increasing potentials are constructed. The eikonal equation for the situation of linearly increasing potentials is studied in detail.

Asymptotics of the bound-state wavefunctions of quantum three-particle systems are well known [1]. However, correct treatment of the non-relativistic quark model [2] requires knowledge of the bound-state eigenfunction for three-body systems with a confinement. In this paper such an asymptotics for a one-dimensional three-particle system is constructed. The dynamics of this system is determined by Schroedinger equation

$$(-\,\partial_\rho^2 - \rho^{-1}\partial_\rho - \rho^{-2}\partial_\varphi^2 + V(\rho,\varphi))\Psi = E\,\Psi , \tag{1}$$

where

$$V(\rho,\varphi) = \rho^n\alpha(\varphi) = \rho^n\sum_{k=1}^{3}\lambda_k|\sin(\varphi - \varphi_k)|^n, \quad \lambda_k > 0,\ n > 0, \tag{2}$$

is the potential energy and the angles φ_k depend on the masses of
the particles only.

Coordinate asymptotics of the discrete-spectrum
eigenfunction Ψ when $\rho \rightarrow \infty$ will be constructed by the eikonal
method [3]. The eikonal $L(\rho,\varphi)$ is defined in the form of
expansion:

$$L(\rho,\varphi) = \sum_{k=0}^{[1/2+1/n]+1} \rho^{\omega_k} \Phi_k(\varphi), \qquad \omega_k = \omega_0 - kn, \quad \omega_0 = 1 + \frac{n}{2}. \tag{3}$$

where the function $\Phi_k(\varphi)$ is defined in terms of the functions
$\{\Phi_m\}_{m=0}^{k-1}$ by the following recurrsive relation

$$\Phi_k(\varphi) = \exp(-\mu(\varphi)\omega_0\omega_k) \int^{\varphi} (\frac{F_k}{2\Phi_0}) \exp(\mu(s)\omega_0\omega_k)ds, \tag{4}$$

where $\mu(\varphi) = \int^{\varphi} \Phi_0(s)(\Phi_0'(s))^{-1}ds$, and F_k can be calculated for
every k (one has, for example, $F_1 = -E$). The function $\Phi_0(\varphi)$ has
the form

$$\Phi_0(\varphi) = \omega_0^{-1} \alpha^{1/2}(\varphi) \sin u(\varphi), \tag{5}$$

where $u(\varphi)$ obeys the equation

$$u' = \omega_0 - \frac{1}{2} \alpha' \bar{\alpha}^{1} \, tg \, u, \tag{6}$$

together with the boundary conditions $\Phi_0(\varphi) > 0$, $\Phi_0(\varphi) = \Phi_0(\varphi+2\pi)$, $\varphi \in S^1$.

The eikonal amplitude $A(\rho,\varphi)$ has the factorized form

$$A(\rho,\varphi) = C \rho^{-\omega_0/2} (\Phi_0')^{-1/2}, \qquad C = const \tag{7}$$

everywhere except at the vicinities of selected directions, $\{\varphi^{\bullet} \in$

$S^1 : \Phi_0'(\varphi^*) = 0 \}$.

Studying of eq. (7) is the main purpose of this paper. It is equivalent to investigation of trajectories of some analytical dynamical system on a 2-dimensional torus. The vector field determining such a system can be constructed by analytical continuation for the function $\alpha(\varphi)$ (see [4])

$$|\sin(\varphi-\varphi_k)| \longrightarrow (\sin^2(\varphi-\varphi_k))^{1/2} \tag{8}$$

The following table contains the characteristics of Riemann surfaces of the function $\alpha(\varphi)$ in the cases when the degree $n=p/q$ of the potential growth is rational.

	p odd	p even q = 1	p even q ≠ 1
s	6q	1	3q
g	6(2pq-p-q)+1	0	3(pq-p-q)+1

The problem of obtaining periodic solutions of the eq.(7) can be reformulated as looking for a single closed geodesics on the Riemann surface R_g^s with the corresponding metrics. If $g=0$, there exists, in general, only a finite number of closed geodesics, while for $g \geq 1$, these geodesics are dense in R_g^s in some sense. All these cases are analyzed in detail in [5].

Not every closed geodesics, however, is a solution to eq.(7) and, in general, the choice of the corresponding geodesics represents a difficult problem. In some special cases the problemcan be dealt with using analysis of the phase-space picture for this dynamical system. In the most interesting case [2] of a linearly increasing potential the trajectories corresponding to the solutions of (7) can be written in the analytical form

$$u = \varphi - \varphi_0, \quad u = \varphi - \varphi_0 + \pi, \tag{9}$$

where φ is determined by the masses of the particles and the sheet of the Riemann surface R_g^s. Both trajectories (9) generate the same function $\Phi_0(\varphi)$.

Large-distance asymptotics for the wave function of a three-body system with increasing interactions in one dimension are found. Complete asymptotical expansion of the eikonal is obtained. The Riemann surface generating the solution of eikonal equation is constructed. For the linear potential the complete analysis of the trajectories of the system on the Riemann surface is done and the solution is constructed in the analytical form.

REFERENCES

1. Merkuriev S.P., Sov.J.Nucl.Phys. 19 (1974) 222.
2. Kuperin Yu.A., Kvitsinsky A.A., Merkuriev S.P., Yarevsky E.A., Proc.of Int.Sem. on High Energy Phys., JINR 1,2-86-668, vol.1, Dubna 1987, p.212.
3. Merkuriev S.P., Faddeev L.D., *Quantum Scattering Theory for Few-Body Systems*, Moscow, Nauka, 1985 (in Russian).
4. Fiziev P.P., Fizieva Z.I., Preprint JINR, R4-86-339, 1986, Dubna.
5. Klingenberg W. *Lectures on Closed Geodesics*. Springer-Verlag, Berlin 1978.

Department of Mathematical & Computational Physics,
Institute for Physics, Leningrad University,
Leningrad 198904 , USSR .

Operator Theory:
Advances and Applications, Vol. 46
© 1990 Birkhäuser Verlag Basel

SURPRISES OF QUANTUM TUNNELING (SOMETHING ABOUT THE VELOCITY OF THE SUB-BARRIER MOTION)

Zakhariev B.N., Olhovsky V.S., Shilov V.M.

We discuss the time evolution in sub-barrier motion to investigate the unexpected effect discovered by Fletcher, namely the independence of the time delay on the barrier length for sufficiently long barriers. It is shown that the velocity of a wave packet under the barrier is exponentially decreasing from the beginning to the end of the barrier so the packet spends the main part of time penetrating the last part of the barrier.

The barrier tunnelling is one of the clearest manifestations of specific quantum mechanical features. It is also of principal importance for the nuclear fission and fusion, for most nuclear reactions and processes of the atomic physics. The famous Josephson effect and the tunnelling microscope are other examples of this phenomenon. And though it was investigated from the very beginning of the quantum theory, some interesting and incomprehensible questions still remain. It is especially true for the multi-channel [1] and many-particle [2] aspects of the phenomenon, but even in the simplest one-dimensional case of a single-particle motion in an external field there are some mysterious moments. For instance, Fletcher [3] has discovered that the time t_A spent by a wave packet under the rectangular potential barrier is linearly increasing as a function of the barrier width A for small values of A, but becomes independent of A for sufficiently wide barriers (see Fig.1 and the solid lines in the Fig 3).

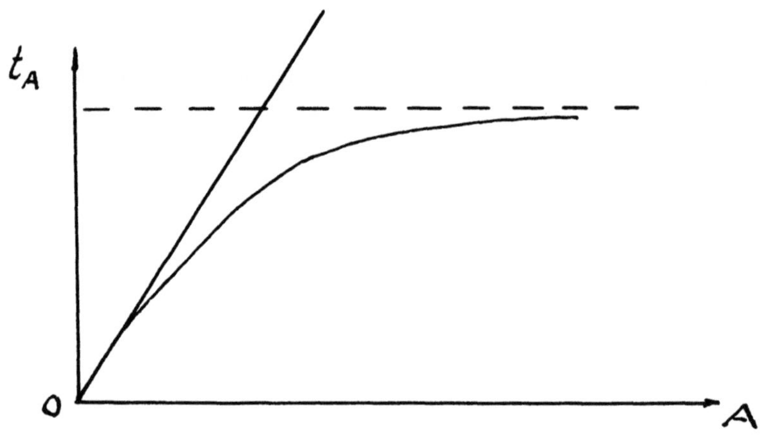

Fig.1. Barrier crossing time t_A as a function of barrier thickness A (solid curve). The dashed horizontal line indicates the asymptotic time delay $t_{A=\infty}$.   The sloped line corresponds to the constant (independent of A) velocity of the tunneling.

It turns out that the mean velocity of sub-barrier motion is increasing with A. However, Fletcher has not investigated the time evolution of wave packets inside the barriers. Has the motion in different parts of the barrier a constant velocity or is it slowing down (or accelerating) during the tunneling ? The determination of the sub-barrier velocity as a function of the coordinate x may shed more light on this intriguing problem.

In the present paper we have calculated the times t_x when the "center of mass" of the packet is at different points x of the sub-barrier interval. It turns out that the packet is exponentially slowing down from the beginning to the end of the barrier where its velocity becomes equal to the velocity of the free motion. Thus the last part of the barrier is mostly responsible for the time delay.

STATIONARY SOLUTIONS

First we shall consider the simplest case of a rectangular barrier. The solution of the Schroedinger equation in front of (x < 0), under (0 < x < A) and behind the barrier (x > 0) has the known form (h = m = 1) :

$$\Psi(x)= \exp(ikx) + R \exp(-ikx) \; ; \; x \leq 0 \; ; \; k = (2E)^{1/2} \qquad (1)$$

$$\Psi(x)= \alpha \, \exp(-\kappa x) + \beta \, \exp(\kappa x) \; ; \; 0 \leq x \leq a \; ;$$
$$\kappa = \{2(V-E)\}^{1/2} \qquad (2)$$

$$\Psi(x)= T \, \exp(ikx) \; ; \; x \geq 0 \qquad (3)$$

The current

$$J = (i/2m) \; [\Psi(x) \, \Psi^{*'}(x) - \Psi^{*}(x) \, \Psi'(x)] \qquad (4)$$

is a constant on the whole axis x. In front of the barrier it is equal to the sum of the currents of the incoming and reflected waves (two components of the linear combination of free solutions in (1)). It is less known that inside the barrier the currents for separate components of the solution (exponentially increasing and decreasing) are zero. Only their interference provides the conservation of J.

Another interesting fact is that the decreasing

component of the solution under the barrier becomes comparable with the increasing component at x = A   (the coefficient α is modulo exp(2κA) times greater than β) :

$$\alpha = \frac{2k}{D_+} \ \frac{k + i\kappa}{(k^2 - \kappa^2) D_- / D_+ + 2ik\kappa}$$

$$\beta = \frac{2k}{D_+} \ \frac{-k + i\kappa}{(k^2 - \kappa^2) D_- / D_+ + 2ik\kappa} \ \exp(-2\kappa a) \tag{5}$$

where $D_\pm = 1 \pm \exp(-2\kappa a)$,   so that the increasing solution is exponentially decreasing in the direction to the beginning of the barrier from the least value of the decreasing component at the end point x = A.

CONSTRUCTION OF THE WAVE PACKET

We shall make up the packet of the solutions Ψ given by (1-3) integrating them with a weight g(k) :

$$\Psi_{pack}(t,x) = \int dk\ g(k)\ \Psi(k,x)\ \exp(-iEt) \tag{6}$$

It is known that if the factor g(k) is chosen of a Gaussian   form

$$C \exp [-b(k-k_\circ)^2]\ ,$$

then in the configuration space also the part of the packet constructed of the incoming waves, will be of a Gaussian form. The integral of the reflected and the transferred waves in eq.(6) must give the corresponding retarded wave packets. But the question may arise whether the energy dependence of coefficients R(k) and T(k) will result into appearance of non-negligible packet components, for example, behind the barrier even before the incoming packet approaches the barrier. This would hinder physical interpretation of our results. To check it, we have calculated the packet form in the configurational space for different weight functions g(k) at the time before the packet of incoming waves exp(ikx) has approached the barrier: a particular case is shown in the Fig.2a. For a more illustrative presentation the part of the packet inside the barrier was multiplied by the increasing exponential exp $[(V-k_o^2/2)^{1/2} x]$, and behind the barrier by the exp $[(V-k_o^2/2)^{1/2} A]$ in order to compensate the exponential decrease of waves under the barrier. Then everything which is much less than the packet in front of the barrier may be neglected. Fig.2 shows that it is possible to construct the incoming packet of the solutions (1-3). No significant packet components appearing "prematurely", even with exponential amplification, can be seen in Fig.2a.

In addition, it is desirable to have also the over-barrier part of the packet which takes no part in tunneling negligible in order not to distort the picture of sub-barrier penetration.

The mean time t_x for the packet transition through the point x is naturally determined as [4] :

$$t_x = \left\{ \int_{-\infty}^{\infty} t \, |\Phi_{pack}(t,x)|^2 \, dt \right\} / \left\{ \int_{-\infty}^{\infty} |\Phi_{pack}(t,x)|^2 \, dt \right\}. \qquad (7)$$

It is especially easy to calculate these integrals with the weight function g(k) for which

$$g(k)^2 = \delta(k^2-k_o^2) \equiv \delta(E-E_o). \qquad (8)$$

Indeed, in this case substituting eqs.(6),(8) into eq.(7), we get

$$t_x = \partial \, \arg \, [\Phi_{pack}(t,x)] \, / \, \partial E . \qquad (9)$$

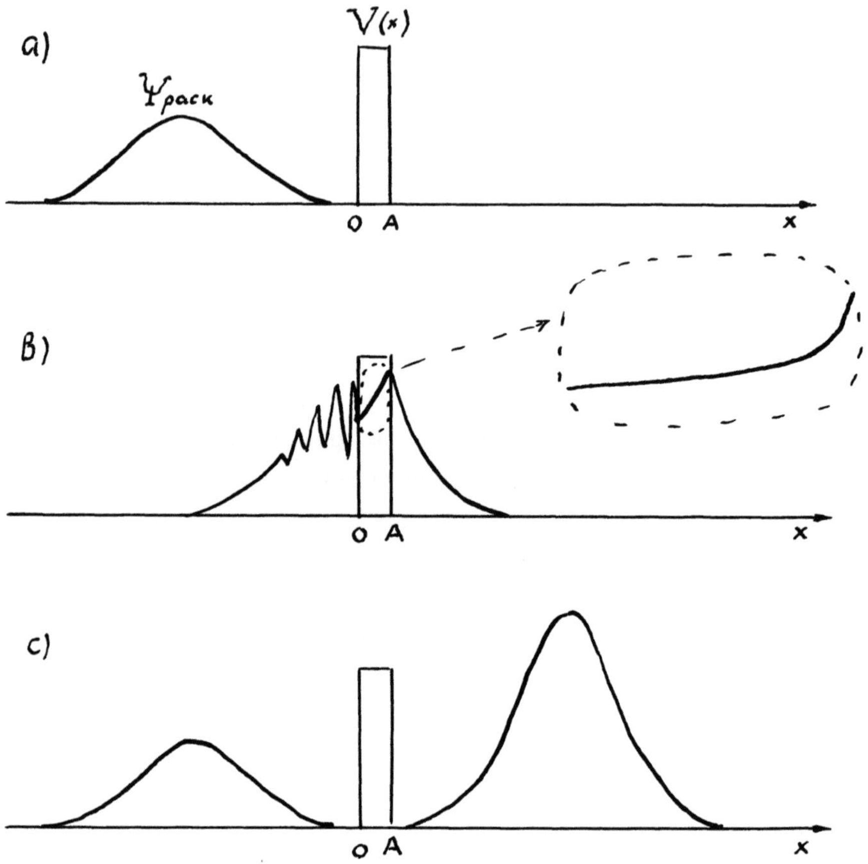

Fig.2. The shape of the wave packet with the weight function $g(k) = \exp[-b(k-k_o)^2]$ with the cut-off of energy values outside of the interval $0 < E < V$. The following values of parameters were chosen : $E_o = 0.5$; $A = 5$; $b = 40$ and different times a) $t = -30$; b) $t = -5$; c) $t = 30$. The parts of the packets for $x > 0$ are given with exponential amplification. The sub-barrier part of the packet in the Fig.b surrounded by the dashed line, is shown enlarged.

In the derivation of eq.(9) we have changed the time t in the numerator of eq.(7) to the derivative

$$t \, \exp(-iEt) \; \longrightarrow \; i \, \partial \, \exp(-iEt) \, / \, \partial E$$

and integrated by parts.

Just the same eq.(9) is obtained by the stationary phase method provided we follow the motion of the packet maximum only.

It is also possible to calculate the time t_x averaging the currents $j^{\pm}_{pack}$

$$j^{\pm}_{pack} = (i/2) \; [\Phi^{\pm}_{pack}(x,t) \; \Phi^{\pm *\prime}_{pack}(x,t) - \Phi^{\pm *}_{pack}(x,t) \; \Phi^{\pm \prime}_{pack}(x,t)] \; :$$

$$t^{\pm}_x = \left\{ \int_{-\infty}^{\infty} t \, j^{\pm}_{pack}(x,t) \; dt \right\} \Big/ \left\{ \int_{-\infty}^{\infty} j^{\pm}_{pack}(x,t) \; dt \right\}. \qquad (10)$$

where the signs $\pm$ correspond to the positive- and negative-momentum Fourier components.

NUMERICAL RESULTS

The x-dependence of t_x for barriers with different A is given in Fig.3. Furthermore, in Fig.4 the time dependence of the average currents $j_{pack}(x,t)$ is given in different points x under the barrier. The negative parts of the currents shifted according the advanced positive parts are the first direct evidence for existence of the incident and reflected components of packets under barriers.

All the curves in Fig.3 start from the origin. This is so only if t_x and t_A are counted not from the moment when the center of the incoming packet arrives at the edge of the barrier but from the time t_0 when the center of the whole packet is at the point $x = 0$ (with both incident and reflected components, the latter having a significant time delay of its own).

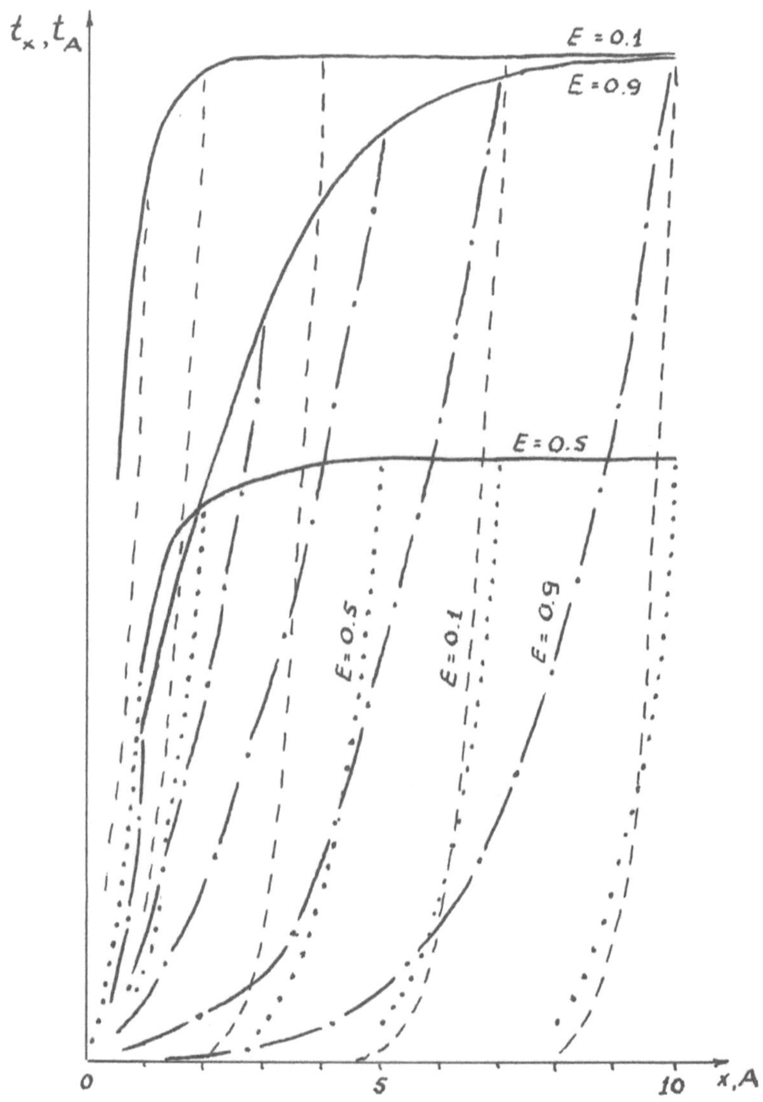

Fig.3   The time t_x of the packet "centre of mass" transition through the point x as a function of the position of x under the barrier   ($0 < x \leq A$)   for different values of the packet mean energy value E_o : E_o = 0.1 (-----) ; E_o= = 0.5 (....) ; E_o = 0.9 (——.——.——) .

The time delay t_A of the packet by the barrier as a function of the barrier width A (————) .

The steep rise of the curves t_x at the end of the barrier corresponds to the decrease of the "packet velocity under the barrier" during the motion from $x = 0$ to $x = A$.

Our results are in agreement with the ones by Fletcher [1]. The time delay t_A for sufficiently wide barriers does not depend on A, and these limiting delay times have a minimum at the mean packet energy equal to one half of the barrier height, namely

$$\lim_{large\ A} t_A = h\ [E(V-E)]^{-1/2}\ .$$

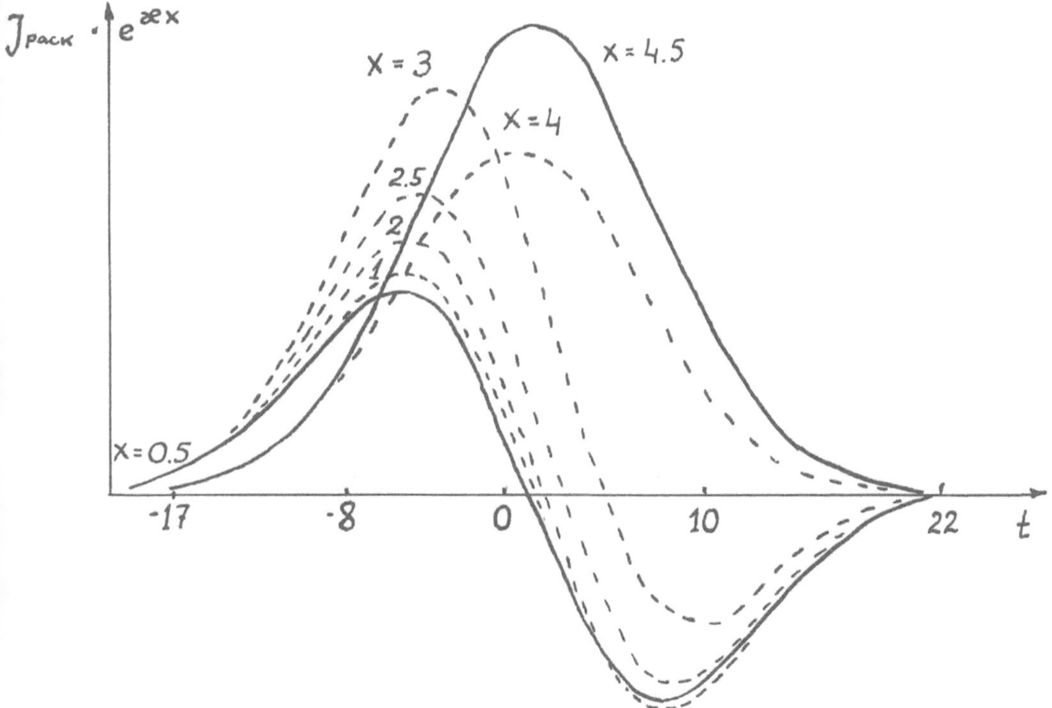

Fig.4. The time dependence of the average flux in different points under the barrier (A = 5) with the factor exp(κx). The "movement" of the maximum corresponds to the slowing down of the wave packets at the end of the barrier.

ACKNOWLEDGMENTS

The authors are grateful to Professsors V.G.Kadyshevsky, V.L.Liuboshits, V.P.Permyakov, A.Radosh, M.I.Shirokov and L.G.Zastavenko for useful discussions.

References

1   Tarakanov A.V., Shilov V.M. Sov.J.Nucl.Phys.$\underline{48}$, 108 (1988) (in Russian).
2   Zakhariev B.N., Suzko A.A. *Potentials and quantum scattering. Direct and Inverse Problems.* Energoatomizdat, Moscow.1985; a revised English edition is due to appear in Springer Verlag, Heidelberg 1990
3   Fletcher J.R., J.Phys.$\underline{C18}$, L55 (1985).
4   Olhovsky V.S., Sov.J.Elem.Part.and Nucl. $\underline{15}$, 289 (1984) (in Russian).

Laboratory of Theoretical Physics
Joint Institute for Nuclear Research
141980 Dubna, USSR

and

Institute for Nuclear Research
Ukrainian Academy of Sciences
Kiev, USSR

Part 2

POINT AND CONTACT INTERACTIONS, SELF-ADJOINT EXTENSIONS

Operator Theory:
Advances and Applications, Vol. 46
© 1990 Birkhäuser Verlag Basel

LIFSHITZ-TAILS AND NON-LIFSHITZ-TAILS FOR ONE-DIMENSIONAL RANDOM POINT INTERACTIONS

Werner Kirsch, <u>Frank Nitzschner</u>

We investigate the integrated density of states (henceforth abbreviated by i.d.s.) of a simple model of continuous random Hamiltonians and give a description of the asymptotic behaviour at the band edges of the spectrum. In particular, a new kind of asymptotic behaviour (polynomial decay at some band edges, in contrast to exponential one at other band edges) of the i.d.s. of random Schrödinger operators is established. Finally, the extension of the proof of (internal) Lifshitz tails to more general kinds of one-dimensional continuous models is discussed.

1. THE MODEL

We consider operators on $L^2(\mathbb{R})$ formally given by $H = -\frac{d^2}{dx^2} + \sum_{i \in \mathbb{Z}} \lambda_i \delta(\cdot - i)$, i.e. H describes δ-interactions of strength λ_i centered at $i \in \mathbb{Z}$. These Kronig-Penney-type operators are seen to be selfadjoint operators e.g. for real-valued bounded sequences $(\lambda_i)_i$ by the method of quadratic forms (see [4]) or by extension of the restriction of $-\frac{d^2}{dx^2}$ to the set of smooth functions with compact support contained in $\mathbb{R} \setminus \mathbb{Z}$. The most important description for our purposes is given by

$$(1) \quad \begin{cases} D(H) = \{\varphi \in H^1(\mathbb{R}) \cap H^2(\mathbb{R} \setminus \mathbb{Z}) : \varphi'(i+) - \varphi'(i-) = \lambda_i \varphi(i) \quad \forall i \in \mathbb{Z}\}, \\ H\varphi = -(\varphi\!\restriction\!(\mathbb{R} \setminus \mathbb{Z}))'', \qquad \varphi \in D(H), \end{cases}$$

where $H^k(\Omega)$=Sobolev space, ψ'=derivative of ψ in distributional sense, and $\restriction$ denotes restriction. There is an extensive discussion of these operators in the literature, see e.g. the monograph [2].

The model under consideration is given by

$$(2a) \qquad H_\omega = -\frac{\mathrm{d}^2}{\mathrm{d}x^2} + \sum_{i\in\mathbb{Z}} \lambda_i(\omega)\delta(\cdot - i), \qquad \omega \in \Omega,$$

where $(\lambda_i)_{(i\in\mathbb{Z})}$ is a family of independent and identically distributed random variables on some probability space $(\Omega, \mathcal{A}, I\!P)$ such that

$$(2b) \qquad 0 \leq \lambda_{\min} := \inf \mathrm{supp} I\!P_{\lambda_0} < \lambda_{\max} := \sup \mathrm{supp} I\!P_{\lambda_0} < +\infty.$$

It will be instructive to regard H_ω as a random perturbation of

$$(3) \qquad H_{\underline{\lambda}} := -\frac{\mathrm{d}^2}{\mathrm{d}x^2} + \lambda \sum_{i\in\mathbb{Z}} \delta(\cdot - i) \qquad \text{with } \lambda = \lambda_{\min}.$$

2. "A REVIEW" ABOUT THE I.D.S.

When restricting (2) to some bounded interval $I_L = [-L-\frac{1}{2}, L+\frac{1}{2}]$ and imposing boundary condition b.c., we obtain operators

$$(4) \qquad H_\omega^{L,\mathrm{b.c.}} = \left(-\frac{\mathrm{d}^2}{\mathrm{d}x^2}\right)^{L,\mathrm{b.c.}} + \sum_{i=-L}^{L} \lambda_i(\omega)\delta(\cdot - i) \qquad \text{on } L^2(I_L),$$

which can be rigorously defined by the same methods as H_ω. In the following the most important boundary condition will be b.c.=D,N,per,ap, which will denote Dirichlet, Neumann, periodic and antiperiodic boundary conditions respectively.

The description by quadratic forms yields, that each $H_\omega^{L,\mathrm{b.c.}}$ has discrete spectrum

$$(5) \qquad E_1 \leq E_2 \leq \ldots \leq E_n \leq E_{n+1} \leq \ldots \to +\infty,$$

where E_j is counted according to multiplicity (cf. [10]).

We may now define the i.d.s. ρ of (H_ω) by

$$(6) \qquad \rho(E) = \lim_{L\to\infty} \frac{1}{2L+1} \# \left\{ E_n(H_\omega^{L,\mathrm{b.c.}}) \leq E \right\} \qquad I\!P\text{-a.s.}, E \in \mathbb{R},$$

where $\#\{\cdot\}$ denotes the number of elements of the set $\{\cdot\}$.

In fact, one may follow the proof of [5] in order to obtain the existence of the limit (6) if one notes that each $H_\omega^{L,\text{b.c.}}$ is a perturbation of $\left(-\frac{d^2}{dx^2}\right)^{L,\text{b.c.}}$ of rank $2\cdot(2L+2)$ in the sense that this holds for the resolvents. Furthermore, $\rho(E)$ is independent of the b.c. chosen, since $H_\omega^{L,D}$ is a perturbation of $H_\omega^{L,N}$ of rank 2 (for $\omega \in \Omega$ fixed) in the same sense. As a corollary we obtain the so-called Dirichlet-Neumann bracketing

$$(7) \quad \begin{cases} \dfrac{1}{2L+1}\int \#\left\{E_n(H_\omega^{L,D})\leq E\right\}\,\mathbb{P}(d\omega) \leq \rho(E) \leq \\[2ex] \qquad\qquad \leq \dfrac{1}{2L+1}\int \#\left\{E_n(H_\omega^{L,N})\leq E\right\}\,\mathbb{P}(d\omega) \qquad \forall L\in \mathbb{N}. \end{cases}$$

As usual (see e.g. [3]; we note that Feynman-Kač-type formulas for our model may be found in [9]) we see that ρ is continuous (as a function of E) and that the points of non-constancy of ρ, i.e. the set $\Sigma = \{E \in \mathbb{R} : \rho(E+\varepsilon) - \rho(E-\varepsilon) > 0 \ \ \forall \varepsilon > 0\}$, coincides with "the" spectrum of H_ω in the sense that $\Sigma = \sigma(H_\omega)$ $\mathbb{P}$-a.s. On the other hand, it is well-known that

$$(8) \quad \sigma(H_\omega) = \sigma(H_{\underline{\lambda_{\min}}}) = \bigcup_{k=1}^{\infty}\left[B_k(\lambda_{\min}), k^2\pi^2\right] \quad \mathbb{P}\text{-a.s.},$$

where $\lambda_{\min}$ is given by (2) and $B_k(\cdot)$ is a *strictly* monotone increasing function on $[0,\infty)$. Thus there are infinitely many gaps in the spectrum of H_ω if $\lambda_{\min} > 0$. We emphasize that the upper band edges $k^2\pi^2$ are independent of $\lambda_{\min}, \lambda_{\max}$; this will have a major impact on the asymptotic properties of ρ.

3. A LEMMA

The following lemma will be crucial for Theorem 2 below:

LEMMA $\quad \rho(k^2\pi^2) = k, \ \ k \in \mathbb{N}$.

PROOF: If $\varphi_L(x) = \sin(k\pi x), |x| \leq L+\frac{1}{2}$, and k is odd, it is easy to check that $\varphi_L \in D(H_\omega^{L,N})$ for any $\omega \in \Omega$ (note (1) and $\varphi_L(i) = 0$ for $i = -L,\ldots,+L$), $H_\omega^{L,N}\varphi_L = k^2\varphi^2$ and that φ_L has exactly $k\cdot(2L+1)$ zeros in the interval I_L. By Sturm oscillation theory [1]

$$\rho(k^2\pi^2) = \lim_L \frac{1}{2L+1}\#\left\{E_n(H_\omega^{L,N}) \leq k^2\pi^2\right\}$$
$$= \lim_L \frac{1}{2L+1}k(2L+1) = k.$$

For k even, the same test function φ_L works, but one has to switch to Dirichlet b.c. ∎

4. THE NOTION OF LIFSHITZ TAILS

On the basis of physical arguments, Lifshitz [7] predicted that the i.d.s. of general random Hamiltonians H_ω behaves at band edges E_∂ like

$$(9) \qquad C_1 e^{-c_2 |E - E_\partial|^{-\frac{d}{2}}} \qquad \text{as } E \to E_\partial,$$

where $E \in \Sigma \setminus \{E_\partial\}$, Σ is "the" spectrum of H_ω and d is the dimension of the underlying Euclidean space. For some models and/or some E_∂ Lifshitz tails have been proven rigorously in the weaker form

$$(10) \qquad \lim_{E \to E_\partial} \frac{\ln(-\ln|\rho(E) - \rho(E_\partial)|)}{\ln|E - E_\partial|} = -\frac{d}{2}$$

(see e.g. [6],[8]). In the following, we will be concerned with the inequality "$\leq$" in (10) since it is *this* inequality that distinguishes the random case from the periodic one.

5. LIFSHITZ TAILS AT (SOME) LOWER BAND EDGES

If ρ denotes the i.d.s. (6) of model (2), then the lemma allows us to estimate as follows: For $E > B_k = B_k(\lambda_{\min})$,

$$\rho(E) - \rho(B_k) \leq \frac{1}{2L+1} \int \# \left\{ E_n(H_\omega^{L,N}) \leq E \right\} \, \mathbb{P}(d\omega) - (k-1),$$

by Dirichlet-Neumann-bracketing and the fact that $\rho(B_k) = \rho((k-1)^2 \pi^2)$. Letting $K = K(L) = (2L+1) \cdot (k-1) + 1$ and using $H_\omega^{L,N} \geq H_{\lambda_{\min}}^{L,N}$, we see that

$$(11) \qquad \begin{cases} \rho(E) - \rho(B_k) \leq \dfrac{1}{2L+1} \displaystyle\int_{E_K(H_\omega^{L,N}) \leq E} \# \left\{ E_n(H_\omega^{L,N}) \leq E \right\} \, \mathbb{P}(d\omega) \\[2ex] \qquad\qquad \leq \dfrac{1}{2L+1} \cdot \# \left\{ E_n(H_{\lambda_{\min}}^{L,N}) \leq E \right\} \cdot \mathbb{P} \left\{ E_K(H_\omega^{L,N}) \leq E \right\}. \end{cases}$$

Since $\frac{1}{2L+1} \cdot \#\{\ldots\}$ is bounded (for E bounded), we have to bound $E_K(H_\omega^{L,N})$ from below, when we want to prove

THEOREM 1   With the above notations,

$$\varlimsup_{E \searrow B_k} \frac{\ln\{-\ln(\rho(E) - \rho(B_k))\}}{\ln(E - B_k)} \leq -\frac{1}{2}$$

holds if k is odd.

<u>Main ingredients for the lower bound of</u> $E_K(H)$, <u>where</u> $H = H_\omega^{L,N}$:

I)   First of all, we note the following generalization of Temple's inequality to non-ground state energies:

Let $H \geq -M$ be self adjoint, $\mu \leq E_{n+1}(H), \varphi \in D(H)$ normalized such that $\langle \varphi, H\varphi \rangle < \mu$. Then

$$(12) \qquad E_n(H) \geq \langle \varphi, H\varphi \rangle - \frac{\langle H\varphi, H\varphi \rangle}{\mu - \langle \varphi, H\varphi \rangle}.$$

II)   In order to apply the above generalized Temple's inequality, we have to select a "good" test function φ. To this end, let ψ_0 denote the eigenfunction on $I_0 = [-\frac{1}{2}, +\frac{1}{2}]$, which "produces" the lower band edge. Since k is odd, ψ_0 is periodic. By explicit knowledge of ψ_0, it satisfies Neumann boundary conditions and

$$(13) \qquad \psi_0(0) \neq 0.$$

Let ψ_L denote the normalized periodic extension of ψ_0 to I_L. Since $\psi_L \notin D(H_\omega^{L,N})$ (cf. (1),(13)), we choose some $\mathcal{Z} \in C^0(\mathbb{R}) \cap C^\infty(\mathbb{R} \setminus \{0\})$ such that

- $\mathcal{Z}(0) = 0$ (for simplicity)
- $\mathcal{Z}'(0\pm) = \pm\frac{1}{2}$ (in order to obtain the right behaviour at integer points in (1))
- $\text{supp}\mathcal{Z} \subset (-\frac{1}{2}, +\frac{1}{2})$ (in order to retain Neumann b.c. of ψ_0)

and let

$$(14) \qquad \varphi = \varphi_{\omega,L} = \left\{ \psi_0 + \sum_{i=-L}^{L} \eta_i(\omega)\mathcal{Z}(\cdot - i) \right\} \restriction I_L.$$

Then $\varphi \in D(H_\omega^{L,N})$ if $\eta_i(\omega) \sim \lambda_i(\omega) - \lambda_{\min}$.

III)   We, still, have to ensure $\langle \varphi, H\varphi \rangle < \mu \leq E_{K+1}(H)$ before we can apply (12). In fact, one can achieve this by switching to some $\tilde{H} \leq H$ if one knows that

$$(15) \qquad E_{K+1}(H^{L,N}_{\underline{\lambda_{\min}}}) - E_K(H^{L,N}_{\underline{\lambda_{\min}}}) \geq \frac{c}{(L+\frac{1}{2})^2}.$$

This can be seen by reduction to the well-known behaviour of the i.d.s. of the periodic operator $H_{\underline{\lambda_{\min}}}$.

Now, I,II,III (and the innocent-looking (13)!) can be used to imply the following crucial proposition, where $L = L(\beta, \varepsilon) = \left[\frac{1}{2} \left(\frac{1}{\sqrt{\beta\varepsilon}} + 1 \right) \right]$:

PROPOSITION   There are constants $\alpha, \kappa, \beta_0, \varepsilon_0 > 0$ (*not* dependent on L) such that the following implication holds for $\omega \in \Omega, \beta \geq \beta_0, 0 < \varepsilon \leq \varepsilon_0$:

$$(16) \qquad \begin{cases} E_K(H^{L,N}_\omega) \leq B_k + \varepsilon \\[2mm] \qquad \Rightarrow \quad \# \{|i| \leq L : \lambda_i(\omega) - \lambda_{\min} < \alpha \cdot \varepsilon\} \geq \kappa \cdot (2L+1). \end{cases}$$

The proposition implies Theorem 1 by some standard large deviations arguments (see e.g. [6]). ∎

6. EXTENSIONS AND REMARKS CONCERNING THE PROOF OF LIFSHITZ TAILS

1)   The additional assumption "k odd" in Theorem 1 should be redundant for the *result* (cf. Sect. 6.2 below), but it is not for the *proof* of Theorem 1 as sketched in Section 5. Indeed, we need *Neumann* eigenfunctions to make use of (11).

2)   In the statement of (generalized) Temple's inequality, the expression $\langle \varphi, H\varphi \rangle$ appears twice. On the one hand, this expression has to be *small* (so that we may apply (12); cf. Sect. 5.III). On the other hand, we need $\langle \varphi, H\varphi \rangle$ to be *large* in order to get an effective lower bound of $E_K(H)$, because it is the leading term in (12). In the case of the lower band edges, this latter fact is ensured by (13).

When we try to prove the analogue of Theorem 1 for upper band edges (or at least for those which are connected to Neumann boundary conditions; cf. Sect. 6.1 above) we realize that the whole proof goes through without *any* change if we had (13) as well for the upper band edges. But we know that this is wrong from the proof of the lemma. In physical terms: the particle doesn't feel the interaction. This is very much related to the fact that $B_k(\cdot)$ is moving with the coupling constant whereas the upper band edges are *not*[*]. Thus the proof of Lifshitz tails at the upper band edges fails only slightly, but nevertheless the behaviour of the i.d.s. changes dramatically, since we have:

7. POLYNOMIAL BEHAVIOUR OF THE I.D.S. AT UPPER BAND EDGES

THEOREM 2   Let $k, \lambda_{\min}, \lambda_{\max}$ be given. Then there are positive constants C_+, C_- such that

$$C_- \sqrt{\varepsilon} \leq \rho(k^2 \pi^2) - \rho(k^2 \pi^2 - \varepsilon) \leq C_+ \sqrt{\varepsilon} \qquad \text{for } 0 \leq \varepsilon \leq \varepsilon_0.$$

PROOF: Obviously

$$H^{L,N}_{\underline{\lambda_{\min}}} \leq H^{L,N}_{\omega} \leq H^{L,N}_{\underline{\lambda_{\max}}}$$

(in the sense of quadratic forms) for any $L \in \mathbb{N}, \omega \in \Omega$. Thus by definition (6) of ρ we obtain

$$\rho_{\lambda_{\min}}(E) \geq \rho(E) \geq \rho_{\lambda_{\max}}(E) \qquad \forall E \in \mathbb{R},$$

where ρ_λ denotes the i.d.s. of the periodic operator $H_{\underline{\lambda}}$. Now, letting E be of the form $k^2 \pi^2 - \varepsilon$, we deduce from the lemma

$$\rho_{\lambda_{\min}}(k^2 \pi^2) - \rho_{\lambda_{\min}}(k^2 \pi^2 - \varepsilon) \leq \rho(k^2 \pi^2) - \rho(k^2 \pi^2 - \varepsilon) \leq \rho_{\lambda_{\max}}(k^2 \pi^2) - \rho_{\lambda_{\max}}(k^2 \pi^2 - \varepsilon).$$

By the well-known behaviour of the l.h.s. and the r.h.s., we arrive at Theorem 2. ∎

---

[*] There are similar structures of the spectrum for one-dimensional δ'-interactions and three-dimensional δ-interactions, see [2].

REFERENCES

[1]   S. Albeverio, J.E. Fenstad, R. Høegh-Krohn, *Singular perturbations and nonstandard analysis*, Trans. Amer. Math. Soc. **252** (1979), 275-295

[2]   S. Albeverio, F. Gesztesy, R. Høegh-Krohn, H. Holden, **Solvable models in quantum mechanics**, Springer, Berlin (1988)

[3]   W. Kirsch, *Random Schrödinger operators*, pp. 264-370 in: **Schrödinger operators**; edt. by H. Holden, A. Jensen; Lect. Notes Phys. **345**, Springer, Berlin (1989)

[4]   W. Kirsch, F. Martinelli, *On the spectrum of Schrödinger operators with a random potential*, Comm. Math. Phys. **85** (1982), 329-350

[5]   W. Kirsch, F. Martinelli, *On the density of states of Schrödinger operators with a random potential*, J. Phys. **A15** (1982), 2139-2156

[6]   W. Kirsch, B. Simon, *Lifshitz tails for periodic plus random potentials*, J. Statist. Phys. **42** (1986), 799-808

[7]   I. Lifshitz, *Energy spectrum structure and quantum states of disordered condensed systems*, Sov. Phys. Usp. **7** (1965), 549-573

[8]   G. Mezincescu, *Bounds on the integrated density of electronic states for disordered Hamiltonians*, Phys. Rev. **B32** (1985), 6272-6277

[9]   S. Nakao, On the spectral distribution of the Schrödinger operator with random potential, Japan. J. Math. **3** (1977), 111-139

[10] M. Reed, B. Simon, **Methods of modern mathematical physics, IV: Analysis of operators**, Academic Press, New York (1977)

Werner Kirsch, Frank Nitzschner
Institut für Mathematik
Ruhr-Universität Bochum
Postfach 102148
D-4630 Bochum 1
FRG

Operator Theory:
Advances and Applications, Vol. 46
© 1990 Birkhäuser Verlag Basel

POINT INTERACTIONS WITH AN INTERNAL STRUCTURE
AS LIMITS OF NONLOCAL SEPARABLE POTENTIALS

S.E.Cheremshantsev, K.A.Makarov

A one-parameter family of self-adjoint extensions of the symmetric operator $h_0 = -\Delta$ in $L_2(R^3)$ acting on the space of smooth functions which vanish in the vicinity of the origin serves as a rigorous definition for one-particle point interaction Hamiltonian [1]. The resolvents $R^{(\alpha)}$ of this family $h^{(\alpha)}$ can be given explicitly and in the p-representation they have the form

$$R^{(\alpha)}(z) = R_0(z) - t(z)K(z) \ , \quad \text{Im } z \neq 0 \ , \tag{1}$$

where $R_0(z) = (p^2 - z)^{-1}$ is the resolvent of the Laplace operator, $K(z)$ is given by the integral kernel

$$K(p,k,z) = (p^2 - z)^{-1}(k^2 - z)^{-1},$$

and $t(z) = (2\pi^2)^{-1}(-\sqrt{-z} + \alpha)^{-1}$, $\text{Re}(\sqrt{-z}) \geq 0$ for $z<0$, plays the role of the t-matrix if $\alpha \in R$; on the other hand, $\alpha = \infty$ corresponds to the free operator $-\Delta$.

It is well known [1] that the above mentioned family $h^{(\alpha)}$ can be obtained as a limit in the norm resolvent sense of a sequence of Hamiltonians h_n with nonlocal rank-one potentials

$$\int_{1}^{\infty} \frac{d\mu^{(2)}(R)}{R^2 + 1} < \infty \ .$$

Let $\mu^{(1)} = \mu_{ac}^{(1)} + \mu_{sing}^{(1)}$ be the decompositon to the corresponding absolutely continuous and singular part, with $f^{(1)} \in L_1([0,1])$ and $f^{(2)}(R)(R^2+1)^{-1} \in L_1([1,\infty])$ being the densities for $\mu_{ac}^{(1)}$.

In view of the definition of the singular measure there exists a sequence of open sets $e_n^{(1)}$, $i=1,2$, such that $e_n^{(1)} = \bigcup_{k=1}^{\infty} \Delta_k^{(1)}(n)$, with $\Delta_k^{(1)}$ being open intervals, $\Delta_k^{(1)}(n) \cap \Delta_j^{(1)}(n) = \varnothing$, $k \neq j$, mes$(e_n^{(1)}) < 1/n$, $i=1,2$, and for every n, $\mu_{sing}^{(1)}([0,1] \setminus e_n^{(1)}) = 0$, $\mu_{sing}^{(2)}([1,n] \setminus e_n^{(2)}) = 0$. Notice that if $0 \in \Delta_m^{(1)}(n)$ for some m, we should choose $\Delta_m^{(1)}(n)$ such that

$$\frac{\text{mes } (\Delta_m^{(1)}(n) \cap \mathbb{R}_-)}{\text{mes } (\Delta_m^{(1)}(n))} < 1/n.$$

Let $F_n^{(1)}(R) = 0$ for $R \notin e_n^{(1)}$ and

$$F_n^{(1)}(R) = \frac{\mu_{sing}^{(1)}(\Delta_k^{(1)}(n))}{\text{mes } (\Delta_k^{(1)}(n))}$$

for $R \in \Delta_k^{(1)}$. The sought sequence u_n has the form

$$u_n(p) = u_n^{(1)}(p)\chi(|p|<1) + u_n^{(2)}(p)\chi(|p|\geq 1)\chi(|p|\leq n) \ ,$$

$$u_n^{(1)}(p) = |p|^{-1}(\sqrt{f^{(1)}(|p|)} \exp(in|p|) + \sqrt{F_n^{(1)}(|p|)}) \ ,$$

$$u_n^{(2)}(p) = \sqrt{1+qp^2/n} + \sqrt{f^{(2)}(|p|)} \exp(in|p|) + \sqrt{F_n^{(2)}(|p|)} \ ,$$

where $q = \pi E_2/2$ (see (3)).

$$h_n = p^2 + c_n(*, u_n) u_n \quad , \tag{2}$$

where $u_n(p) = \chi(|p| < n)$ is the characteristic function of the ball of radius n and $(2\pi)^{-1} \lim_{n \to \infty} ((\pi c_n)^{-1} + 4n) = \alpha$.

Self-adjoint extensions of h_0 with extension to some wider Hilbert space seem to provide a much richer structure. One can treat such Hamiltonians as models of a quantum mechanical particle with an internal structure [2]. The corresponding resolvents have a natural block-structure and the part corresponding to the "channel" $L_2(R^3)$ (the Krein's generalized resolvent) still has the form (1). The only difference is that $t(z)$ is now parametrised by an arbitrary R-function $g(z)$ [3,4],

$$t(z) = (2\pi^2)^{-1} (-\sqrt{-z} + g(z))^{-1} \quad , \quad \text{Im } z \neq 0 \quad , \tag{3}$$

Another extension of the scheme is possible when the the norm resolvent approximation is replaced by the weak one ; a simple realization of this idea can be found in [5]. Our main result is the following

THEOREM 1. Let $R^{(g)}(z)$ be the generalized resolvent of the type (1) with $t(z)$ given by (4) and $g(z)$ being an R-function, with the support of the spectral measure lying on the nonnegative axis. Then $R^{(g)}(z)$ is the weak resolvent limit for a sequence of the Hamiltonians h_n of the type (2) with suitable $u_n \in L_2(R^3)$ and $c_n \in R$.

Let us describe the construction.

LEMMA. Every R-function $g(z)$ satisfying the conditions of Theorem 1 admits the represantation

$$g(z) = E_1 + E_2 z + \int_0^1 \frac{d\mu^{(1)}(R)}{R^2 - z} + z \int_1^\infty \frac{d\mu^{(2)}(R)}{R^2 - z} \tag{4}$$

where $\mu^{(i)}$, $i = 1,2$, are Borel measures and

REFERENCES

1. S.Albeverio, F.Gesztesy, R.Hoegh-Krohn, H.Holden. Solvable
   Models in Quantum Mechanics. Springer-Verlag, 1988.
2. B.S.Pavlov. Teor. Mat. Fiz.$\underline{59}$ (1984),No. 3, 345-353.
3. F.V.Atkinson. Discrete and Continuous Boundary Problems.
   Academic Press, 1964.
4. M.G.Krein. Dokl.Akad.Nauk SSSR $\underline{87}$ (1952), 881-884.
5. Yu.G.Shondin. Teor.Mat.Fiz.$\underline{64}$ (1985),No.3, 432-441.

Leningrad Branch of the Steklov Institute
Fontanka 27, 191011 Leningrad
and
Department of Mathematical and Computational Physics
Leningrad State University
198904 Leningrad, USSR

Operator Theory:
Advances and Applications, Vol. 46
© 1990 Birkhäuser Verlag Basel

GREEN'S FUNCTION FOR THE AHARONOV-BOHM EFFECT WITH A NON-ABELIAN GAUGE GROUP

Pavel Šťovíček

In the case of more than one solenoid, the computation of the Green's function for the Aharonov-Bohm effect becomes rather complicated. In this paper, we discuss two possible approaches to this problem. We start with the Schulman's approach based on the Feynman path integral. Applying this technique, recently having been generalized by several authors to non-Abelian gauge groups, we are able to derive some explicit formulae in the two-solenoid case. On the other hand, the interpretation of the problem in the framework of the theory of self-adjoint extensions allows us to apply the Krein's formula. But in this case, some open problems still remain.

1. INTRODUCTION

In this paper, we consider a non-relativistic particle moving in an external and generally non-Abelian gauge field with vanishing field strength. The problem here is discussed from the mathematical point of view. More precisely, the aim is to compute the Green's function. Provided the underlying configuration space is multiply connected, the corresponding Schrödinger operator need not be equivalent to a multiple of the Laplace-Beltrami operator. In fact, this is the meaning of the Aharonov-Bohm effect [1] on the theoretical level.

The generalization of the magnetic field to the non-Abelian gauge field was proposed by Wu and Yang [2]. One way how to attack this problem is to use the Feynman path integral. This approach was elaborated by Schulman [3] for the U(1) gauge group and then it was generalized to the non-Abelian case by Sundrum and Tassie [4] and by Oh, Soo and Lai [5]. But all these deliberations, sketched here in Sec.2, can be concentrated into the unique relation which we shall refer to as the Schulman's Ansatz. The simplest example providing the truly non-Abelian effect is the two-solenoid with the doubly punctured plane as the configuration space [4,5,6]. In this concrete case, one is able to derive more explicit formulae [7] which are presented here without some details in Sec.3. On the other hand, the problem can be considered as belonging to the theory of self-adjoint extensions. The possible application of the Krein's formula is discussed in Sec.4.

2. THE SCHULMAN'S ANSATZ

Let M be connected but generally multiply connected Riemannian manifold and A_ν be a gauge field with the gauge group $U(N)$ and such that the field strength $F_{\mu\nu} = d_\mu A_\nu - d_\nu A_\mu + [A_\mu, A_\nu]$ vanishes on M. It can be shown that, up to equivalence, all such gauge fields, i.e., all flat connections on M are in one-to-one correspondence with unitary representations U of the fundamental group $\Gamma = \pi_1(M, x_{ref})$. Provided a reference point $x_{ref} \in M$ is fixed, the unitary representation is given by the parallel transport. The inverse procedure is also clear. Let $\bar{M}$ designate the universal covering space which is again a Riemannian manifold. We accept the formalism when Γ acts on $\bar{M}$ from the left and the group multiplication in Γ is defined as follows: $[\gamma_1][\gamma_2] = [\gamma_1 \circ \gamma_2]$, where $\gamma_1 \circ \gamma_2$ means that the curve γ_2 follows the curve γ_1. Then the quotient $\Gamma \backslash \bar{M}$ coincides with M and $\bar{M}$ is a principal bundle over M with the structure group Γ. Since $\dim \bar{M} = \dim M$, in this principal bundle there exists only one connection and it is necessarily flat. With the help of the representation U , one can construct the associated vector bundle together with the connection and this connection is again flat.

The corresponding Schrödinger operator can be introduced in at least three unitarily equivalent ways. The most usual one is to replace the derivatives in the Laplace-Beltrami operator by covariant derivatives. Maybe a more convenient formulation is the other one making use of the existence of the universal covering space. The Hamiltonian is $H_U = - (\hbar^2/2\mu)\, \Delta_{LB}$ acting in the Hilbert space consisting of U - equivariant vector functions on $\bar{M}$. The integration in the Hermitian product is over any fundamental domain in $\bar{M}$ (with respect to the action of Γ). The third possibility is to fix a simply connected fundamental domain $D \subset \bar{M}$. This domain can be identified with a set $M \setminus L$ with L being a cut in M of codimension 1 . Now the Hamiltonian H is a self-adjoint extension of the differential operator $X = - (\hbar^2/2\mu)\, \Delta$ defined on $C_o(M \setminus L) \otimes \mathbb{C}^N \subset L^2(M,\mathbb{C}^N,dV)$. The self-adjoint extension is specified by the boundary conditions on L . The explicit results presented here will be given in this third formulation.

The heuristic path integral approach leads to an expression for the propagator in the first formulation.

$$K_t^A(x,x_o) = \int D_M[\gamma] \; P\exp\left(- \int_\gamma A_\nu \, dx^\nu \right) \; \exp[\; i\, S(\gamma)\;] \; ,$$

where the integration is over all paths $\gamma: [0,t] \to M$ such that $\gamma(0) = x_o$, $\gamma(t) = x$, $S(\gamma)$ is the free-particle action. The path-ordered integral appearing in the integrand actually depends only on the homotopic class $[\gamma]$. Let the bar indicate that the corresponding object has been pulled-back from M to $\bar{M}$. Then we get

$$\bar{K}_t^A(x,x_o) = \sum_{g \in \Gamma} \int_{\substack{\gamma(0)=x_o \\ \gamma(t)=g\circ x}} D_{\bar{M}}[\gamma] \; P\exp(- \int_\gamma \bar{A}_\nu \, dx^\nu) \quad x$$

$$x \quad \exp[\; i \; S(\gamma) \; / \; \hbar \;] \; =$$

$$= \sum_{g \in \Gamma} P\exp(- \int_{x_o}^{g.x} \bar{A}_\nu \, dx^\nu) \; \bar{K}_t(g\circ x,x_o) \quad .$$

$\bar{K}_t(\circ,\circ)$ designates the free-particle propagator on $\bar{M}$ which is necessarily Γ - invariant. It is worth to stress that in the case of the free particle on $\bar{M}$,wave functions and the propagator are scalar and the integration in the Hermitian product is over the whole space $\bar{M}$. Since $\bar{M}$ is simply connected, one can express $\bar{A}_\nu = - V^{-1} \partial_\nu V$, $V(x) = P \exp(- \int^x \bar{A}_\nu \, dx^\nu)$. Hence $P \exp(- \int_\gamma \bar{A}_\nu \, dx^\nu) = V(g\circ x) V(x_o)^{-1} = V(x) U(g^{-1}) V(x_o)^{-1}$ and $K_t^U(x,x_o) = V(x)^{-1} \bar{K}_t^A(x,x_o) V(x_o)$ is the propagator in the second formulation. Finally, one arrives at the Schulman's Ansatz:

$$K_t^U(x,x_o) = \sum_{g \in \Gamma} U(g^{-1}) \bar{K}_t(g\circ x,x_o) \quad . \tag{1}$$

Manipulating formally the (possibly infinite) series in (1), K_t^U is easily checked to fulfil all the basic properties:

$$(\; i\hbar \frac{\partial}{\partial t} + \frac{\hbar^2}{2\mu} \Delta_{LB}) \; \vartheta(t) \; K_t^U(x,x_o) = i\hbar \; \delta(t) \; \delta^U(x,x_o) \; ,$$

where $\vartheta(t)$ is the Heaviside step function and the Dirac-type generalized function $\delta^U(x,x_o) = \sum_{g\in\Gamma} U(g^{-1}) \, \delta_{\bar{M}}(g\circ x,x_o)$ plays the role of the unit-operator kernel in the second formulation,

$$\int_D K^U_{t_1}(x,y)\, K^U_{t_2}(y,x_o)\, dV(y) \;=\; K^U_{t_1+t_2}(x,x_o) \quad ,$$

$$K^U_t(x,x_o)^* \;=\; K^U_{-t}(x_o,x) \quad .$$

The propagator in the third formulation is obtained as a restriction to the fixed fundamental domain: $K_t(x,x_o) = K^U_t(x,x_o) \mid D \times D$. It appears to be more convenient to treat the Green's function rather than the propagator. But, of course , the both objects are closely related by the Laplace transformation $G(z) = (i/\hbar)\, {}_0\!\int \exp(izt)\, K(t)\, dt$, Im $z > 0$.

3. THE TWO-SOLENOID

Let us now apply this procedure to the two-solenoid. In the idealized setup, $M = \mathbb{R}^2 \setminus \{a,b\}$, $a = (0,0)$, $b = (\rho,0)$, $\rho = |a - b| > 0$. The cut L is chosen to be a union $L_a \cup L_b$ of two halflines lying on the x - axis: $L_a = \{ (x,0) ; x < 0 \}$, $L_b = \{ (x,o) ; x > \rho \}$. We shall need two polar coordinate systems centred at the points a and b . The angles are counterclockwise oriented and $\varphi_a, \varphi_b \in (-\pi,\pi)$. Γ is the free group with the generators g_a , g_b corresponding to two simple positively oriented curves winding round the point a (respectively, b) and with the point b (respectively, a) lying in the outside. The unitary representation U is determined by two unitary matrices $U(g_a) = \exp(2\pi i\alpha)$, $U(g_b) = \exp(2\pi i\beta)$, $0 \leq \alpha,\beta < 1$. The Hermitian matrices α , β are not constrained by any other condition. In the third formulation, the Hamiltonian is defined : $H = - (\hbar^2/2\mu)\, \Delta$ on $\mathbb{R}^2 \setminus L$ together with the boundary conditions on the halflines L_a , L_b:

$$\psi\big|_{\varphi_a=\pi} = e^{2\pi i\alpha}\, \psi\big|_{\varphi_a=-\pi} \quad , \quad \partial_{\varphi_a}\psi\big|_{\varphi_a=\pi} = e^{2\pi i\alpha}\, \partial_{\varphi_a}\psi\big|_{\varphi_a=-\pi} .$$

$$\psi\big|_{\varphi_b=\pi} = e^{2\pi i\beta}\, \psi\big|_{\varphi_b=-\pi} \quad , \quad \partial_{\varphi_b}\psi\big|_{\varphi_b=\pi} = e^{2\pi i\beta}\, \partial_{\varphi_b}\psi\big|_{\varphi_b=-\pi} .$$

The universal covering space $\bar{M}$ results from the infinite process of patching together countably many copies of the typical sheet (fundamental domain) $D = \mathbb{R}^2 \setminus L$. The boundary ∂D consists of four halflines (two sides of L_a and two sides of L_b) and of four points: a , b and two times (reached from the upper and from the lower halfplane). Each sheet is patched together along four halflines with four other sheets. It is reasonable to complete $\bar{M}$ with the boundary points correeesponding to a and b . So the points we have added constitute a union $\mathcal{A} \cup \mathcal{B}$; the countable set $\mathcal{A}$ (respectively, $\mathcal{B}$) is projected into the point a (respectively, b). Let the function $\chi(x,x_o)$ being equal to 1 (respectively, 0) indicate whether two points $x,x_o \in \bar{M} \cup \mathcal{A} \cup \mathcal{B}$ can (respectively, cannot) be connected by a finite geodesic and put

$$Z_t(x,x_o) = \vartheta(t)\, \chi(x,x_o)\, \frac{\mu}{2\pi i\hbar t}\, \exp[\,\frac{i\mu}{2\hbar t}\, \text{dist}^2(x,x_o)\,] \qquad (2)$$

Hence Z_t has the form of the free propagator on the plane. Further, for three points $x_1,x_2,x_3 \in \bar{M} \cup \mathcal{A} \cup \mathcal{B}$ such that $\chi(x_1,x_2) = \chi(x_2,x_3) = 1$ and for two positive times t_1 , t_2 we put

$$V\begin{pmatrix} x_3,x_2,x_1 \\ t_2,t_1 \end{pmatrix} = \frac{i\hbar}{\mu}\left(\frac{1}{\theta - \pi + iu} - \frac{1}{\theta + \pi + iu} \right) , \qquad (3)$$

where $\theta = \text{angle}(x_1, x_2, x_3)$ (the angle is oriented), $u = \ln(t_2 r_1 / t_1 r_2)$ and $r_1 = \text{dist}(x_1, x_2)$, $r_2 = \text{dist}(x_2, x_3)$.

The experience with the one-solenoid case enables us to guess the form of $\bar{K}_t$ ($t > 0$, c.f. [7]) :

$$\bar{K}_t(x, x_o) = \sum_{\gamma, n \geq 0} W_\gamma(t; x, x_o) \tag{4}$$

$$W_\gamma(t; x, x_o) = \int_{\mathbb{R}^{n+1}} dt_n \ldots dt_0 \;\; \delta(t_n + \ldots + t_0 - t) \;\; V\binom{\gamma}{t_n, \ldots, t_0} \;\; x$$

$$x \;\; Z_{t_n}(x, C_n) \;\; Z_{t_{n-1}}(C_n, C_{n-1}) \;\; \ldots \;\; Z_{t_0}(C_1, x_o) \;\; ,$$

$$V\binom{\gamma}{t_n, \ldots, t_0} = V\binom{x, C_n, C_{n-1}}{t_n, t_{n-1}} \; V\binom{C_n, C_{n-1}, C_{n-2}}{t_{n-1}, t_{n-2}} \ldots \; V\binom{C_2, C_1, x_o}{t_1, t_0} \; ,$$

$V = 1$ for $n = 0$ and the sum runs over all piecewise geodesics $\gamma : x \leftarrow C_n \leftarrow \ldots \leftarrow C_1 \leftarrow x_o$, such that $C_1, \ldots, C_n \in \mathcal{A} \cup \mathcal{B}$ and $\text{dist}(C_j, C_{j+1}) = \rho$, $1 \leq j \leq n-1$. To simplify notation we put where necessary $C_0 = x_o$, $C_{n+1} = x$. Treating formally the infinite sum one can again verify the basic properties of the propagator.

Now one can make use of the Schulman's Ansatz and apply the Laplace transformation to get the expression for the Green's function (in the third formulation, $z \notin \mathbb{R}_+$) :

$$G_z(x, x_o) = G_z^a(x, x_o) + G_z^b(x, x_o) - G_z^0(x, x_o) +$$

$$+ \frac{\mu}{\pi \, \hbar^2} \sum_{\gamma, n \geq 2} (-1)^n \int_{\mathbb{R}^n} d^n s \;\; K_0[w \, R_\gamma(s)] \;\; S_\gamma(s; \varphi, \varphi_o) \; , \tag{5}$$

where G^a (respectively, G^b) is the one-solenoid Green's function centred at the point a (respectively, b), G^0 is the free-particle Green's function, K_0 is the Macdonald function.

$$w = (- 2\mu z)^{1/2} / \hbar , \quad \mathrm{Re}\, w > 0 , \tag{6}$$

$$R_\gamma(s)^2 = (r_o + r_1 e^{s_1} + \ldots + r_n e^{s_1+\ldots+s_n}) \times$$

$$\times (r_o + r_1 e^{-s_1} + \ldots + r_n e^{-s_1-\ldots-s_n}) , \tag{7}$$

$$S_\gamma(s;\varphi,\varphi_o) = \frac{\sin \pi \sigma_n}{\pi} \frac{e^{-\sigma_n(s_n-i\varphi)}}{1 + e^{-s_n+i\varphi}} \frac{\sin \pi \sigma_{n-1}}{\pi} \frac{e^{-\sigma_{n-1} s_{n-1}}}{1 + e^{-s_{n-1}}}$$

$$\ldots\ldots \frac{\sin \pi \sigma_2}{\pi} \frac{e^{-\sigma_2 s_2}}{1 + e^{-s_2}} \frac{\sin \pi \sigma_1}{\pi} \frac{e^{-\sigma_1(s_1+i\varphi_o)}}{1 + e^{-s_1-i\varphi_o}} , \tag{8}$$

$r_j = |c_{j+1} - c_j|$, $0 \le j \le n$, and (r_n,φ) (respectively, (r_0,φ_o)) are the polar coordinates of x (respectively, x_o) with respect to the center c_n (respectively, c_1) and the sum runs over all finite sequences $\gamma = (c_n,\ldots,c_1)$, $c_j \in \{a,b\}$, $c_j \ne c_{j+1}$ and $\sigma_j = \alpha$ (respectively, β) provided $c_j = a$ (respectively, b). Again $c_0 = x_o$, $c_{n+1} = x$.

Now using the asymptotics of the Macdonald function and the estimate $R_\gamma(s) \ge r_0 + r_n + (n-1)\rho$, one is able to treat the infinite series (5) quite rigorously.

4. THE KREIN'S FORMULA

It is reasonable to state the problem in a more general formulation. Let X be the differential operator $-\Delta$ defined on $D(X) = C_o(\mathbb{R}^2 \setminus L) \otimes \mathbb{C}^N \subset L^2(\mathbb{R}^2, \mathbb{C}^N, dx\,dy)$, with L being the x - axis in $\mathbb{R}^2$. Let $\nu : \mathbb{R} \to \mathbb{C}^{N,N}$ be a piecewise smooth function satisfying $\nu(x)^* = \nu(x)$, $0 \leq \nu(x) < 1$, and H be the self-adjoint extension of X specified by the boundary conditions on L :

$$\psi(x,0_-) = e^{2\pi i\,\nu(x)}\,\psi(x,0_+) \ , \ \frac{\partial}{\partial y}\,\psi(x,0_-) = e^{2\pi i\,\nu(x)}\,\frac{\partial}{\partial y}\,\psi(x,0_+) \ .$$

$$(9)$$

The free Hamiltonian H_o is another self-adjoint extension of the operator X (corresponding to $\nu(x) \equiv 0$ in (9)). The Krein's formula enables to compare the resolvents of the both self-adjoint extensions (c.f. [8] for a brief review).

Let $N(z)$ designate the deficiency subspace for X corresponding to the spectral parameter z , $\mathrm{Im}\,z \neq 0$. Let $(\varphi_n(z))$ be a basis of $N(z)$ analytically depending on z and $W(z): N(z) \to N(z)$ be the mapping determining the self-adjoint extension $H : D(H) = D(X) \oplus (1 + W(z))N(z)$, $H(\psi + \varphi + W(z)\varphi) = X\psi + z\varphi + \bar{z}W(z)\varphi$. $W_o(z)$ is defined analogously. Then

$$(H - z)^{-1} = (H_o - z)^{-1} + \sum_m \sum_n \Gamma_{mn}(z) \ |\varphi_m(z)> <\varphi_n(\bar{z})| \ , (10)$$

where
$$\sum_m \sum_n <\varphi_j(z)|\varphi_m(z)> \Gamma_{mn}(z) <\varphi_n(\bar{z})|\varphi_k(\bar{z})> \ =$$

$$= \frac{<\varphi_j(z)|W(\bar{z}) - W_o(\bar{z})|\varphi_k(\bar{z})>}{\bar{z} - z}$$

$$(11)$$

It is possible to consider a generalized basis of $N(z)$ and replace the sums in (10,11) by integrals.

The deficiency indices of our operator X are (∞,∞) and the generalized basis of $N(z)$ can be chosen to be $\{\varphi_{\pm,\kappa,j}(z;x,y) = (1/\sqrt{2\pi})\exp[i\kappa x - (\kappa^2 - z)^{1/2}|y|]\,\vartheta(\pm y)\otimes e_j$; $\kappa \in \mathbb{R}$ and $e_1,\ldots,e_N$ is the standard basis in $\mathbb{C}^N$ $\}$. It holds

$$\langle\varphi_{+,\kappa,j}(z)|\varphi_{-,\lambda,k}(z)\rangle = 0 \ ,$$

$$\langle\varphi_{+,\kappa,j}(z)|\varphi_{+,\lambda,k}(z)\rangle = \langle\varphi_{-,\kappa,j}(z)|\varphi_{-,\lambda,k}(z)\rangle =$$

$$= \left(2\,\mathrm{Re}\,(\kappa^2 - z)^{1/2}\right)^{-1}\,\delta(\kappa - \lambda)\,\delta_{jk}$$

Now the operator $\Gamma(z)$ can be obtained after a straightforward but formal computation. Let $\Lambda(z)$ and Ω be the operators with the generalized kernels

$$\Lambda(z)(\kappa,\lambda) = (\kappa^2 - z)^{1/2}\,\delta(\kappa - \lambda)\otimes 1_N$$

$$\Omega(\kappa,\lambda) = \frac{1}{2\pi}\int_{-\infty}^{\infty} e^{2\pi i\,v(x)\,-\,ix(\kappa - \lambda)}\,dx \ . \tag{12}$$

Then

$$\Gamma(z) = \begin{pmatrix} \Gamma_{++}(z) & \Gamma_{+-}(z) \\ \Gamma_{-+}(z) & \Gamma_{--}(z) \end{pmatrix} =$$

$$\begin{pmatrix} [\Lambda(z)\Omega + \Omega\Lambda(z)]^{-1}\,\Omega - \Lambda(z)^{-1}, & [\Lambda(z)\Omega + \Omega\Lambda(z)]^{-1} - \Lambda(z)^{-1} \\ \Omega\,[\Lambda(z)\Omega + \Omega\Lambda(z)]^{-1}\Omega - \Lambda(z)^{-1}, & \Omega\,[\Lambda(z)\Omega + \Omega\Lambda(z)]^{-1} - \Lambda(z)^{-1} \end{pmatrix}$$

But it still remains much to do. It is necessary to prove the existence of the inversion $[\Lambda(z)\Omega + \Omega\Lambda(z)]^{-1}$ in a convenient sence and, hopefully, to find an at least theoretical way how to compute this inverse operator. The Krein's formula approach can be tested on the well known one-solenoid example when $N = 1$ and $\nu(x) = \alpha\,\vartheta(x)$, $\alpha \in [0,1)$. But even in this case the resulting formula for $[\Lambda(z)\Omega + \Omega\Lambda(z)]^{-1}$ is not very lucid.

REFERENCES

[1]   Y. Aharonov and D. Bohm, Phys.Rev. **115** (1959) 485.
[2]   T.T. Wu and C.N. Yang, Phys.Rev.D **12** (1975) 3845.
[3]   L.S. Schulman, Phys.Rev. **176** (1968) 1558.
[4]   R. Sundrum and L.J. Tassie, J.Math.Phys. **27** (1986) 1566.
[5]   C.H. Oh, C.P. Soo and C.H. Lai, J.Math.Phys. **29** (1988) 1154.
[6]   L.S. Schulman, J.Math.Phys. **12** (1971) 304.
[7]   P. Šťovíček, JINR preprint E5-89-369, to appear in Phys. Lett.A
[8]   S.Albeverio et al., Solvable Models in Quantum Mechanics (Springer Verlag, Berlin-Heildelberg 1988) 357.

Laboratory of Theoretical Physics
Joint Institute for Nuclear Research
141 980 Dubna, USSR

A MODEL OF ZERO-WIDTH SLITS AND
THE REAL DIFFRACTION PROBLEM

I.Yu.Popov

A model of zero-width slits based on the extension theory and the real diffraction problems are compared.

Let Ω^{in} a the bounded domain in $\mathbb{R}^m$, $m = 2,3$, with a smooth boundary $\partial\Omega$, and consider a point $x_0 \in \partial\Omega$. Suppose that $\Delta_0^{in,ex}$ are the Laplace operators in $\Omega^{in,ex}$ with Neumann boundary condition defined on the set of all functions vanishing in the vicinity of the point x_0. The operators $\Delta_0^{in,ex}$ are symmetric, and their deficiency elements are the Green's functions $G^{in,ex}(x,x_0,\lambda_0)$, $\text{Im } \lambda_0 > 0$. Any self-adjoint extension of the operator $-\Delta_0 = - (\Delta_0^{in} \oplus \Delta_0^{ex})$ is our model operator. This operator has a simple structure, and one can obtain the exact expression for its resolvent if the Neumann problems in Ω^{in} an Ω^{ex} are solved. The S-matrix for the model operator (provided the conservation of flux through the opening is assumed) has the form [1,2]:

$$S(\omega,\upsilon,\lambda) = S^{ex}(\omega,\upsilon,\lambda) + ik(2\pi)^{-1} (D^{in}(\lambda) + D^{ex}(\lambda))^{-1}$$

$$\times \phi^{ex}(x_0,\omega,\lambda) \; \phi^{ex}(x_0,\upsilon,\lambda).$$

Here S^{ex} and ϕ^{ex} are the S-matrix and the scattered wave for the unperturbed problem,

$$D^{in,ex}(\lambda) = (G^{in,ex}(x,x_0,\lambda) - \mathrm{Re}\; G^{in,ex}(x,x_0,\lambda_0))\big|_{x=x_0}.$$

If the extension parameter λ_0 satisfies the algebraic equation

$$D^{in}(\lambda) + D^{ex}(\lambda) = (G^{in} + G^{ex})(s,x,\lambda)$$

$$-\left(\int_{\Gamma_d} dx \int_{C^{ex}} ds\; G''_{n,n}(x,s,0)\right)^{-1},$$

where $C^{ex} = \{\, x: x \in \Omega^{ex},\; |x - x_0| = d\,\}$, G is the Green function for the Laplace operator in the ball and $s_0 \in C^{ex}$, then the Green function of the model coincides with the leading term of the asymptotic of the Green function of the real problem with an opening Γ_d of radius d (the accuracy is $O((kd)^{5-m})$, $k^2 = \lambda$).

One of the consequences of this fact is that the resonances of our model problem, which can be easily calculated, approximate the resonances of the complicated real problem.

References

1. B.S.Pavlov, Sov.Math.Uspekhi 42(6), 99-131 (1987).
2. I.Yu.Popov, Lecture Notes in Physics 324, 218-230 (1989).

USSR, 197101 Leningrad, Sablinskaya St. 14,
Leningrad Institute of Fine Mechanics and Optics (LITMO),
Department of Mathematics.

Operator Theory:
Advances and Applications, Vol. 46
© 1990 Birkhäuser Verlag Basel

CONSTRUCTION OF AN INELASTIC SCATTERER IN
NANOELECTRONICS BY THE EXTENSION-THEORY METHODS

I.Yu.Popov

A model of inelastic scatterer which is analogous to the Buttiker's model is constructed by the extension theory methods.

A model of inelastic scatterer has been suggested in [1]. It looks as follows : the electron wave passes through a T-like plane waveguide andone branch of this waveguide is connected to the reservoir containing a set of electrons. If the electron comes to the reservoir, another electron with a random phase leaves it.Let t_{nm} and r_{nm} be the averaged transmission and reflection coefficient of m-th mode to n-th mode of waveguide, respectively.Inelastic scattering is characterised by the absence of mixing of modes, $| t_{nm} | = \delta_{nm}$.

Let us construct an analogous model using the extension theory. For simplicity we shall consider the case of two propagating modes only. Suppose we have the following system of "wires": two semiinfinite wires $[M,\infty)$ (numbered 1,3), two semiinfinite wires $[N,\infty)$ (wires 2,4) and two finite wires $[M,Q]$, $[N,Q]$ (numbered 01,02). Let H or H , respectively, be the operator $-d^2/dx^2$, acting in the space L^2 (referring to the particular wire) with the domain $W_2^2 \cap \{f: f(M)=f(N)=f(Q)=0\}$. The wires 1, 3, 01 and 2, 4, 02 correspond to the modes with the wavenumber k_1 and k_2, respectively. The operator $H = H_1 \oplus H_2 \oplus H_3 \oplus H_4 \oplus H_{01} \oplus H_{02}$ is symmetric and have the deficiency indices (8,8) - cf.[2]. We choose a class of self-adjoint extensions of the operator H whose domains consist of the functions from the domain of H^* satisfying the conditions

$$f_1(0)= f_3(0)=f_{01}(0), \quad f_1'(0)+f_3'(0)+f_{01}'(0) = Cf_3(0) \text{ at the point M,}$$

$$f_2(0)=f_4(0)=f_{02}(0), \quad f_2'(0)+f_4'(0)+f_{02}'(0) = Cf_4(0) \quad \text{at N,}$$

$$\begin{pmatrix} f_{01}'(1) \\ f_{01}(1) \end{pmatrix} = D \begin{pmatrix} f_{02}'(1) \\ f_{02}(1) \end{pmatrix} \text{ at Q, where D is a random Hermitian matrix}$$

Now one can obtain the transmission and reflection coefficients for this model problem. If D has zero mean value, the averaged coefficients are $r_{jj} = (C+k_j)^{-1}(2ik_j-k_j-C)$, $t_{jj} = (2ik_j)^{-1}(C+k_j)$, $j = 1,2$, $r_{1j} = t_{1j} = 0$ for $l \neq j$, i.e. the modes are not mixed (a peculiar property of inelastic scattering) in the model under discussion. Using the well-known relation between the transmission and reflection coefficients and the resistance (the Landauer formula - cf.[3]), we obtain the temperature dependence of resistance for the case of inelastic electron scattering.

References

1. M.Büttiker, IBM J.Res.Dev.32, No 3, 317-334 (1987).
2. P.Exner, P.Šeba, P.Šťovíček, Lect.Notes in Phys.324, 257-266 (1989).
3. U.Sivan, Y.Imry, Phys.Rev.B33, 551-558 (1986).

USSR, 197101 Leningrad, Sablinskaya St. 14,
Leningrad Institute of Fine Mechanics and Optics (LITMO),
Departments of Mathematics

Operator Theory:
Advances and Applications, Vol. 46
© 1990 Birkhäuser Verlag Basel

ON BOUNDARY THEORY FOR SCHRÖDINGER OPERATORS AND STOCHASTIC PROCESSES

J.F. Brasche, W. Karwowski

0. INTRODUCTION

Non–standard Schrödinger operators have proved to be a powerful tool for modelling a wide variety of physical phenomena, cfr. [1], [2], [3], [4] and references given therein (a non–standard Schrödinger operator H is, by definition, a self–adjoint operator on $L^2(\mathbb{R}^d, dx)$ (dx denotes the Lebesgue measure) with following properties:

(i) there exists a closed subset Γ of $\mathbb{R}^d$ with Lebesgue measure zero such that $C_0^\infty(\mathbb{R}^d \backslash \Gamma)$, the space of infinitely differentiable functions with compact support away from Γ, is in the domain $D(H)$ of H,

(ii) $H \equiv -\Delta$ on $C_0^\infty(\mathbb{R}^d \backslash \Gamma)$ where $-\Delta$ denotes the free Hamiltonian (cfr. [5], §IX.7) but

(iii) $H \neq -\Delta$).

Unfortunately the standard Schrödinger theory cannot be applied to study these operators since, of course, there exists no operator V such that $-\Delta + V$ satisfies each of above conditions (i), (ii) and (iii). Consequently one is faced with a lot of new (often amazing) mathematical problems if one wants to develop a theory of non–standard Schrödinger operators. Among others the question arises how to construct such operators. In this paper we discuss a method of construction which can be applied whenever the set Γ is not too small in the sense that its classical capacity is strictly positive. We refer to [1] – [4],

[6] – [9], and references given therein for other methods and a detailed analysis of certain classes of non–standard Schrödinger operators.

The construction we have in mind is as follows. One takes a positive Radon measure μ on Γ and a closable positive quadratic form $\mathcal{E}'$ with domain $D(\mathcal{E}')$ on $L^2(\Gamma, \mu)$ such that $f|\Gamma$, the restriction of f on Γ, is in $D(\mathcal{E}')$ and $\mathcal{E}'(f|\Gamma, f|\Gamma) \neq 0$ for some $f \in C_0^\infty(\mathbb{R}^d)$. Then one considers the quadratic form $\mathcal{E}$ on $L^2(\mathbb{R}^d, dx)$, defined by

$$(0.1a) \qquad \mathcal{E}(f, g) := (-\Delta f, g) + \mathcal{E}'(f|\Gamma, g|\Gamma)$$

$((\,,\,)$ denotes the scalar product on $L^2(\mathbb{R}^d, dx))$ for all f and g in the domain

$$(0.1b) \qquad D(\mathcal{E}) := \{f \in C_0^\infty(\mathbb{R}^d) : f|\Gamma \in D(\mathcal{E}')\}$$

of $\mathcal{E}$. In general the form $\mathcal{E}$ is not closable. But if $\mathcal{E}$ is closable then the positive self–adjoint operator H on $L^2(\mathbb{R}^d, dx)$ uniquely associated with the closure $\bar{\mathcal{E}}$ of $\mathcal{E}$ (in the sense that $D(H^{1/2}) = D(\bar{\mathcal{E}})$, $(H^{1/2}f, H^{1/2}g) = \bar{\mathcal{E}}(f, g) \,\forall f, g \in D(\bar{\mathcal{E}}))$ is a non–standard Schrödinger operator, as is easily verified. Thus one is led to the problem to characterize those Γ, μ and $\mathcal{E}'$, for which the form $\mathcal{E}$, defined by (0.1), is closable on $L^2(\mathbb{R}^d, dx)$. This problem is also very important in the theory of stochastic processes. Namely if $\mathcal{E}$ is closable and Markovian and $C_0^\infty(\mathbb{R}^d) \subset D(\mathcal{E})$ then the closure $\bar{\mathcal{E}}$ of $\mathcal{E}$ is a regular Dirichlet form (cfr. [10], §1.1 for the basic definitions and Theorem 2.1.1 for the proof). But each regular Dirichlet form on $L^2(\mathbb{R}^d, dx)$ yields in a canonical way a (sub–) Markov process $\mathbf{M}$ on $\mathbb{R}^d$ (cfr. [10], chapter 6). In particular, $\bar{\mathcal{E}}$ yields a Markov process $\mathbf{M}_\mathcal{E}$ on $\mathbb{R}^d$. It follows easily from general results (cfr. [10], chapter 4) that away from the set Γ the process $\mathbf{M}_\mathcal{E}$ coincides with Brownian motion. Thus the question for which Γ, μ and $\mathcal{E}'$ the form $\mathcal{E}$ is closable is closely related to the boundary theory of Brownian motion.

Actually it is desirable to study closability of the form $\mathcal{E}$ within a somewhat more general framework. First of all one wants to replace the expression $(-\Delta f, g)$ in (0.1a) by $c((-\Delta)^\alpha f, g)$, $\alpha, c > 0$.

The reason is, of course, that the quadratic form $\mathcal{E}_{\alpha,c}$, defined by

$$D(\mathcal{E}_{\alpha,c}) := C_0^\infty(\mathbb{R}^d),$$

$$(0.2) \qquad \mathcal{E}_{\alpha,c}(f,g) := c((-\Delta)^\alpha f, g) \;\; \forall f, g \in C_0^\infty(\mathbb{R}^d)$$

is closable on $L^2(\mathbb{R}^d, dx)$ and (for $\frac{1}{2} < \alpha \le 1$) its closure is a regular Dirichlet form which yields in a canonical way a symmetric stable process with index α. Moreover it is natural to replace the perturbation term $\mathcal{E}'$ in (0.1) by a (not necessarily countable) superposition of perturbation terms $\mathcal{E}_\theta$. Thus one is led to the problem to discuss closability of quadratic forms $\mathcal{E}$ on $L^2(\mathbb{R}^d, dx)$ of the kind

$$D(\mathcal{E}) := \{ f \in C_0^\infty(\mathbb{R}^d) : f|\Gamma_\theta \in D(\mathcal{E}_\theta) \text{ for } \nu - \text{a.e. } \theta \in \Theta,$$

$$\mathcal{E}_\theta(f|\Gamma_\theta, f|\Gamma_\theta) \text{ is } \mathcal{A} - \text{measurable in } \theta,$$

$$(0.3) \qquad \int_\Theta \mathcal{E}_\theta(f|\Gamma_\theta, f|\Gamma_\theta)\nu(d\theta) < \infty \}$$

$$\mathcal{E}(f,g) := \mathcal{E}_{\alpha,c}(f,g) + \int_\Theta \mathcal{E}_\theta(f|\Gamma_\theta, g|\Gamma_\theta)\nu(d\theta), \;\; \forall f, g \in D(\mathcal{E}).$$

Here $(\Theta, \mathcal{A}, \nu)$ denotes an auxiliary measure space and, for each $\theta \in \Theta$, Γ_θ is a closed subset of $\mathbb{R}^d$ and $\mathcal{E}_\theta$ a closable positive quadratic form on $L^2(\Gamma_\theta, \mu_\theta)$ (where μ_θ is a positive Radon measure on Γ_θ).

For smooth manifolds Γ_θ closability of the form $\mathcal{E}$, defined by (0.3), has been discussed already a long time ago (cfr. [10], §2.1, cfr. [11], [12] for more recent results and an extension to a more general class of forms). However with a view to various applications in quantum field theory, polymer physics and other fields it is necessary to consider also "irregular" sets Γ_θ like, e.g., a "typical path of Brownian particle" (cfr. [1], §7.5). For this reason we give results on closability of the form $\mathcal{E}$, defined by (0.3), which include "irregular sets" Γ_θ. Our results are new even if specialized to the form $\mathcal{E}$, defined by (0.1). In Theorem 1.7 we give a sufficient condition on the measures μ_θ in order that the form $\mathcal{E}$ (defined by (0.3)) is closable on $L^2(\mathbb{R}^d, dx)$ for each admissible family $\{\mathcal{E}_\theta\}_{\theta \in \Theta}$, i.e. for each family $\{\mathcal{E}_\theta\}_{\theta \in \Theta}$ such that $\mathcal{E}_\theta$ is a closable positive quadratic form on $L^2(\Gamma_\theta, \mu_\theta)$ for ν−almost every

$\theta \in \Theta$. This condition is fairly weak in the sense that, in contradistinction to previous approaches, we do not exclude "irregular" sets Γ_θ. Nevertheless for some applications our assumptions in Theorem 1.7 are too strong. Thus in Theorem 1.1 we give an even weaker condition on the measure μ_θ which ensures that the form $\mathcal{E}$ is closable for each admissible family $\{\mathcal{E}_\theta\}_{\theta \in \Theta}$ with the additional property that $\mathcal{E}_\theta$ is coercive (i.e. there exists $c_\theta > 0$ such that

$$\mathcal{E}_\theta(f, f) \geq c_\theta(f, f)_{L^2(\Gamma_\theta, \mu_\theta)} \forall f \in D(\mathcal{E}_\theta))$$

for $\nu-$a.e. $\theta \in \Theta$. In a certain sense our condition in Theorem 1.1 is optimal, see remark 1.2. Finally we discuss the closure of the form $\mathcal{E}$ (proposition 1.8 and proposition 1.9). In this short contribution we only present general results. The proofs, illustrating examples and applications will be given elsewhere.

1. THE RESULTS

Througout this section the capacity c_α ($\alpha > 0$ is arbitrary but fixed) will play an important rôle. It is defined as follows. For each compact set K

$$c_\alpha(K) := \inf\{(f, f)_{\bar{\mathcal{E}}_{\alpha,c}} : f \in C_0^\infty(\mathbb{R}^d), \ f(x) \geq 1 \ \forall x \in K\}$$

$(f, g)_{\bar{\mathcal{E}}_{\alpha,c}} := \bar{\mathcal{E}}_{\alpha,c}(f, g) + (f, g)$ for all f and g in the Sobolev space $W_\alpha(\mathbb{R}^d)$ and for an arbitrary subset A of $\mathbb{R}^d$

$$c_\alpha(A) := \sup c_\alpha(K)$$

where the supremum is taken over all compact subsets K of A.

The following theorem gives a very weak condition in order that the form $\mathcal{E}$, defined by (0.3), is closable on $L^2(\mathbb{R}^d, dx)$. In fact, in a certain sense, this condition is optimal (cf. Remark 1.2).

THEOREM 1.1: *Let $(\Theta, \mathcal{A}, \nu)$ be an auxiliary measure space. For $\nu-$a.e. $\theta \in \Theta$ let Γ_θ be a closed subset of $\mathbb{R}^d$, μ_θ a positive Radon measure on Γ_θ and $\mathcal{E}_\theta$ a closable coercive quadratic form on $L^2(\Gamma_\theta, \mu_\theta)$. Suppose that none of the measures μ_θ charges a set with $c_\alpha-$capacity zero. Then the form $\mathcal{E}$, defined by (0.3), is closable on $L^2(\mathbb{R}^d, dx)$.*

REMARK 1.2: Let Γ be a closed subset of $\mathbb{R}^d$ and μ be a positive Radon measure on Γ charging a set with c_α−capacity zero. Then the form $\mathcal{E}$, defined by

$$D(\mathcal{E}) := C_0^\infty(\mathbb{R}^d)$$

$$\mathcal{E}(f,g) := ((-\Delta)^\alpha f, g) + \int_\Gamma f(x)g(x)\mu(dx) \ \forall f,g \in D(\mathcal{E})$$

is not closable on $L^2(\mathbb{R}^d, dx)$.

REMARK 1.3: By a result of Fuglede [13] and Meyers [14] a positive Radon measure charges a set B with $c_\alpha(B) = 0$ if and only if there exists a compact set $K \subset B$ such that

$$\int_K \int_K \int_{\mathbb{R}^d} J_\alpha(x - y) J_\alpha(y - z) dy\mu(dx)\mu(dz) = \infty.$$

(J_α denotes the convolution kernel of the operator $(-\Delta + 1)^{-\alpha/2}$ in $L^2(\mathbb{R}^d, dx)$, i.e.

$$\int_{\mathbb{R}^d} J_\alpha(x - y)(-\Delta + 1)^{\alpha/2} f(y)dy = f(x) \ \forall x \in \mathbb{R}^d$$

for each f in the Schwartz space $\mathcal{S}(\mathbb{R}^d)$ of functions of rapid decrease). Since the properties of the Bessel potentials J_α are well known (cfr. [15], §V.3) this gives a convenient tool in order to check whether the hyphothesis of Theorem 1.1 is satisfied.

In the following we admit closable quadratic forms $\mathcal{E}_\theta$ which are not coercive but only positive. We need, however, stronger conditions on the measures μ_θ. First we state a general theorem on closability of quadratic forms defined on different Hilbert spaces.

THEOREM 1.4: *Let H be a (real) Hilbert space and $(\mathcal{E}_0, D(\mathcal{E}_0))$ a closable positive quadratic form on H. Let $(\Theta, \mathcal{A}, \nu)$ be an auxiliary measure space and for each $\theta \in \Theta$ let H_θ be a Hilbert space, $P_\theta : D(\mathcal{E}_0) \to H_\theta$ a linear transformation and $(\mathcal{E}_\theta, D(\mathcal{E}_\theta))$ a closable positive quadratic form on H_θ. Suppose that*

$$(P_\theta f, P_\theta f)_{H_\theta} \leq a\{\mathcal{E}_0(f, f) + (f, f)_H\}$$

$\forall f \in D(\mathcal{E})$ for ν−a.e. $\theta \in \Theta$ and some $a \in \mathbb{R}$ independent of f and θ. Then the form $\mathcal{E}$ on H, defined by

$$D(\mathcal{E}) := \{f \in D(\mathcal{E}_0) : P_\theta f \in D(\mathcal{E}_\theta) \text{ for } \nu - a.e \; \theta \in \Theta \,,$$

$$\mathcal{E}_\theta(P_\theta f, P_\theta f) \text{ is } \mathcal{A} - \text{measurable in } \theta \in \Theta \,,$$

$$\textstyle\int_\Theta \mathcal{E}_\theta(P_\theta f, P_\theta f)\nu(d\theta) < \infty\}$$

$$\mathcal{E}(f,g) := \mathcal{E}_0(f,g) + \textstyle\int_\Theta \mathcal{E}_\theta(P_\theta f, P_\theta g)\nu(d\theta) \;\; \forall f,g \in D(\mathcal{E})\,,$$

is closable on H.

We want to apply this theorem in the special case when $H = L^2(\mathbb{R}^d, dx)$, $H_\theta = L^2(\Gamma_\theta, \mu_\theta)$, $P_\theta f = f_\theta := f|\Gamma_\theta$ and $\mathcal{E}_0 = \mathcal{E}_{\alpha,c}$. Thus we need criteria in order that $\int_\Gamma f(x)^2 \mu(dx) \leq a\{\mathcal{E}_{\alpha,c}(f,f) + (f,f)\}$ $\forall f \in C_0^\infty(\mathbb{R}^d)$ and some $a \in \mathbb{R}$. Such a criterion is given by the following proposition.

PROPOSITION 1.5: *Let* Γ *be a closed subset of* $\mathbb{R}^d$ *and* μ *a positive Radon measure on* Γ. *Suppose*

$$c_{\mu,\alpha} := \textstyle\int_\Gamma \int_\Gamma \{\int_{\mathbb{R}^d} J_\alpha(x-y)J_\alpha(y-z)dy\}^2 \mu(dx)\mu(dy) < \infty\,.$$

Then

$$\textstyle\int_\Gamma f(x)^2 \mu(dx) \leq a(f,f)_{\bar{\mathcal{E}}_{\alpha,c}} \;\; \forall f \in \mathcal{S}(\mathbb{R}^d)$$

for some $a \in \mathbb{R}$ *independent of* f.

REMARK 1.6: The condition that $c_{\mu,\alpha} < \infty$ is sufficient but not necessary in order that the assertion of proposition 1.5 holds. Other sufficient conditions are well known from the theory of trace operators in Sobolev spaces. However within this theory one requires that the set Γ satisfies certain smoothness conditions. Thus, in contradistinction to above proposition, the results of this theory cannot be applied to "irregular" sets like, e.g., fractal sets, "typical paths of a Brownian particle" and so on.

In the following theorem we give a fairly weak condition in order that the form $\mathcal{E}$, defined by (0.3), is closable on $L^2(\mathbb{R}^d, dx)$ for each admissible family $\{\mathcal{E}_\theta\}_{\theta \in \Theta}$. (In contradistinction to Theorem 1.1 the forms $\mathcal{E}_\theta$ are not necessarily coercive).

THEOREM 1.7: *Let* $(\Theta, \mathcal{A}, \nu)$ *be an auxiliary measure space. For each* $\theta \in \Theta$ *let* Γ_θ *be a closed subset of* $\mathbb{R}^d$, μ_θ *a positive Radon mea-*

sure on Γ_θ and $\mathcal{E}_\theta$ a closable quadratic form on $L^2(\Gamma_\theta, \mu_\theta)$. Suppose that

$$\nu - \operatorname*{ess\ sup}_{\theta \in \Theta} \int_{\Gamma^\theta} \int_{\Gamma^\theta} \{\int_{\mathbb{R}^d} J_\alpha(x-y)J_\alpha(y-z)dy\}^2 \mu_\theta(dx)\mu_\theta(dz) < \infty.$$

Then the form $\mathcal{E}$, defined by (0.3), is closable on $L^2(\mathbb{R}^d, dx)$.

Finally we would like to discuss the closure of the form $\mathcal{E}$. First of all we have the following information on the closure $\bar{\mathcal{E}}$ of $\mathcal{E}$.

PROPOSITION 1.8: *Both under the hypothesis (and with the notation) of Theorem 1.1 and under the hypothesis of Theorem 1.7 the closure $\bar{\mathcal{E}}$ of the form $\mathcal{E}$, defined by (0.3), satisfies*

$$D(\bar{\mathcal{E}}) \subset \{f \in W_\alpha(\mathbb{R}^d) : \tilde{f}^\theta := \tilde{f}|\Gamma_\theta \in D(\bar{\mathcal{E}}_\theta)$$

$$\text{for } \nu - a.e. \ \theta \in \Theta, \ \bar{\mathcal{E}}_\theta(\tilde{f}^\theta, \tilde{f}^\theta) \text{ is } \mathcal{A} - \text{measurable in } \theta,$$

$$\textstyle\int_\Theta \bar{\mathcal{E}}_\theta(\tilde{f}^\theta, \tilde{f}^\theta)\nu(d\theta) < \infty\},$$

$$\bar{\mathcal{E}}(f,g) = \bar{\mathcal{E}}_{\alpha,c}(f,g) + \textstyle\int_\Theta \bar{\mathcal{E}}_\theta(\tilde{f}^\theta, \tilde{g}^\theta)\nu(d\theta) \ \forall f,g \in D(\bar{\mathcal{E}})$$

($\tilde{f}$ denotes an arbitrary quasi-continuous representant of f, cfr. [16] for the definition of quasi-continuity).

In the remaining part of this section, for the sake of simplicity, we concentrate on the case that Θ is a singleton, i.e. we consider the quadratic form $\mathcal{E}$, given by

$$D(\mathcal{E}) := \{f \in C_0^\infty(\mathbb{R}^d) : f|\Gamma \in D(\mathcal{E}')\},$$

$$(1.1) \qquad \mathcal{E}(f,g) := \mathcal{E}_{\alpha,c}(f,g) + \mathcal{E}'(f|\Gamma, g|\Gamma) \ \forall f,g \in D(\mathcal{E}),$$

where Γ is a closed subset of $\mathbb{R}^d$ and $\mathcal{E}'$ a closable positive quadratic form on $L^2(\Gamma, \mu)$ (μ a positive Radon measure on Γ) such that $\mathcal{E}$ is closable on $L^2(\mathbb{R}^d, dx)$. By proposition 1.8, $\{\tilde{f}|\Gamma : f \in D(\bar{\mathcal{E}})\} \subset D(\bar{\mathcal{E}}')$. It is a sophisticated question whether conversely each $u \in D(\bar{\mathcal{E}}')$ is the restriction on Γ of a suitable chosen $f \in D(\bar{\mathcal{E}})$, i.e. whether $\{\tilde{f}|\Gamma : f \in D(\bar{\mathcal{E}})\} = D(\bar{\mathcal{E}}')$. We discuss this question in proposition 1.9 below. First we introduce some notations. We denote by

$N_{\alpha,c}$ the closure in the Hilbert space $(W_\alpha(\mathbb{R}^d), (\, , \,)_{\bar{\mathcal{E}}_{\alpha,c}})$ of the space $\{f \in C_0^\infty(\mathbb{R}^d) : f \equiv 0 \text{ on } \Gamma\}$ and by P_0 the orthogonal projection in $W_\alpha(\mathbb{R}^d)$ on $N_{\alpha,c}$. We set $P := I - P_0$ and $D_{\alpha,c} := PC_0^\infty(\mathbb{R}^d)$.

PROPOSITION 1.9: *Let Γ be a closed subset of $\mathbb{R}^d$, μ a positive Radon measure on Γ charging no set with c_α-capacity zero and $\mathcal{E}'$ a closable positive quadratic form on $L^2(\Gamma, \mu)$ such that the form $\mathcal{E}$, given by (1.1), is closable on $L^2(\mathbb{R}^d, dx)$. Suppose that $\{\tilde{f}|\Gamma : f \in D_{\alpha,c}, \tilde{f}|\Gamma \in D(\bar{\mathcal{E}}')\}$ is a core for $\bar{\mathcal{E}}'$ and that there exists an $a > 0$ such that*

$$\bar{\mathcal{E}}(\tilde{f}|\Gamma, \tilde{f}|\Gamma) + \int_\Gamma \tilde{f}(x)^2 \mu(dx) \geq a(f,f)_{\bar{\mathcal{E}}_{\alpha,c}} \ \forall f \in D_{\alpha,c}.$$

Suppose further that for each $f \in C_0^\infty(\mathbb{R}^d)$ either $f|\Gamma \in D(\mathcal{E}')$ or $f|\Gamma \notin D(\bar{\mathcal{E}}')$. Then $\{\tilde{f}|\Gamma : f \in D(\bar{\mathcal{E}})\} = D(\bar{\mathcal{E}}')$.

ACKNOWLEDGEMENT

It is a pleasure to thank J. Dittrich, P. Exner, H. Neidhardt and P. Šeba for the invitation to the conference, and very pleasant stays in Dubna as well as many stimulating discussions. The second named author is also indebted to SFB 237 for supporting his visits to the Institute of Mathematics of the Ruhr–Universität Bochum where part of the research has been done. We are both grateful to S. Albeverio for interest in our work and helpful discussions.

REFERENCES

[1] Albeverio, S., Fenstad, J.E., Høegh–Krohn, R., Lindström, T.: "Non-standard methods in Stochastic Analysis and Mathematical Physics", Academic Press, New York – San Francisco – London, 1986

[2] Albeverio, S., Gesztesy, F., Høegh–Krohn, R., Holden, H.: "Solvable Models in Quantum Mechanics", Springer Verlag, Berlin – Heidelberg – New York, 1988

[3] Exner, P.,Šeba, P. (Eds]: "Application of self–Adjoint Extensions in Quantum Physics", Springer Verlag, 1989

[4] Exner, P., Šeba, P. (Eds]: "Schrödinger Operators – Standard and Non–Standard", World Scientific Publishing Company, Singapore, 1989

[5] Reed, M., Simon, B.: Methods of Modern Mathematical Physics II, Academic Press, 1975

[6] Albeverio, S., Brasche, J.F., Röckner, M.: "Dirichlet forms and generalized Schrödinger Operators", p. 1-42 in Holden, H, Jensen, A.: "Schrödinger Operators", lecture Notes in Physics 345, Springer, 1989

[7] Brasche, J.F.: "Generalized Schrödinger Operators, an inverse problem in spectral analysis and the Efimov effect", to appear in the Proceedings of the international conference in Ascona–Locaouo on "Stochastic Processes, Physics and Geometry", eds. Albeverio, S., Casati, G., Cattaneo, U., Merlini, K., Moresi, R., World Scientific Publishing Company

[8] Brasche, J.F.: "Perturbations of self–adjoint operators supported by null sets", PhD–Thesis, Bielefeld 1988

[9] Teta, A.: "Singular Perturbations of the Laplacian and Connection with Models of Random Media", PhD–Thesis, Trieste, 1989

[10] Fukushima, M.: "Dirichlet forms and Markov processes", North–Holland/Kodansha, Amsterdam–Oxford–New York, 1980

[11] Davies, E.B.: '1'Heat kernels and spectral theory", Cambridge University Press 1989

[12] Karwowski, W., Marion, J.: "On the Closability of Some Positive Definite Symmetric Differential Forms on $C_0^\infty(\Omega)$", Journal of Functional Analysis, **62** (1985), 266–275

[13] Fuglede, B.: "Applications du théorème minimax à l'étude de diverses capacités" C.R. Acad. Sei. Paris, sér A 266 (1968), 961–923

[14] Meyers, N.G.: "A theory for capacity for potentials of function in Lesbesgue classes", math. Scand. **26** (1970), 255–292

[15] Stein, E.M.: "Singular integrals and differentiability properties of functions", Princeton University Press, Princeton, N.J., 1980

[16] Maz'ja, V.G., Havin, V.P.: "Non–linear potential theory", Uspehi Mat. Nauk **27** (1972), 67–138 English transl. in Russian Mathematical Surveys

J.F. Brasche,
Institut für Mathematik,
Ruhr–Universität Bochum,
4630 Bochum, Germany

W. Karwowski,
Institute of Theoretical Physics,
University of Wroclaw,
ul. Cybulskiego 36,
50–205 Wroclaw, Poland.

Operator Theory:
Advances and Applications, Vol. 46
© 1990 Birkhäuser Verlag Basel

DIRAC HAMILTONIAN WITH COULOMB POTENTIAL
AND CONTACT INTERACTION ON A SPHERE

J.Dittrich, P.Exner and P.Šeba

All rotationally and space-reflection symmetric contact interactions supported by a sphere are constructed for a Dirac particle in the Coulomb potential with the source in the centre of the sphere. Some of their spectral properties and unitary equivalences among them are discussed.

1. INTRODUCTION

Dirac equation with external Coulomb potential is very successful in atomic physics and at the same time belongs to the rather limited number of exactly solvable models in relativistic quantum mechanics. It is also desirable to study Dirac equation with Coulomb potential supplemented by a short-range interaction. Phenomenologically such a situation occurs, e.g., in hadronic atoms. To obtain a simple mathematically manageable model of this kind, we consider a contact interaction on a spherical shell with Coulomb source in its centre. As an example, we can formally write the Hamiltonian

$$-i\vec{\alpha}\cdot\vec{\nabla} + \beta m - \frac{Z\alpha}{r} + g_S\beta\delta(r-R) + g_V\delta(r-R) \tag{1}$$

where R is the radius of the sphere.

2. THE HAMILTONIAN

The contact interactions are constructed by a well-known method [1], namely, the Hamiltonian without contact-interaction terms is restricted to the domain containing only function with supports disjoint with the support of the interaction ; the obtained operator is not self-adjoint in general, and the Hamiltonians with contact interactions are constructed as its self-adjoint extensions.

We are going to perform this procedure for the Dirac Hamiltonian with Coulomb field. We restrict ourselves to the class of Hamiltonians which are symmetric with respect to space rotations and reflections. Then we can decompose the problem into partial waves of a given angular momentum j and parity $(-1)^l$. It is therefore sufficient to consider the radial Hamiltonian for each partial wave.

We start from the radial operator

$$H_0 = \begin{pmatrix} m - \dfrac{Z\alpha}{r} & -\dfrac{d}{dr} + \dfrac{\kappa}{r} \\ \\ \dfrac{d}{dr} + \dfrac{\kappa}{r} & -m - \dfrac{Z\alpha}{r} \end{pmatrix} \tag{2}$$

in the Hilbert space $\mathcal{H} = L^2(\mathbb{R}_+) \otimes \mathbb{C}^2$ with the restricted domain

$$\mathcal{D}(H_0) = C_0^\infty((0,R) \cup (R,\infty)) \otimes \mathbb{C}^2 . \tag{3}$$

Here we have denoted $\kappa = (-1)^{j-l+1/2}(j+1/2)$, m is the mass of Dirac particle, Z is the charge number of Coulomb source and α is the fine-structure constant.

The deficiency indices $d_\pm := \dim \mathrm{Ker}\,(H_0^* \mp i)$ can be calculated by solution of the equations for the elements of deficiency subspaces. Since H_0 is real, $d_+ = d_- \equiv d$ and the self-adjoint extensions exist. The values of d for various Z are given in Table I.

TABLE I. Deficiency indices d.

z	$\lvert Z\rvert \leq (\kappa^2-1/4)^{1/2}\,\alpha^{-1}$	$(\kappa^2-1/4)^{1/2}\,\alpha^{-1} < \lvert Z\rvert < \lvert\kappa\rvert\,\alpha^{-1}$
d	2	3

For $\lvert Z\rvert \leq \sqrt{3}(2\alpha)^{-1} \approx 118.7$ only the first possibility $d=2$ occurs for all j . Self-adjoint extensions H of H_0 are then defined by imposing two boundary conditions on functions from $\mathcal{D}(H)$ at the point R .

For $\sqrt{3}(2\alpha)^{-1} < \lvert Z\rvert < \alpha^{-1}$, the second case $d=3$ occurs for $j=1/2$. Self-adjoint extensions are then defined by two boundary conditions at the point R and one boundary condition at zero. Hence in addition to the shell contact interaction at $r=R$ there could be, in general, also a point contact interaction at $r=0$. However, we are not interested in the latter and we can avoid it choosing a suitable self-adjoint extension in the same way as one does it for Dirac Hamiltonian with Coulomb potential without the shell [2-7].

For simplicity, we restrict ourselves to $\lvert Z\rvert \leq \sqrt{3}(2\alpha)^{-1}$ in the following. Let us denote $\tilde{H}_0$ the formal differential operator of the form (2). The self-adjoint extensions H are defined by relations

$$\mathcal{D}(H) = \left\{ \psi \in \mathcal{H} : \psi \text{ absolutely continuous inside } (0,R)\cup(R,\infty), \right.$$
$$\left. \tilde{H}_0\psi \in \mathcal{H} \, , \; C\psi(R^-) + D\psi(R^+) = 0 \right\} , \tag{4}$$

$$H\psi = \tilde{H}_0\psi \, . \tag{5}$$

The limits $\psi(R^-)$, $\psi(R^+)$ exist for $\psi \in \mathcal{D}(H_0^*)$ and C,D are 2×2 matrices which satisfy the same requirements as in the case of Z=0 [8,9]. The matrices C and D can be transformed into one of the following standard forms : either

$$C = I , \qquad D = e^{i\delta} A \tag{6a}$$

where δ is a real number and A is a real 2×2 matrix, det A = 1, or

$$C = \begin{pmatrix} c_1 & c_2 \\ 0 & 0 \end{pmatrix}, \qquad D = \begin{pmatrix} 0 & 0 \\ d_1 & d_2 \end{pmatrix} \tag{6b}$$

where either $c_1=1$ and c_2 is real or $c_1=0$ and $c_2=1$, and either $d_1=1$ and d_2 is real or $d_1=0$ and $d_2=1$.

3. UNITARY EQUIVALENCES

All self-adjoint extensions of the type (6a) with the same matrix A are unitary equivalent. Let us define the unitary operator $F : \mathcal{H} \longrightarrow \mathcal{H}$ as

$$(F\psi)(r) = \begin{cases} \psi(r) & \text{for } r < R \\ e^{-i\varphi} \psi(r) & \text{for } r > R \end{cases} \tag{7}$$

where φ is a real constant and $\psi \in \mathcal{H}$. If H is a self-adjoint extension of H_0 defined by the boundary condition (6a) then $H'=FHF^{-1}$ is the operator of the same type with the only difference that δ is replaced by

$$\delta' = \delta + \varphi . \tag{8}$$

In particular, operators that differ only by the value of δ have the same spectra. The question whether all unitary equivalences, or more generally, all isospectral transformations between self-adjoint extensions of the operator H_0 are of the form (8) remains open.

4. THE SPECTRUM

Since the essential spectra of all self-adjoint extensions of a symmetric operator with finite deficiency indices are equal [12], we have

$$\sigma_{ess}(H) = \sigma_{ess}(H^{Coulomb}) = (-\infty,-m] \cup [m,\infty) .$$

Here we denoted as $H^{Coulomb}$ the Dirac Hamiltonian with pure Coulomb potential (without the shell interaction). Therefore the spectra $\sigma(H)$ may differ by eigenvalues in $(-m,m)$ only.

The eigenvalues ε in $(-m,m)$ can be looked for in a straightforward way : we have to solve the eigenvalue equation

$$H \psi_\varepsilon = \varepsilon \psi_\varepsilon \tag{9}$$

First we solve the differential equation

$$H_0^{\sim} \psi = \varepsilon \psi \tag{10}$$

in $(0,R)$ and (R,∞), wé select the square integrable solutions, and from them we construct the solution of eq.(9) in $\mathcal{D}(H)$ — cf.(4). We use the substitutions [10]

$$\psi = \left(\begin{array}{c} f \\ g \end{array} \right) \tag{11a}$$

$$f = (m+\varepsilon)^{1/2} e^{-\rho/2} \rho^{\gamma} (Q_1 + Q_2) \tag{11b}$$

$$g = -(m-\varepsilon)^{1/2} e^{-\rho/2} \rho^{\gamma} (Q_1 - Q_2) \tag{11c}$$

and denote

$$\lambda = (m^2 - \varepsilon^2)^{1/2} \ , \qquad \gamma = (\kappa^2 - Z^2\alpha^2)^{1/2} \ , \qquad \rho = 2\lambda r \ ,$$

$$a = \gamma - \frac{Z\alpha\varepsilon}{\lambda} \ , \qquad a_1 = \kappa - \frac{Z\alpha m}{\lambda} \ , \qquad a_2 = \kappa + \frac{Z\alpha m}{\lambda} \ ,$$

$$c = 2\gamma + 1 \ .$$

The square integrable solutions of (10) are given by the functions

$$Q_1(r) = a_1\Phi(a,c;\rho) \ , \qquad Q_2(r) = -a\Phi(a+1,c;\rho) \tag{12}$$

for $r \in (0,R)$, and

$$Q_1(r) = \Psi(a,c;\rho) \ , \qquad Q_2(r) = a_2\Psi(a+1,c;\rho) \tag{13}$$

for $r \in (R,\infty)$. Here Φ is the degenerate hypergeometric function and Ψ is the other solution of the degenerate hypergeometric equation [11].

If ε coincides with one of the purely Coulombic eigenvalues, i.e., if $a = 0,-1,-2,\ldots$ and $(a,a_1) \neq (0,0)$, the functions (13) are proportional to the known Coulombic eigenfunctions (12).

The eigenfunctions of H should be of the form

$$\psi_\varepsilon(r) = u\ \theta(R-r)\ \psi(r) + v\ \theta(r-R)\ \psi(r) \tag{14}$$

where $u,v \in \mathbb{C}$ and ψ is given by formulas (11)-(13). The boundary condition

$$C\ \psi_\varepsilon(R^-) + D\ \psi_\varepsilon(R^+) = 0$$

from (4) have a nontrivial solution (u,v)   iff

$$\det\left[C\psi(R^-) \ , \ D\psi(R^+)\right] = 0 \ .$$

(15)

The eigenvalues ε of our Hamiltonian H are just those solutions of the equation (15) which lead to nonzero eigenvectors ψ_ε of the form (14). The functions ψ_ε which are identicall zero and must be therefore excluded may occur for $a = a_1 = 0$ only.

5. AN EXAMPLE : SCALAR AND VECTOR DELTA SHELL

Let us consider the Hamiltonian (1) which contains contact interaction on a sphere as a scalar and the time component of a vector external fields. The corresponding radial Hamiltonians are then formally

$$\tilde{H} = \tilde{H_0} + \begin{pmatrix} g_V + g_S & 0 \\ 0 & g_V - g_S \end{pmatrix} \delta(r - R) \ .$$

(16)

In order to define operators corresponding to the heuristic expressions (1) and (16), we start with the formal integration of the radial Dirac equation and the formal definition

$$\int_{R-\eta}^{R+\eta} \psi(r) \ \delta(r-R) \ dr = \frac{1}{2}\left[\psi(R^-) + \psi(R^+)\right] \ .$$

It yields a boundary condition of the form (4) with the matrices (cf.[8,9])

$$C_1 = \begin{pmatrix} 1 & x \\ y & 1 \end{pmatrix} , \quad D_1 = \begin{pmatrix} -1 & x \\ y & -1 \end{pmatrix}$$

(17)

where

$$x = \tfrac{1}{2}\left(g_S - g_V\right) , \qquad y = \tfrac{1}{2}\left(g_S + g_V\right) .$$

The relations (4) and (5) with the matrices C_1 and D_1 define a self-adjoint extension of H_0 which is a well-defined realization of the heuristic operator (16).

The matrices C_1 , D_1 are not of the standard form (6). For $xy \neq 1$, i.e., $g_V^2 - g_S^2 + 4 \neq 0$, they can be transformed into the standard form (6a),

$$C = I , \qquad D = \frac{1}{xy - 1} \begin{pmatrix} xy + 1 & -2x \\ -2y & xy + 1 \end{pmatrix} . \tag{18}$$

For $xy = 1$, the matrices C_1, D_1 can be transformed into the standard form (6b),

$$C = \begin{pmatrix} 1 & x \\ 0 & 0 \end{pmatrix} , \qquad D = \begin{pmatrix} 0 & 0 \\ 1 & -x \end{pmatrix} . \tag{19}$$

It can be seen from (18) and (19) that different pairs of coupling constants (g_S, g_V) define actually different operators H . Let us consider the unitary equivalences (8) applied to these operators. Since we allow only real coupling constants g_S and g_V , the only possibility is the transformation

$$D' = -D \tag{20}$$

for the case (18). The transformation (20) is possible only between the operators with $xy \neq 0$, i.e., $g_S^2 \neq g_V^2$, and has the form $x' = y^{-1}$, $y' = x^{-1}$, i.e.,

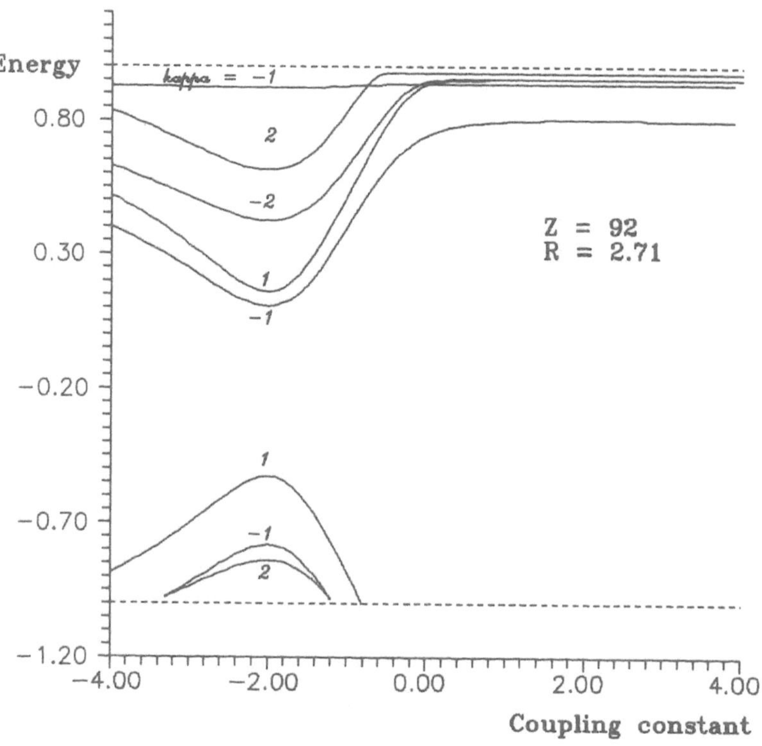

Fig.1a   The dependence of the lowest eigenvalues on the
            strong-shell coupling constant g_S

$$g_S' = \frac{4g_S}{g_S^2 - g_V^2} \quad , \quad g_V' = \frac{4g_V}{g_S^2 - g_V^2} \quad . \tag{21}$$

For $g_S^2 = g_V^2$, the transformation (20) leads again to another
unitarily equivalent self-adjoint extension which is not of the
form (16) now.

For $g_S^2 - g_V^2 = 4$, i.e., in the case (6b) when the
sphere is impenetrable, the transformation (21) reduces to the
identical mapping.

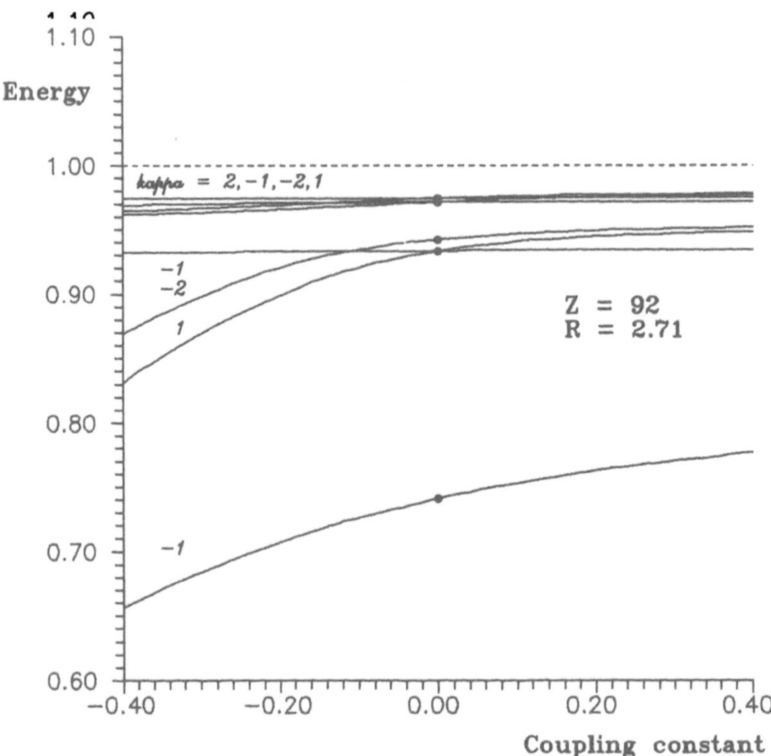

Fig.1b   A detail of the previous picture

In the special case $g_V = 0$ the equations (21) give

$$g'_S = \frac{4}{g_S} \ , \qquad g'_V = 0 \ .$$
(22)

To illustrate these results, we have calculated
numerically from the equation (15) a few "lowest" eigenvalues for
the Hamiltonians (16) with $g_V = 0$, choosing the uranium charge
$Z = 92$ and $R = 2.71$ for the radius of the sphere. This
dimensionless value is in fact related to the Compton wavelength

of the particle under consideration as a unit. Taking into account the empirical formula $R \approx 1.2 \ Z^{1/3}$ fm for the nucleus radius, we see that our choice corresponds roughly to the uranium muonic atom. Of course, a muon does not interact strongly, but we can in this way illustrate the spectral behaviour avoiding at the same time some numerical difficulties that arise for hadronic atoms.

The results are plotted on Figs.1a,b. On the first of them, the dependence of eigenvalues on the scalar shell coupling constant g_S clearly exhibits the symmetry (22). The second picture shows how the eigenvalues behave around $g_S = 0$; for comparison we have plotted here also the well-known eigenvalues [11] for the purely Coulombic case. We see that that the strong interaction removes degeneracy in the sign of κ , i.e., in the orbital momentum or parity, and that for $|g_S|$ large enough it causes a level rearrangement.

REFERENCES

1 S.Albeverio, F.Gesztesy, H.Holden, R.Hoegh-Krohn : *Solvable Models in Quantum Mechanics*. Springer, Berlin, 1988.
2 F.Rellich, Math.Z.<u>49</u> (1943-44),702.
3 G.Nenciu, Helv.Phys.Acta <u>50</u> (1977),1.
4 G. Nenciu, Commun. Math.Phys.<u>48</u> (1976),235.
5 M.Klaus, R.Wüst, Commun.Math.Phys.<u>64</u> (1979),171.
6 M.Klaus, Helv.Phys.Acta <u>53</u> (1980),463.
7 B.Karnarski, J.Operator Theory <u>13</u> (1985),171.
8 J.Dittrich, P.Exner, P.Seba, in *Schroedinger Operators, Standard and Nonstandard* (P. Exner and P. Seba, eds.). World Scientific, Singapore, 1989 ; pp.191-204.
9 J.Dittrich,P.Exner, P.Seba, J.Math.Phys.<u>30</u> (1989), to appear
10 V.B.Beresteckij, E.M.Lifschitz, L.P.Pitajevskij : *Quantum Electrodynamics*. Nauka, Moscow, 1980 (in Russian).
11 H.Bateman, A.Erdélyi : *Higher Transcendental Functions*, Vol.1. Mc Graw-Hill, New York, 1953.
12 H.Behncke, Proc.Am.Math.Soc.<u>72</u> (1978),82.

Nuclear Physics Institute
Czechoslovak Academy of Sciences
25068 Řež near Prague
Czechoslovakia

Operator Theory:
Advances and Applications, Vol. 46
© 1990 Birkhäuser Verlag Basel

SELF-ADJOINT EXTENSIONS OF SCHROEDINGER
OPERATORS WITH SINGULAR POTENTIALS

A. N. Kochubei

Let L_o be a minimal operator in $L_2(0,\infty)$ generated by $l = -\dfrac{d^2}{dt^2} + q$ where q is a real potential with a non-integrable singularity at $t = 0$. For various classes of potentials we present an explicit description of all self-adjoint extensions. The multi-dimensional case is also considered.

1. Let A be a symmetric operator in a Hilbert space H . The boundary value space (BVS) of A is a triplet $(\mathfrak{H},\Gamma_1,\Gamma_2)$ where $\mathfrak{H}$ is another Hilbert space and $\Gamma_j : \mathcal{D}(A^*) \longrightarrow \mathfrak{H}$ are linear mappings such that (i) for each $f,g \in \mathcal{D}(A^*)$

$$(A^*f,g)_H - (f,A^*g)_H = (\Gamma_1 f,\Gamma_2 g)_{\mathfrak{H}} - (\Gamma_2 f,\Gamma_1 g)_{\mathfrak{H}}$$

and (ii) for each $F_1,F_2 \in \mathfrak{H}$ there exists $f \in \mathcal{D}(A^*)$ such that $\Gamma_j f = F_j$, $j = 1,2$.

It is well-known that when a BVS is given one can describe any self-adjoint extension of A as a restriction of A^* to the set of $f \in \mathcal{D}(A^*)$ satisfying

$$(\cos C)\Gamma_2 f - (\sin C)\Gamma_1 f = 0$$

where C is a self-adjoint operator in $\mathfrak{H}$.

2. Suppose that q is locally integrable away from zero and that l is of the limit-point type at $t = \infty$ and the limit-circle at $t = 0$. If $q \in L_1(0,\varepsilon)$, $\varepsilon > 0$, one can take $\mathfrak{H} = \mathbb{C}$, $\Gamma_1 f = f'(0)$, $\Gamma_2 f = -f(0)$ obtaining the well-known BVS for L_0 . This fails when q has a stronger singularity. The general results applicable to this case (see [1] for a review) are not constructive. The problem of finding an explicit description has been studied only for the powerlike behavior $q(t) \sim \text{at}^{-\alpha}$, $t \rightarrow 0$, where $1 \le \alpha \le 2$ [2,3] while for $\alpha > 2$ (and, of course, a < 0) it was stated as unsolved in [4].

In this paper we present a constructive description of the self-adjoint extensions (in fact, explicit formulas for the BVS) for potentials admitting application of the WKB method [5]. This includes a wide class of potentials with $\alpha > 2$. Our approach provides some new results also for $1 \le \alpha \le 2$.

We start with the following basic lemma.

LEMMA. *Let* η_1, η_2 *be real functions such that*
(a) $\eta_1, \eta_2 \in C^1(0,\varepsilon)$,
(b) $[\eta_2,\eta_1]_0 \equiv \lim_{t \to 0} \{\eta_2(t)\eta_1'(t) - \eta_1(t)\eta_2'(t)\} = 1,$
(c) *for any* $f \in \mathcal{D}(L_0^*)$ *there exist finite limits* $[f,\eta_1]_0$, $[f,\eta_2]_0$.
Then a BVS for L_0 *can be obtained as follows:*

$$(1) \qquad \mathfrak{H} = \mathbb{C}, \quad \Gamma_j f = [f,\eta_j]_0, \quad j = 1,2.$$

3. **THEOREM 1.** *Let* $q = -q_1 + q_2$ *where* q_1 *is positive in* $(0,\varepsilon)$ *such that* $q_1 \in C^2(0,\varepsilon)$, $q_1^{-1/4} \in L_2(0,\varepsilon)$ *while* $q_2 \in$

$C(0,\varepsilon)$. *If the condition*

$$\int_0^\varepsilon dt \ \left| [q_1(t)]^{-\frac{1}{4}} \frac{d^2}{dt^2} \left\{ [q_1(t)]^{-\frac{1}{4}} \right\} - [q_1(t)]^{-\frac{1}{2}} q_2(t) \right| < \infty$$

is satisfied, then by (1) a BVS for L_0 *is given defining*

$$(2) \qquad \eta_1(t) = [q_1(t)]^{-\frac{1}{4}} \cos\left\{ \int_t^\varepsilon [q_1(\tau)]^{\frac{1}{2}} d\tau \right\},$$

$$(3) \qquad \eta_2(t) = [q_1(t)]^{-\frac{1}{4}} \sin\left\{ \int_t^\varepsilon [q_1(\tau)]^{\frac{1}{2}} d\tau \right\}$$

for $0 < t < \varepsilon$.

Proof is based on the Lemma and WKB-estimates by Olver [5].

EXAMPLES. 1) Let $q_1(t) = t^{-\alpha} s(t)$ for $0 < t < \varepsilon$ where $\alpha > 2$, $s \in C^2(0,\varepsilon)$ and $0 < s_0 \le s(t) \le s_1$. The conditions of Theorem 1 are satisfied if

$$\int_0^\varepsilon t^{\frac{\alpha}{2}} \{ [s'(t)]^2 + |s''(t)| + |q_2(t)| \} dt < \infty.$$

2) Let $q_1(t) = \exp(\gamma t^{-\beta}) s(t)$ for $0 < t \le \varepsilon$ where $\gamma > 0$, $\beta > 0$, $s \in C^2(0,\varepsilon)$ and $0 < s_0 \le s(t) \le s_1$. The conditions of Theorem 1 are satisfied if

$$\int_0^\varepsilon \exp(-\tfrac{\gamma}{2} t^{-\beta}) \left\{ [s'(t)]^2 + |s''(t)| + |q_2(t)| \right\} dt < \infty.$$

3) Let $q(t) = -t^{-2} s(t) + p(t)$ where $s \in C^2(0,\varepsilon)$, $\tfrac{1}{4} < s_0 \le s(t) \le s_1$, $p \in C(0,\varepsilon)$ and

$$\int_0^{\varepsilon} t \left\{ t^{-1}|s'(t)| + |s'(t)|^2 + |s''(t)| + |p(t)| \right\} dt < \infty.$$

If $q_1(t) = t^{-2}[s(t) - \frac{1}{4}]$, $q_2(t) = -\frac{1}{4} t^{-2} + p(t)$ and $q = -q_1 + q_2$, then the conditions of Theorem 1 are satisfied. The case $s(t) \leq \frac{1}{4}$ will be considered below.

4. THEOREM 2. *Let* $q(t) = t^{-2}s_1(t) + s_2(t)$ *for* $0 < t < \varepsilon$, *where* $s_1 \in C^2[0,\varepsilon]$, $\sigma_0 = s_1(0) \in (-\frac{1}{4}, 0) \cup (0, \frac{3}{4})$ *and* $t^{\beta}s_2 \in L_2(0,\varepsilon)$, $\beta = \frac{1}{2} - \sqrt{\sigma_0 + \frac{1}{4}}$. *Setting* $\sigma_1 = s_1'(0)$ *and*

$$\eta_j(t) = b_0 t^{\beta_j} + b_j t^{\beta_j + 1}, \quad j = 1,2,$$

where $\beta_1 = \frac{1}{2} + \sqrt{\sigma_0 + \frac{1}{4}}$, $\beta_2 = \beta$, $b_0 = (1+4\sigma_0)^{-1/4}$ *and* $b_j = b_0 \sigma_1 [\beta_j(\beta_j + 1) - \sigma_0]^{-1}$, $j = 1,2$, *we give a BVS for* L_0 *by* (1).

THEOREM 3. *Let* $q(t) = t^{-2}s_1(t) + s_2(t)$ *for* $0 < t < \varepsilon$, *where* $s_1 \in C^2[0,\varepsilon]$, $s_1(0) = -\frac{1}{4}$ *and* $(t^{1/2}\log t)s_2(t) \in L_2(0,\varepsilon)$. *Then by* (1) *a BVS for* L_0 *can be given with*

$$\eta_1(t) = (t^{1/2} + \sigma t^{3/2})\log t - 2\sigma t^{3/2},$$

$$\eta_2(t) = t^{1/2} + \sigma t^{3/2}, \quad \sigma = s_1'(0).$$

Next we consider potentials with $1 \leq \alpha < 2$.

THEOREM 4. *Let* $q_1(t) = t^{-1}s_1(t) + s_2(t)$ *for* $0 < t < \varepsilon$, *where* $s_1 \in C^1[0,\varepsilon]$, $s_2 \in L_2(0,\varepsilon)$. *Then by* (1) *a BVS for* L_0 *can be given fixing* η_1 *and* η_2 *by*

$$\eta_1(t) = s_1(0) \; t \log t + 1, \quad \eta_2(t) = -t.$$

THEOREM 5. *Let* $q(t) = t^{-\alpha}s_1(t) + s_2(t)$ *for* $0 < t < \varepsilon$, *where* $1 < \alpha < 2$, $s_1 \in C^2[0,\varepsilon]$, $s_1(0) \neq 0$, $s_2 \in L_2(0,\varepsilon)$. *Then by* (1) *a BVS for* L_0 *can be given defining* η_1 *and* η_2 *as follows: If* N,M *are natural numbers such that*

$$N > (\alpha - \tfrac{3}{2})(2 - \alpha)^{-1}, \quad M > (\alpha - \tfrac{1}{2})(2 - \alpha)^{-1}$$

and if $\sigma_0 = s_1(0)$, $\sigma_1 = s_1'(0)$, *then*

$$\eta_1(t) = \Gamma(1 + (2-\alpha)^{-1}) \sum_{m=0}^{N} \frac{\sigma_0^m (2-\alpha)^{-2m}}{m! \, \Gamma(1+m+(2-\alpha)^{-1})} \, t^{1+m(2-\alpha)}$$

and (if $(2-\alpha)^{-1}$ *is not an integer)*

$$\eta_2(t) = - \sum_{k=0}^{M} \left\{ a_k t^{k(2-\alpha)} + b_k t^{k(2-\alpha)+1} \right\}$$

where

$$a_k = - \frac{\sigma_0^k \, \Gamma(1-(2-\alpha)^{-1})}{(2-\alpha)^{2k} \, k! \, \Gamma(1+k-(2-\alpha)^{-1})},$$

b_0 *is an arbitrary number and* b_k, $k \geq 1$, *are determined recursively by*

$$a_k \sigma_1 + b_k \sigma_0 = b_{k+1}[(k+1)(2-\alpha) + 1](k+1)(2-\alpha).$$

If $(2-\alpha)^{-1} = 1$ *is an integer number, then*

$$\eta_2(t) = - \sum_{k=0}^{M} \left\{ a_k t^{k(2-\alpha)} + b_k t^{k(2-\alpha)+1} \log t + c_k t^{k(2-\alpha)+1} \right\}$$

where $a_o = -1$,

$$a_{k+1}[(k+1)(2-\alpha) - 1](k + 1)(2 - \alpha) = a_k\sigma_o,$$

$$k = 0,1,\ldots,1-2, \quad a_1 = a_{1+1} = \ldots = a_M = 0,$$

$b_o = a_{1-1}\sigma_o$, c_o *is an arbitrary number and* b_k *and* c_k *are determined by the relations*

$$b_{k+1}[(k+1)(2-\alpha) + 1](k+1)(2-\alpha) = b_k\sigma_o,$$

$$c_{k+1}[(k+1)(2-\alpha) + 1](k+1)(2-\alpha) =$$

$$a_k\sigma_1 + c_k\sigma_o - b_{k+1}[2(k+1)(2-\alpha) +1], \quad k = 0,1,\ldots,M-1.$$

5. Finally, let us discuss the minimal operator $\mathcal{L}_o$ in $L_2(\mathbb{R}^n)$, $n \geq 2$, corresponding to $l_n = -\Delta + q(|x|)$, $x \in \mathbb{R}^n$, where q satisfies the conditions of Theorem 1. We assume in addition that q is bounded from below and $r^{-2}[q_1(r)]^{-1/2} \in L_1(0,\varepsilon)$.

Let B be the Laplace-Beltrami operator in $L_2(S^{n-1})$, $0 = \kappa_o > \kappa_1 >\ldots$ be its eigenvalues with eigensubspaces Q_1, $1 = 0,1,\ldots$, which yield the partial wave decomposition

$$L_2(\mathbb{R}^n) = \overset{\infty}{\underset{1=0}{\oplus}} H_1, \quad H_1 = L_2((0,\infty),r^{n-1}dr) \otimes Q_1.$$

We shall use the following version of the Weyl function (a special case of a general Weyl function introduced in [6]). Consider the equation $\lambda_1[y] = iy$ where

$$\lambda_1 = -\frac{d^2}{dr^2} + q(r) + \left[\frac{(n-1)(n-3)}{4} - \kappa_1\right]r^{-2}, \quad 0 < r < \infty.$$

It has two solutions $Y_{1,1}$ and $Y_{1,2}$ such that

$$Y_{1,j} \sim \eta_j(r), \quad r \to 0, \quad j = 1,2$$

(η_j are given by (2) and (3)). There exists a unique number $m_1 \in \mathbb{C}$, $\mathrm{Im}(m_1) > 0$, such that $Y_{1,1} - m_1 Y_{1,2} \in L_2(0,\infty)$. Setting

$$\zeta_{1,1} = (\mathrm{Im}(m_1))^{-1/2}\{\eta_1 - (\mathrm{Re}(m_1)\eta_2\},$$

$$\zeta_{2,1} = (\mathrm{Im}(m_1))^{1/2}\eta_2.$$

we get

THEOREM 6. *For* $\mathcal{L}_0$ *a BVS can be obtained as follows:*

$$\mathfrak{H} = \bigoplus_{l=0}^{\infty} Q_l, \quad \Gamma_j f = \left\{[\tilde{f}_l, \zeta_{j,l}]_o\right\}_{l=0}^{\infty}, \quad j = 1,2,$$

where $\tilde{f}_l(r) = r^{\frac{1}{2}(n-1)} f_l(r)$, f_l *is a component of* f *in* H_l.

REFERENCES

1   Fulton, C.T., *Trans.Amer.Math.Soc.*229 (1977), 51-63.
2   Rellich, F., *Math.Z.* 49 (1943/44), 702-723.
3   Bulla, W.; Gesztesy, F., *J.Math.Phys.*26 (1985), 2520-2528.
4   Burnap, C.; Greenberg, W.; Zweifel, P.F., *Nuovo Cim.* A50 (1979), 457-465.
5   Olver, F.W.J., *Introduction to asymptotics and special functions*, Academic Press, New York, 1974.
6   Derkach, V.A.; Malamud, M.M., *Sov.Math.Dokl.*35 (1987), 393 -398.

Energosetproject
Solomenskaya 5
252 110 Kiev, USSR

Operator Theory:
Advances and Applications, Vol. 46
© 1990 Birkhäuser Verlag Basel

CURRENTS AND THE EXTENSIONS THEORY

Yu.A.Kuperin, E.A.Yarevsky

Formulation of von Neuman's extension theory in terms of boundary
forms is given and equivalence of this theory to the variational
principle for some action functional is proved. It is shown that
the way in which one treats $U(1)$-invariance of this action
selects aselfadjoint extensions and the corresponding Noether's
invariant coincides with the boundary form.

SECTION 1

Let A be a selfadjoint (s.a.) operator in a Hilbert
space $\mathcal{H}$, and suppose that U is its Cayley transform, $\mathfrak{N}_{1}$ and $\mathfrak{N}_{-1}$
are the minimal generating subspaces of A. Let us consider the
class of restrictions A_0,

$$A_0 = A \mid \mathcal{D} (A_0)$$

$$\mathcal{D} (A_0) = (A - iI)^{-1} \mathfrak{N}_1^{\perp} ,$$

(1)

with $\mathfrak{N}_1^{\perp}$ being the orthogonal complement to $\mathfrak{N}_1$. One can show [1]
that

1) A_0 is a symmetric operator on $\mathcal{D}(A_0)$ with deficiency subspaces

$\mathfrak{N}_{\pm 1} = \{Ran(A_0 \mp iI)\}^{\perp}$ and

2) A_0 is densely defined if and only if $\mathcal{D}(A) \cap \mathfrak{N}_{\pm 1} = \{0\}$.

For the minimal $\mathfrak{N}_1$ with the basis $\{\vartheta_s\}$ the vectors $W_s^+ = \frac{1}{2}(U^* + I)\vartheta_s$ and $W_s^- = \frac{1}{2i}(U^* - I)\vartheta_s$ form a basis in $\mathfrak{N} = \mathfrak{N}_{-1} + \mathfrak{N}_1$ and thus each $u \in \mathfrak{N}$ can be decomposed as $u = \sum_s (\xi_s^+ W_s^+ + \xi_s^- W_s^-)$, $\xi_s^\pm \in$ C. The elements $\xi^\pm = \sum_s \xi_s^\pm \vartheta_s$ will be called boundary vectors for the element u.

Define the operator $\hat{A} : \mathfrak{N} \longrightarrow \mathfrak{N}$ by its action on the basis

$$\hat{A} W_s^\pm = \mp W_s^\mp . \tag{2}$$

If A_0 is densely defined, then $\hat{A} = A_0^* | \mathfrak{N}$. The universal symplectic form can de defined in terms of $\hat{A}$ as $\mathfrak{J}(u,v) = \langle\hat{A}u, v\rangle_{\mathfrak{N}} - \langle u, \hat{A}v\rangle_{\mathfrak{N}}$ for $u,v \in \mathfrak{N}$. Since $K_{\pm 1} = \mathfrak{N}_{\pm 1}$ and $\mathfrak{J}(u,v) = \langle\xi^-(u), \xi^+(v)\rangle_{\mathfrak{N}_1} - \langle\xi^+(u), \xi^-(v)\rangle_{\mathfrak{N}_1}$ we call $\mathfrak{J}$ the boundary form of the operator A_0.

For every s.a. operator Γ in $\mathfrak{N}_1$ we define the null-hyperplane $\mathcal{L}_\Gamma = \{\xi^\pm : \xi^- = \Gamma\xi^+ \} \subset \mathfrak{N}$ such that $\mathfrak{J}|\mathcal{L}_\Gamma = 0$. Let $\mathcal{G}$ be the group of all $\mathcal{L}$-invariant transformations in $\mathfrak{N}$. Then the following theorem is valid [1]

THEOREM 1.   *All s.a. extensions* A_Γ *of the operator* A_0 *coincide with the restrictions of* $\hat{A}$ *to the orbit* O_Γ *of the group* $\mathcal{G}$ *containing the null-hyperplane* $\mathcal{L}_\Gamma : A_\Gamma = \hat{A}|O_\Gamma$.

SECTION 2

Let us now reformulate the abstract theory described above in terms of a variational principle. More specifically, for the restrictions A_0 such that $\mathcal{D}(A) \cap \mathfrak{N}_1 = \{0\}$, dim $\mathfrak{N}_1 < \infty$, define the action S by

$$S = \langle (A_0^* - zI)u, u \rangle + \langle u, (A_0^* - zI)u \rangle +$$
$$+ \langle \Gamma^{-1}\xi^-, \xi^- \rangle_{\mathfrak{N}_1} + \langle \Gamma\xi^+, \xi^+ \rangle_{\mathfrak{N}_1} \tag{3}$$

for all $\Gamma : \mathfrak{N}_1 \longrightarrow \mathfrak{N}_1$ with det $\Gamma \neq 0$, $u \in \mathcal{D}(A_0^*)$. Then we have

THEOREM 2. *The condition of stationarity, $\delta S = 0$, of the action (3) is equivalent to the equation of motion*

$$(A_0^* - zI)u = 0 \tag{4}$$

with the boundary conditions

$$\xi^- = \Gamma\xi^+. \tag{5}$$

If we additionally assume that $\Gamma = \Gamma^*$ then the action (3) turns out to be invariant with respect to the gauge transformations $u \longrightarrow u\, e^{i\alpha}$. Such invariance allows to compute the corresponding conserved current. Namely, the infinitesimal gauge transformation has the form $\eta = i\alpha u$, and the action variation δS on the elements satisfying the equation of motion (4) is given by

$$\delta S = -4\alpha \text{ Im } \langle \xi^+(u), \xi^-(u) \rangle_{\mathfrak{N}_1} \tag{6}$$

For the functional spaces the coefficient of $-\alpha$ at the *rhs* of (6) can be interpreted as the integral of $U(1)$-current divergence [2]. Comparing (6) and $\mathfrak{J}(u,v)$,

$$\mathfrak{J}(u,u) = 2i \text{ Im } \langle \xi^+(u), \xi^-(u) \rangle_{\mathfrak{N}}$$

we see that $\delta S = 2i\alpha\mathfrak{J}$, i.e. the coefficient of $-\alpha$ coincides with the boundary form divided by $-2i$. This fact makes the "physical meaning" of the boundary form $\mathfrak{J}$ transparent : it is the abstract version of the $U(1)$-current divergence for differential operators.

Let us summarize the assertions formulated above.

THEOREM 3. *The extensions A_Γ of the operator A_0 defined by the stationarity condition $\delta S = 0$ are selfadjoint for any $U(1)$-invariant action S.*

SECTION 3

Let us now generalize the $U(1)$-invariance property of the action introduced in Sec.2 to an arbitrary group G. Let $G(\omega)$, $\omega = (\omega_1,\omega_2...\omega_n) \in \mathbb{R}^n$ be a representation of an n - parameter Lie group supported by the subspaces $\mathfrak{N}_1$ and $\mathfrak{N}_{-1}$ and normalized by the condition $G(0) = I$.

Let 1_k , $k = 1,...,n$, be generators of a representation of Lie algebra of the group G by linear operators acting in $\mathfrak{N}_{\pm 1}$.

Varying the action (3) on the elements $u \longrightarrow u + \eta$, $u \in \mathcal{D}(A_0^*)$, $\eta \in \mathcal{D}(A_0)$ we have again, in analogy with Section 2, the boundary conditions (5). The invariance condition for the action with respect to the group $G(\omega)$ leads to additional relations.

In particular, let us calculate the variation of interaction in (3) on the elements $u \longrightarrow u + \eta_k^\pm$, where $\eta_k^\pm = G(\omega_k)\xi^\pm(u) - G(0)\xi^\pm(u) = 1_k\xi^\pm(u)\cdot\omega_k$. It gives

$$\delta S_k^{int} = i\omega_k \langle(1^*\Gamma^{-1} + \Gamma^{-1}1)\xi^-,\xi^-\rangle_{\mathfrak{N}_1} -$$
$$- i\omega_k \langle(1^*\Gamma + \Gamma 1)\xi^-,\xi^-\rangle_{\mathfrak{N}_1} . \tag{7}$$

This relation shows that the variation of action vanishes provided

$$1_k^*\Gamma + \Gamma 1_k = 0 , \qquad \Gamma 1_k^* + 1_k \Gamma = 0 , \qquad k = 1,2 \ldots n. \tag{8}$$

Using (8), the total variation of the action can be written in the form

$$\delta S_k = 2\,\mathrm{Re}\,\omega_k \langle\xi^+(1_k u),(\Gamma - \Gamma^*)\xi^+(u)\rangle_{\mathfrak{N}_1} . \tag{9}$$

If invariance of the action (3) with respect to the group G is required, one has to demand $\delta S_k = 0$.

For a unitary representation of the group G one has $\overset{\bullet}{1_k} + 1_k = 0$, in which case (8) is equivalent to the commutation of the operators Γ and 1_k ,

$$[\Gamma,\ 1_k] = 0\ ;   \tag{10}$$

this relation replaces (7).

If $G \simeq U(1)$, $n = 1$, then $1_1 = I_{\mathfrak{N}_{\pm 1}}$. Consequently, (10) is an identity in this case, and the condition $\delta S = 0$ can be rewritten as $\mathrm{Im}\ \langle \xi^+(u),(\Gamma - \Gamma^*)\xi^+(u)\rangle_{\mathfrak{N}_1} = 0$. The latter is equivalent to self-adjointnees of Γ , $\Gamma = \Gamma^*$. If $U(1)$ is a subgroup of G, then we again have $\Gamma = \Gamma^*$ and thus it is necessary to check the condition (10) only. In accordance with the results of section 2, for s.a. extensions A_Γ eq.(10) provides the validity of the invariance condition.

REFERENCES

1. B.S.Pavlov, The Theory of Extensions and an Explicitly Solvable Model, Sov. Math. Surveys <u>42</u> (1987),127-168.

2. B.S. DeWitt, *Dynamical Theory of Groups and Fields*. Gordon and Breach, New York 1965.

3. Yu.A.Kuperin, K.A.Makarov, S.P.Merkuriev, A.K.Motovilov, B.S.Pavlov, Teor. Mat. Fiz. <u>75</u> (1986) 431; ibid <u>76</u> (1988) 242.

Department of Mathematical & Computational Physics, Institute for Physics, Leningrad University, Leningrad 198904 , USSR .

Part 3

QUANTUM CHAOS

Operator Theory:
Advances and Applications, Vol. 46
© 1990 Birkhäuser Verlag Basel

CHAOTIC QUANTUM BILLIARDS

Petr Šeba *+#

We discuss the properties of quantized classically chaotic systems which are illustrated on a solvable singular biliard model.

1.INTRODUCTION

Quantum behavior of classically chaotic systems is a topic which has attracted lot of interest during the last ten years. The progress in this field can be roughly divided into two groups:

The first one contains works dealing with time dependent systems. Here typically one-dimensional time periodic systems are investigated and questions concerning their long-time behavior are answered. The most prominent model from this group is the kicked quantum rotator [1].

The second group concerns time-independent systems. Here one has to investigate multidimensional models. (The dimension of the configuration space must be greater then one. The point is that all classical stationary one-dimensional models are integrable and therefore not interesting from the chaotic point of view). The most prominent members of this group are the so called billiards , i.e., models consisting of a free two-dimensional particle bouncing inside enclosure of a given shape. Depending on its shape the classical billiard can range from being integrable (rectangular billiard) to being completely chaotic (the stadium of Bunimovich). From this point of view the billiards represent an ideal laboratory for testing the influence of the classical chaos on the corresponding quantum system.

In the present note we restrict ourselves solely on the two-dimensional billiards. We give first a short review of their classical/quantum features (Section 2) which will be than illustrated on a solvable model (Section 3).

2.A SHORT BILLIARD REVIEW

By a billiard we understand a classical/quantum particle bouncing inside a flat compact two-dimensional enclosure Ω with perfectly elastic boundary $\delta\Omega$ (we will assume $\delta\Omega$

---

* Alexander von Humboldt Fellow

+ On leave of absence from the Nuclear Physics Institute, Czechoslovak Academy of Sciences, Řež near Prague; Czechoslovakia

to be smooth, i.e. $\delta\Omega \in C$). In what follows we will focus our attention primarily to the behavior of the quantized billiards the Hamiltonian of which is given as

$$H = -\Delta$$

with Dirichlet boundary conditions on $\delta\Omega$

$$f(x,y) = 0; \qquad (x,y) \in \delta\Omega$$

a) THE INTEGRABLE CASE.

Let us first discuss the more simple case when the classical dynamics is integrable, i.e., when there is (in addition to the Hamiltonian) an independent integral of motion. The phase-space trajectory of the billiard is in this case confined to a two-dimensional torus [2] and is easily described using the action-angle variables. Only a limited number of billiards belong, however, to this category since in this case the boundary $\delta\Omega$ has to fulfill some special requirements. For instance all billiards which are connected with discrete crystallographic groups (equilateral triangle, square,...) [3] as well as the circle and elliptic billiards [4] are integrable.

The quantum case can be solved using the Einstein-Brilouin-Keller (EBK) quantization [5]. According to it the quantum eigenvalues are obtained as

$$E_{n_1 n_2} = H\left(I_{n_1}, I_{n_2}\right) \tag{1}$$

where $H(I_1, I_2)$ is the classical Hamiltonian in the action-angle representation and

$$I_{n_1} = \frac{1}{2\pi} \int_{\Gamma_1} \vec{p}d\vec{q} = \left(n_1 + \frac{\alpha}{4}\right) \tag{2a}$$

$$I_{n_2} = \frac{1}{2\pi} \int_{\Gamma_2} \vec{p}d\vec{q} = \left(n_2 + \frac{\alpha}{4}\right) \tag{2b}$$

n_k are integers, α is the corresponding Maslov index and Γ_1, Γ_2 are two different invariant cycles of the torus. ($\hbar = 1$). The corresponding wavefunction is easily expressed as

$$\Psi(\Theta_1, \Theta_2) = \frac{1}{\sqrt{2\pi}} exp(i(\Theta_1 I_{n_1} + \Theta_2 I_{n_2})) \tag{3}$$

In the usual x-representation we get

$$\Psi(x,y) = \sum_j \sqrt{P_j} exp(iS_j) \tag{4}$$

where the sum runs over all independent directions of the classical orbit through the point (x,y); P_j is the classical probability distribution and S_k is the k-th branch of the classical action.

THE EIGENVALUE STATISTICS. One of the basic questions connected with the quantized billiard is the statistical analysis of its eigenvalues. Let E_n be the eigenvalues of the billiard ordered with respect to the magnitude

$$E_n \leq E_{n+1}$$

and $s_i = E_{i+1} - E_i$ is the spacing between two such levels. The simplest statistical characterization of the spectrum is the level spacing distribution P(s): P(s)ds is the probability of finding spacing s_i with $s_i \in [s, s + ds]$.

It is well known that for numbers E_i being independently distributed with density one P(s) is the Poisson distribution

$$P(s) = e^{-s} \tag{5}$$

Berry and Tabor [6] shoved that the unfolded spectrum of a typical integrable system is Poisson distributed. The level statistic of an general integrable billiard is therefore expected to be Poissonian. (The word general means that we are not working with highly degenerate models like the square billiard etc.) As an example we quote the rectangular billiard with incommensurate sites. Its spectrum is given by

$$E_{n,m} = (an)^2 + m^2; \qquad n, m = 1, 2, ... \tag{6}$$

with a being a positive irrational number. This case has been extensively studied by Casati,Chirikov and Guarneri which have computed the first 10^5 eigenvalues and showed that the corresponding level spacing distribution is fairly close to the Poisson one [7].

We can summarize this part as follows:*eigenvalues of a typical integrable quantum billiard are uncorelated leading to Poisson distribution and therefore to expressive level clustering (degeneraties).*

EIGENFUNCTION TOPOGRAPHY. The next question which has to be answered is how a typical wavefunction of an integrable billiard looks like. (By wavefunction topography we mean the localization of its nodal lines, maxima, minima etc.) From the semiclassical quantization we learn that there is a *one to one correspondence between the eigenfunctions and the quantized classical tori.* Roughly speaking we can say that *the wavefunction structure is determined by the x-space projection of the corresponding torus.*(This statement is strictly valid only for billiards with quadratic angular-action representation, since the classical density P is constant in this case and the argument S of the wavefunction (4) is nothing but the classical trajectory winding around the corresponding torus, see [8] for details.) The structure of the wavefunction is expected to be regular. The topography of the positive part of the eigenfunction in a equilateral triangle and ist solution in terms of periodic trajectories is presented on Fig.1.

TIME EVOLUTION. Let us close the discussion on the integrable case describing the evolution of the wavepacket. We will investigate the autocorrelation function $f_\Psi(t)$ defined as

$$f_\Psi(t) = |\langle \Psi(0)|\Psi(t)\rangle|^2 \tag{7}$$

Here $\Psi(0)$ is the initial wavepacket and $\Psi(t)$

$$\Psi(t) = exp(-iHt)\Psi(0) \tag{8}$$

gives the wavepacket after time t when evolving inside the billiard. We decompose first $\Psi(0)$ into the billiard basis

$$|\Psi(0)> = \sum_n a_n |n> \tag{9}$$

where $|n>$ are the billiard eigenfunctions

$$H|n> = E_n|n> \tag{10}$$

Hence

$$f_\Psi(t) = \left| \sum_{n=1}^{\infty} |a_n|^2 e^{-iE_n t} \right|^2 \tag{11}$$

and we find immediately that f_Ψ is a quasiperiodic function of time. This is a direct consequence of the discreteness of the spectrum of H and reflects the quantum recurrence. The discrete nature of the spectrum is, however, resolved by the system only after some time Δt

$$\Delta t \Delta E \geq 1 \tag{12}$$

where ΔE denotes the mean distance between eigenvalues. For $t < \Delta t$ is the wavepacket not influenced by the individual levels but rather by the corresponding spectral density. Our question is now how the autocorrelation function looks like for a typical integrable billiard if $t < 1/\Delta E$

For this purpose we replace the individual levels by the mean values and define a "mean" autocorrelation function $\widetilde{f_\Psi}$ as

$$\widetilde{f_\Psi(t)} = \left| \sum_n |a_n|^2 \int e^{-iEt} P_n(E)dE \right|^2 \tag{13}$$

where $P_n(E)$ is the probability density that the n-th level of the billiard will be equal to E. We know, however, than the level spacing distribution is Poissonian.

$$P(s) = \begin{cases} e^{-s} & \text{for } s \geq 0 \\ 0 & \text{for } s < 0 \end{cases}$$

The densities $P_n(s)$ are obtained from P(s) by subsequent convolution

$$P_1(E) = P(E)$$

$$P_2(E) = \int P_1(x)P(E-x)dx$$

$$\vdots \tag{14}$$

$$P_{n+1}(E) = \int P_n(x)P(E-x)dx$$

$$\vdots$$

and thus

$$\widetilde{f_\Psi(t)} = \left| \sum_n |a_n|^2 \left(\widehat{P(t)}\right)^n \right|^2 \tag{15}$$

where $\widehat{P(t)}$ is the Fourier transform of P(s):

$$\widehat{P(t)} = \int e^{-ist}P(s)ds = \frac{1}{1+it} \tag{16}$$

We get as a result that the *typical decrease of f_Ψ is in the integrable case given by*

$$f_\Psi(t) = \frac{c(\Psi)}{1+t^2} + \sum_{n=3} \frac{c_n}{t^n}; \qquad t < \Delta t \tag{17}$$

(see Fig.3.)

b) THE CHAOTIC CASE

We come now to the second type of billiards, namely to those which are classically completely chaotic. We mention here two representatives of this family: The stadium of Bunimovich and the billiard of Sinai.

The phase-space trajectories of those billiards do not lie on tori but they trace out ergodically the whole energy surface. The semiclassical EBK quantization does not apply in this case making the corresponding quantum properties more delicate.

THE EIGENVALUE STATISTICS The quantized billiard is expected to be "chaotic".

The major recent finding about quantized classically chaotic systems is that their spectra obey the same rules as the spectra of random matrices.

The random matrix theory has been developed 30 years ago [8] to describe the spectra of very complicated quantum systems like heavy nuclei, complicated molecule etc.[9-10]. The apparent difference between those systems and the chaotic billiards is that the billiards (although chaotic) are in fact described by very simple quantum Hamiltonians. Hence the coincidence between their spectral properties and those of the random matrices must be connected with the fact that the simple and purely deterministic billiard Hamiltonians *self-generate the randomness during the time evolution.*

The random matrix theory predicts a level spacing distribution given by the Wigner surmise

$$P(s) = \frac{\pi}{2}s\, exp\left(-\frac{\pi}{4}s^2\right) \tag{18}$$

This distribution displays a clear level repulsion , i.e., the probability to find two eigenvalues close to each other tends to zero as $s \to 0$ (we remember that this probability tends to one in the integrable case)

Numerical calculations performed on chaotic billiards demonstrated a clear agreement between the level spacing statistics and the predictions of the random matrix theory. A phenomenological description based on the Gutzwiler quantization has been given by Berry [11]. See also [19] for interesting connection between the level spacing statistic of the billiard models and the specific heat of small metallic particles.

THE EIGENFUNCTION TOPOGRAPHY. We will now describe the typical shape of the eigenfunctions corresponding to highly excited levels. Heuristically it is reasonable to suppose that the wavefunction is obtained as a superposition of plane waves with a given wavelength (corresponding to the energy) and propagating along the path traced by the classical particle. (This is the standard connection between the classical "ray" orbit and the associated wavefront [12].) The classical path is, however, chaotic and wavefunction is therefore assumed to be *given by a random superposition of plane waves* :

$$\Psi(x,y) = \sum_n exp(i(k_1(n)x + k_2(n)y + \varphi(n))) \tag{19}$$

where $k_1(n); k_2(n)$ *and* $\varphi(n)$ are random numbers connected through

$$k_1(n)^2 + k_2(n)^2 = E \tag{20}$$

with E being the particle energy.

Such a random plane-wave superpositions have been intensively studied. Berry showed that the resulting eigenfunction should be a Gaussian random function with Bessel spatial correlation

$$C(\vec{x}; \vec{x} + \vec{\delta}) = \int \Psi(\vec{x})\Psi(\vec{x} + \vec{\delta})d^2x = const \, J_0\left(\sqrt{E}|\vec{\delta}|\right) \tag{21}$$

Numerical investigations have been performed by Heller, O'Connor and Gehler [13]. They demonstrated, however, that the resulting function has a surprising internal structure consisting of network of long ridges. We will see later that the existence of such ridges is typical for the wavefunctions of classically chaotic billiards.

It is clear that the random plane wave superposition describes well only the local structure of the wavefunction. Globally long-range correlations are expected. First of all the eigenfunction must possess all discrete symmetries of the billiard. The numerical results showed also that the eigenstates have the tendency to organize their ridges into definite patterns which are connected with classical (unstable) periodic orbits inside the billiard. This patterns (denoted as "scars" in the literature) served as a surprise when discovered by numerical calculations [13]. It was found that they have the following properties:

short periodic orbits produce heavier scars

the scars become narrower as energy increase and/or $\hbar \to 0$

It is not difficult to understand why scars must exist. We give here a short heuristic explanation.

It is known from the mathematical literature that the solution of the time-dependent Schroedinger equation is (for high enough energies) localized along the classical trajectory [14]. The resulting asymptotic behavior is uniform in the whole time interval and can be expressed as follows:

$$\overline{(x - X(t))^2} = \int [x - X(t)]^2 \, |\Psi(x,t)|^2 \, d^2x = o(\hbar) \tag{22}$$

where $X(t)$ is the solution of the classical equations corresponding to initial conditions

$$X_k(0) = \int x_k \, |\Psi(x,0)|^2 \, d^2x \tag{23a}$$

$$X'_k(0) = i \int \overline{\Psi(x,0)} \frac{\partial \Psi(x,0)}{\partial x_k} d^2x \tag{23b}$$

and $\Psi(x,t)$ is the solution of the Schroedinger equation.

Let us now assume that the classical trajectory is periodic. Then the evolving wavepacket will after some time come back to its original position and its shape will not change too much for orbits with short period. This procedure will finally develop a "standing wave" localized along the periodic orbit (the scar). The localization of this "standing wave" will be better for smaller $\hbar$ (or for higher energy) leading to narrower scars. On the other hand scars cannot occur along all periodic orbits. Some interplay between the energy of the wavefunction and between the length of the orbit is necessary for the development of standing wave being possible.

It is now clear that the scaring phenomenon is not restricted only to chaotic billiard models. It is expected to occur in all models where periodic orbits exist and the correspondence (22) is satisfied. (Let us mention for instance the well known eigenfunction concentration in a neighborhood of closed geodesics in the Riemannian space [15])

TIME EVOLUTION. To compute the time behavior of the autocorrelation function we use again the formula (15) but now with the level-spacing distribution given by (18). It can be easily seen that

$$\widehat{P(t)} \approx \frac{1}{t^2} \tag{24}$$

and thus

$$\widetilde{f_\Psi(t)} \approx \frac{c(\Psi)}{t^4} \tag{25}$$

Hence f_Ψ decreases faster that in the integrable case. This means that the spreading of the wavepacket is in a typical chaotic billiard faster than in the integrable one. We demonstrate the difference on Fig.2a for both the integrable and the chaotic case. The corresponding spectra were generated by random number generator. The approximated autocorrelation functions (15) are plotted on Fig.2b.

3. EXAMPLES

a) Integrable model. We will demonstrate the above discussed features on a solvable model.

Let us start with an rectangle Ω

$$\Omega = [0, \pi/a] \times [0, \pi]$$

and define the quantum Hamiltonian as

$$H = -\Delta \tag{26}$$

$$D(H) = \left\{ f \in L^2(\Omega);\ p^2 \widehat{f(p)} \in L^2(\Omega)\ and\ f(x,y) = 0\ for\ x \in [0, \pi/a]; y \in [0, \pi] \right\}$$

The number a is assumed to be irrational (we will take $a = \sqrt{5} - 1$ in all numerical calculations).

The quantum dynamics inside this billiard is trivially solvable. The spectrum of H is given by numbers

$$E_{n,m} = a^2 n^2 + m^2; \quad n, m = 1, 2... \tag{27}$$

This eigenvalues have been investigated by Casati,Chirikov and Guarneri [7] which showed that they posses Poisson level spacing distribution.

The corresponding wavefunctions are

$$f_{n,m}(x,y) = \frac{2\sqrt{a}}{\pi} sin(nax)sin(my) \tag{28}$$

Later we will use also a one-index notation with eigenvalue equation written as

$$H|n> = E_n|n> \tag{29}$$

where E_n are the numbers (27) ordered into an increasing sequence

$$E_{n+1} > E_n$$

and $|n>$ are the corresponding wavefunctions (28). In order to check the formula (15) we will compute the autocorrelation function f_Ψ for two different initial wavepackets $|\Psi(0)>$. The first of them will be strongly localized in the basis $|n>$ and given by (see Fig.3)

$$|\Psi(0)> = \sum_n \frac{1}{\sqrt{n!}}|n> \tag{30}$$

The autocorrelation function $\widetilde{f_\Psi}$ can be in this case calculated explicitly leading to

$$\widetilde{f_\Psi(t)} = \left| \sum_{n=1}^{\infty} \frac{\left(\widehat{p(t)}\right)^n}{n!} \right|^2 = 1 + exp\left(\frac{2}{1+D^2t^2}\right) - 2exp\left(\frac{1}{1+D^2t^2}\right) cos\left(\frac{Dt}{1+D^2t^2}\right)$$

where D is the mean level spacing. The behavior of $\tilde{f}_\Psi$ is compared with the exact auto-correlation function f_Ψ and is plotted on Fig.4. The second example illustrates an opposite situation. The function $|\Psi(0)>$ is now very weakly localized in the basis $|n>$

$$|\Psi(0)> = \sum_{n=1}^{100} \frac{1}{\sqrt{n}}|n>$$

This function which has a peak localized in the upper left corner of the billiard is plotted on Fig.5. The autocorrelation function $\tilde{f}_\Psi$ can be again calculated and is approximately given by

$$\tilde{f}_\Psi(t) = \left(arctg\left(\frac{1}{Dt}\right)\right)^2 + \frac{1}{4}\left(ln\left(\frac{t^2D^2+1}{t^2D^2}\right)\right)^2$$

The result compared with the exact calculation of f_Ψ is ploted on Fig.6. In both cases we observe rather good agreement for times $t \leq 2\Delta t$.

b) CHAOTIC MODEL. We will now construct a model which meets two seemingly contrawise features: this model will be chaotic and solvable at the same time.

Our strategy goes back to the work of Sinai. He proved that if one takes an integrable rectangular billiard and puts into its centre a rigid elastic circle one obtains a model which is classically ergodic (the circle scatters the classical trajectories.) We replace the circle by a point scatterer. From the quantum point of view it is reasonable to expect that this change will not modify the physical situation too much.

We start with the Hamiltonian H defined by (26). In order to add the point scatterer at point $(x_0, y_0) \in \Omega$ we follow the strategy used in [16]. We first remove this point from Ω restricting H to an operator H_0

$$H_0 = H \mid D_0 \tag{31}$$

$$D_0 = \{ f \in D(H); \quad f = 0 \ \text{in some neighborhood of } (x_0, y_0)\}$$

The operator H_0 is symmetric but it is not self-adjoint. The desired Hamiltonians are obtained as its self-adjoint extensions. At this point we have to use an abstract mathematical theory developed by J.von Neumann. We assume that the reader is familiar with this part of mathematics. For a comprehensive review see [17].

It can be easily seen that the operator H_0 has deficiency indices (1,1). This means that there exist exactly one-parameter family of its self-adjoint extension:

$$H_\alpha = -\Delta \tag{32}$$

$$D(H_\alpha) = \{f \in L^2(\Omega); \quad f = 0 \ on \ \delta\Omega \ and \ L_0(f) = \alpha L_1(f)\}$$

with

$$L_0(f) = \lim_{\rho\to 0} \frac{f(x,y)}{\ln \rho}$$

$$L_1(f) = \lim_{\rho\to 0} [f(x,y) - L_0(f)\ln\rho]$$

$$\rho = \sqrt{(x - x_0)^2 + (y - y_0)^2}$$

The parameter α describes the coupling constant of the scatterer localized at the point (x_0, y_0). For $\alpha = 0$ we get, of course, the original Hamiltonian H

$$H_{\alpha=0} = H$$

The self-adjoint extension theory is able to describe also the resolvent of H_α. From the famous Krein formula we get

$$(H_\alpha - z)^{-1} = (H - z)^{-1} - \frac{2\pi\alpha}{1 + 2\alpha\xi(z)} |g_z(x, y, x_0, y_0) >< g_z(x_0, y_0, x, y)| \qquad (33)$$

where $g_z(x, y, x', y')$ is the Green function of H

$$\left((H - z)^{-1} f\right)(x, y) = \int_\Omega g_z(x, y, x', y') f(x', y') dx' dy' \qquad (34)$$

and $\xi(z)$ is a meromorphic function given by

$$\xi(z) = \sum_{n=1}^{\infty} \sum_{m=1}^{\infty} \left(\frac{4a \sin(nax_0)^2 \sin(my_0)^2}{\pi (n^2 a^2 + m^2 - z)} - \frac{1}{2m} \right) \qquad (35)$$

The spectrum of H_α can be easily calculated from the pole structure of $(H_\alpha - z)^{-1}$. It is given as a solution of the transcendental equation

$$1 + 2\alpha\xi(z) = 0 \qquad (36)$$

The eigenfunctions are obtained from the corresponding residui of $(H_\alpha - z)^{-1}$ and are given by

$$f_{E_n}(x, y) = g_{E_n}(x, y, x_0, y_0) \qquad (37)$$

where E_n is the solution of (36). Since the Green's function g_z is explicitly known we get finally

$$f_{E_n}(x, y) = \frac{4a}{\pi^2} \sum_{n=1}^{\infty} \sum_{m=1}^{\infty} \frac{\sin(nax) \sin(my) \sin(nax_0) \sin(my_0)}{n^2 a^2 + m^2 - E_n} \qquad (38)$$

We show now that what we get in such a way fulfills all requirements placed on a quantum chaotic system.

THE EIGENVALUES. In order to make the live easy we will place the scatterer in the centre of the rectangle Ω

$$(x_0, y_0) = \left(\frac{\pi}{2a}, \frac{\pi}{2} \right) \qquad (39)$$

The Hamiltonian H_α posses in this case discrete symmetries due to reflection with respect to the symmetry axes of the rectangle

$$[H_\alpha, P_1] = 0$$

$$[H_\alpha, P_2] = 0$$

for all α. Here

$$(P_1 f)(x,y) = f\left(\frac{\pi}{a} - x, y\right)$$

$$(P_2 f)(x,y) = f(x, \pi - y)$$

are the corresponding parity operators. Since the parity is conserved during the time evolution we will work only in the subspace corresponding to parity (1,1), i.e. in the subspace defined by

$$P_1 f = f$$

$$P_2 f = f$$

The eigenvalues belonging to it are determined by the equation (36) with simplified function $\xi(z)$:

$$\xi(z) = \sum_{n,m=1}^{\infty} \left(\frac{4a}{\pi \left(a^2(2n-1)^2 + (2m-1)^2 - z\right)} - \frac{1}{2n} \right) \tag{40}$$

(the eigenvalues of H_α belonging to other parity subspaces coincide with those of the operator H. The reason for this behavior is that the corresponding wavefunctions have in this case nodal lines which go through the point (x_0, y_0) and do not feel the point interaction.)

The level spacing distribution P(s) for a quantum chaotic system is expected to follow the Wigner surmise. We will now prove that the distribution corresponding to H_α has this property. In what follows we show that P(s)

$$P(s) \approx ks \; ; \quad s \to 0$$

for all $\alpha \neq 0$ with k being some constant which depends only on α.

For this reason we rewrite the function ξ as

$$\xi(E) = \sum_{n=1}^{\infty} \left(\frac{4a}{\pi} \frac{1}{E_n - E} - \frac{1}{2n} \right) \tag{41}$$

where E_n are the eigenvalues of H. The function $\xi(E)$ is meromorphic with poles having a Poisson distribution. Let us take for simplicity the coupling constant $\alpha = \infty$. Then the eigenvalues of H_α are given by zeros of $\xi(E)$. The level spacing distribution for H_α can be obtained from the following theorem:

THEOREM: Let $\xi(E)$ be a meromorphic function of the form (41) with poles which have a Poisson spacing distribution. Then the probability density P(s) of finding two roots of $\xi(E)$ with distance s fulfills

$$\lim_{s \to 0} \frac{P(s)}{s} = k \tag{42}$$

with k being some constant.

The proof of this theorem will be given elsewhere [18]. We give, however, an heuristic argument which leads directly to this result:

Let us suppose that the function $\xi(E)$ has two poles $E_n; E_{n+1}$ clustered together and all the remaining poles localized apart from this particular cluster

$$|E_n - E_m| > K \quad ; \quad m \neq n, n+1$$

with K being some positive constant. It is then clear (see Fig.7) that the zeros of the function ξ are not clustered in the neighborhood of E_n. Hence *clusters of two poles do not lead to clusters of zeros.*Therefore a cluster of zeros can be born only by a cluster of three and more poles (see Fig.8). The poles E_n have, however, a Poisson distribution. Thus the probability $P_n(s)$ to find n poles within an interval with length s fulfills

$$\lim_{s \to 0} \frac{P_n(s)}{s^{n-2}} = \frac{1}{(n-2)!} \tag{43}$$

(We suppose that the mean distance between the poles equals to 1.) The spacing distribution for roots of $\xi(E)$ is therefore governed by the distribution $P_3(s)$ which leads directly to the result. A typical plot of the level-spacing distribution for our model is given on Fig.9.

EIGENFUNCTIONS. The eigenfunctions are given by the formula (38). In the simplest case of a centered scatterer (39) we get for the k-th eigenfunction $|\Psi_k >$

$$|\Psi_k >= \sum_n a_{n,k}|n > \tag{44}$$

with $a_{n,k}$ given by

$$|a_{n,k}| = \frac{1}{|E_n(0) - E_k(\alpha)|} \tag{45}$$

Here $E_k(\alpha)$ are the eigenvalues of H_α and $E_n(0)$ are the eigenvalues of the unperturbed Hamiltonian H. Using the fact that

$$E_n(0) \approx \frac{4\pi}{a}n \quad as\ n \to \infty$$

we find

$$\sum_n |a_{n,k}| = \infty \ for\ all\ k \tag{46}$$

This means that the "chaotic" eigenfunctions $|\Psi_k >$ are very badly "localized" in the "regular" basis $|n >$ and have therefore a good chance to be wild. The topography of a typical eigenfunction is ploted on Fig.10 and is compared with the nearest "regular" wavefunction $|n >$ (Fig.11) (as nearest we denote the function $|n >$ with the maximal coefficient $|a_{n,k}|$). The function $|\Psi_k >$ has a shape which is typical for chaotic systems and which has been observed also in the stadium of Bunimovich: It consists of rather long and narrow ridges which are snaked in a complicated way. As already mentioned such a structure is typically displayed by a random superposition of plane waves. (A superposition of 500 plane waves which have the same energy as the wavefunction of Fig.10 and propagate in random directions is plotted on Fig.12.) We would like, however, to stress that in our case the superposition leading to $|\Psi_k >$ is purely deterministic.

In order to compare the complexity of $|n >$ and $|\Psi_k >$ we can introduce a "wavefunction entropy" ($\|\Psi\| = 1$)

$$S(\Psi) = -\int_\Omega |\Psi(x,y)|^2 \ln |\Psi(x,y)|^2 dx dy \qquad (47)$$

This quantity is expected to characterize quantitatively the "irregularity" of Ψ. A direct calculation shows that the entropy of $|\Psi >$ is in all cases substantially higher then the entropy of the nearest $|n >$. (This difference is especially clearly manifested in a toroidal billiard (i.e. in a rectangular billiard with periodic boundary conditions) where the entropy of the "regular" functions $|n >$ equals to 0.)

The last phenomenon we would like to illustrate is the existence of scars. The concentration of the wavefunctions $|\Psi_k >$ along some periodic orbits is illustrated on Fig.13. We would like to mention, however, that there are also wavefunction which do not display scars.

Summarizing we can say that the above model has all features of a "true" quantum chaotic system. It has, however, also the advantage of being extremely simple. It represents therefore an ideal laboratory for testing the features of stationary quantum chaos.

REFERENCES

1. G.Casati at al: Phys.Rev.A 34 (1986) 1413
2. B.Eckhardt: Phys.Rep. 163 (1988) 207
3. G.B.Shaw: J.Phys.A 7 (1974) 1537
4. A.J.S.Traiber at al: J.Phys.A 22 (1989) L 365
5. J.B.Keller: Ann.Phys.4 (1958) 180
6. M.V.Berry, M.Tabor: Proc.R.Soc.Lond. A349 (1976) 101
7. G.Casati,B.V.Chirikov,I.Guarneri: Phys.Rev.Lett. 54 (1985) 1350
8. M.L.Mehta: Random matrices, Academic Press, New York 1967
9. I.I.Gurevich,M.I.Pevsner: Nucl.Phys. 2 (1957) 575
10. T.Zimmerman at al: Phys.Rev.Lett.61 (1988) 3
11. M.V.Berry: Proc.R.Soc.Lond. A 400 (1985) 229
12. E.J.Heller,P.W.O'Connor: Nucl.Phys.B (Proc.Suppl.) 2 (1987) 201
13. P.W.O'Connor,J.N.Gehlen E.J.Heller: Phys.Rev.Lett. 58 (1987) 1296
14. V.M.Babich,Yu.P.Danilov :Mathematical Problems in Wave Propagation Theory II;Consultant Bureau, New York 1971
15. V.M.Babich: Mathematical Problems in Wave Propagation Theory I; Consultant Bureau, New York 1970
16. S.Albeverio, F.Gesztesy,R.Hoegh-Krohn, H.Holden: Solvable models in quantum mechanics, Springer Verlag, New York 1988.

17. M.Reed,B.Simon: Methods of Modern Mathematical Physics, vol.2,
Academic Press, New York 1978
18. I.Ya.Goldsheid, P.Seba: In preparation
19. J.Barojas at al: Ann.Phys. 107 (1977) 95

Institute of Mathematics, Ruhr University Bochum; D-4630 Bochum, F.R.G.

FIGURE CAPTIONS.

Fig.1: The positive part of a eigenfunction inside a equilateral triangle
billiard compared with the corresponding periodic classical
trajectories

Fig.2a: The autocorrelation function f_Ψ for an abstract solvable and
chaotic billiard. The corresponding quantum spectra has been
generated by random number generator.

Fig.2b: The approximated function $\widetilde{f_\Psi}$ corresponding to the
situation of Fig.2a

Fig.3: The shape of the function (30)

Fig.4: The exact and approximated autocorrelation functions of the
wavepacket (30)

Fig.5: The shape of the weakly localized wavepacket

Fig.6: The exact and approximated autocorrelation functions.

Fig.7: The plot of the meromorphic function $\xi(z)$ near the cluster
of two poles

Fig.8: The plot of the meromorphic function $\xi(z)$ near the cluster
of three poles

Fig.9: The level spacing distribution of first 900 eigenvalues of H_α
with $\alpha = 100$

Fig.10: The positive part of the eigenfunction of the 411-th eigenvalue
of H_α with $\alpha = 100$. The scatterer is localized at
$(0.55\pi/a, 0.65\pi)$

Fig.11: The nearest regular function corresponding to Fig.10

Fig.12: The positive part of function obtained as a superposition of 500 plane waves with random phases and directions. The energy of the waves is the same as the energy of the eigenfunction on Fig.10

Fig.13: The plot of the eigenfunction of the 412 state of H_α. Only the part of the function larger then 0.3 has been plotted.

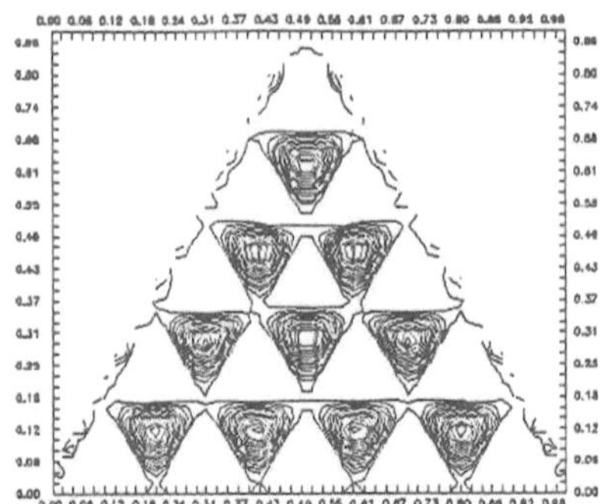

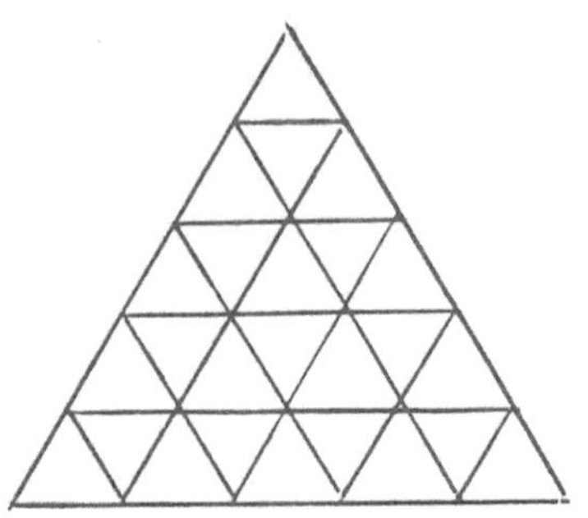

Fig.1

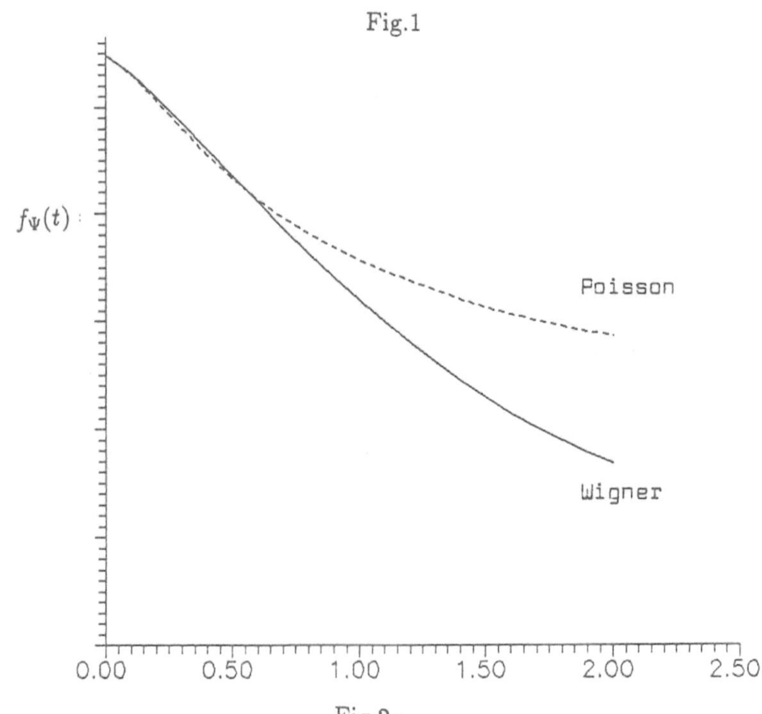

Fig.2a

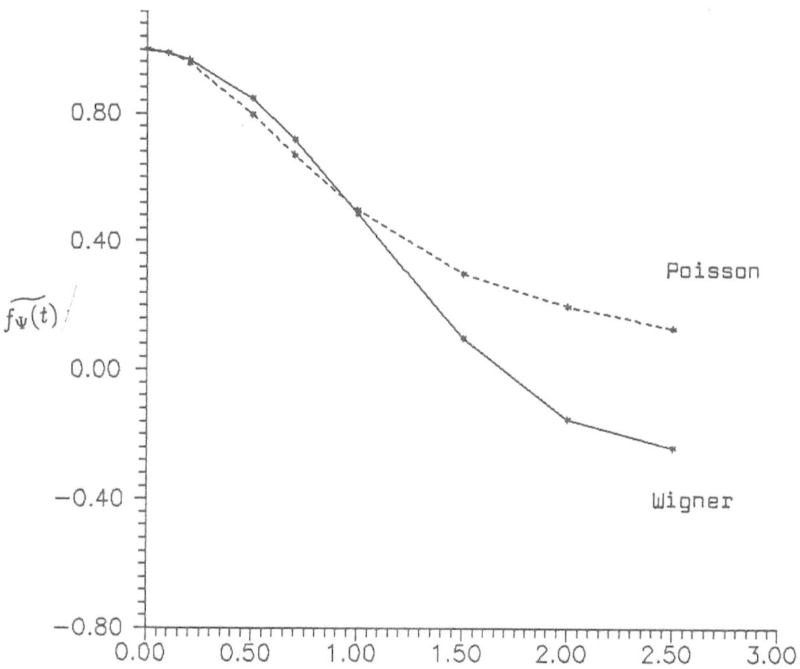

Fig.2 b

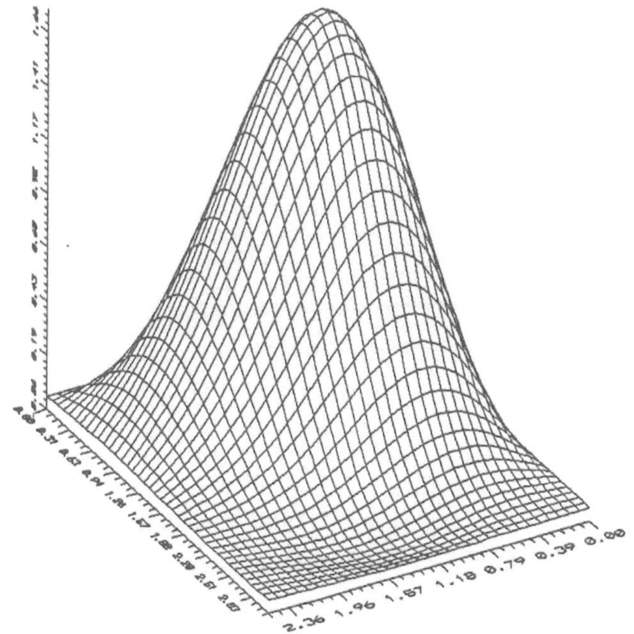

Fig.3

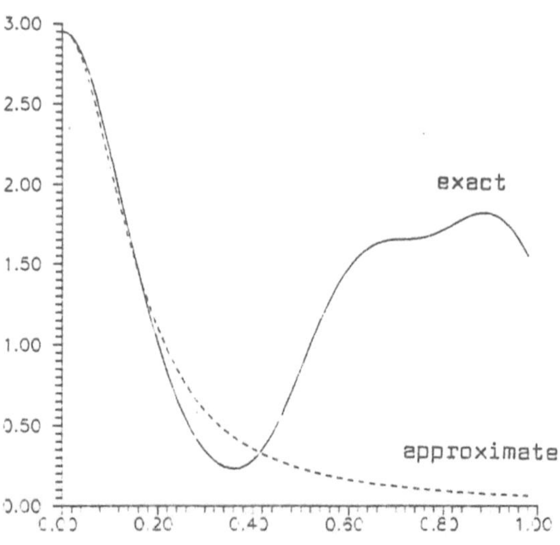

Fig.4

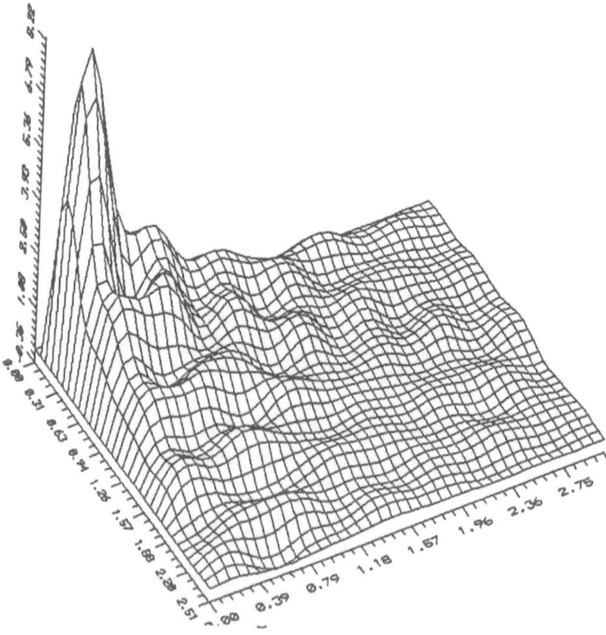

Fig.5

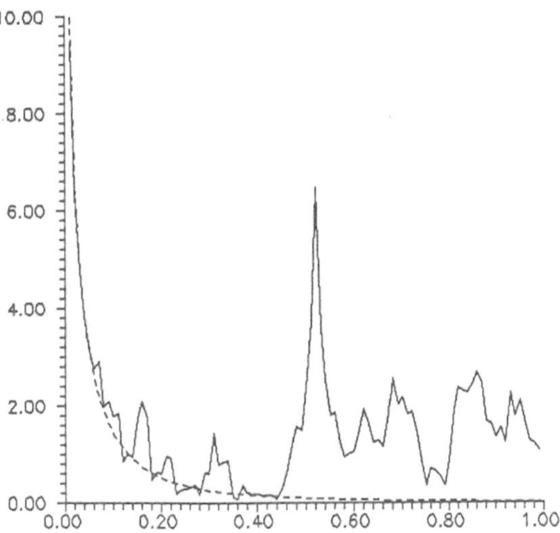

Fig.6

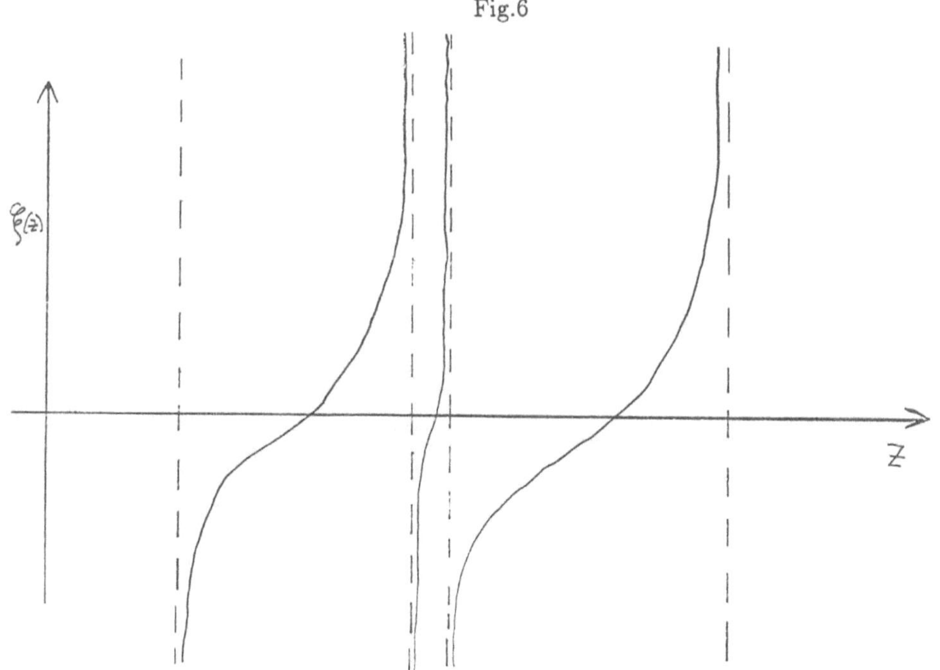

Fig.7

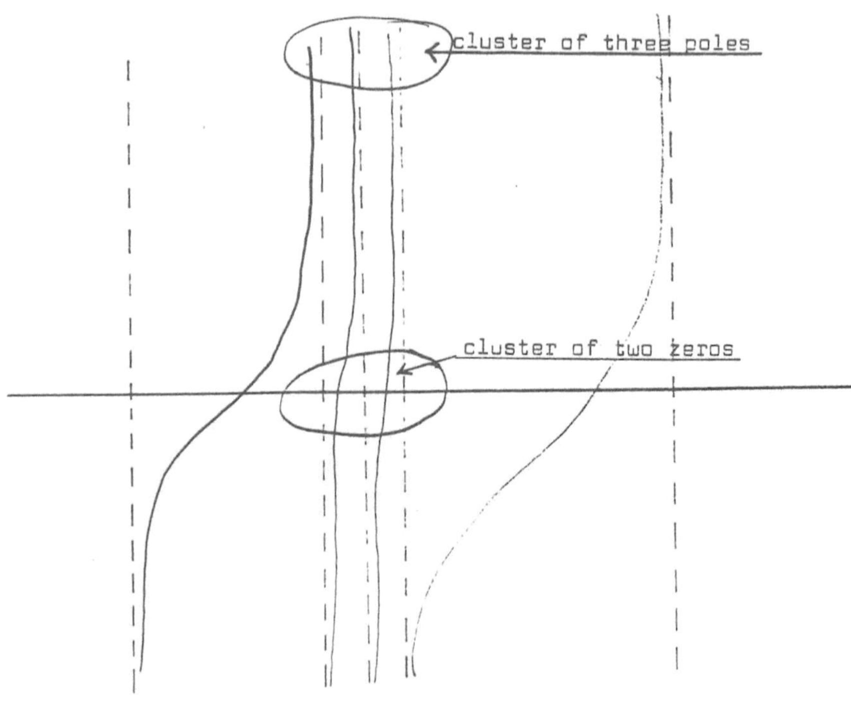

Fig.8

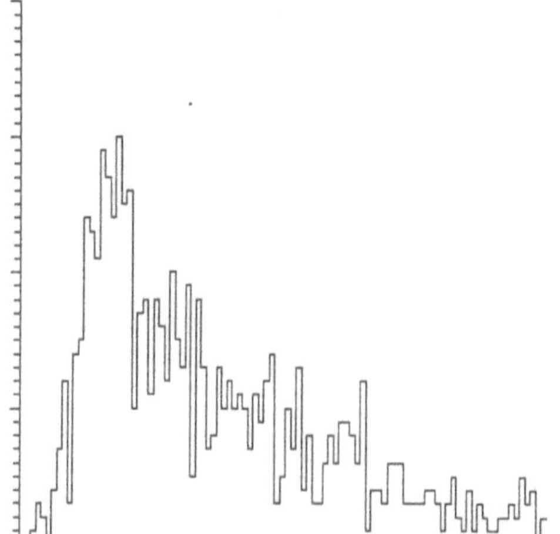

Fig.9

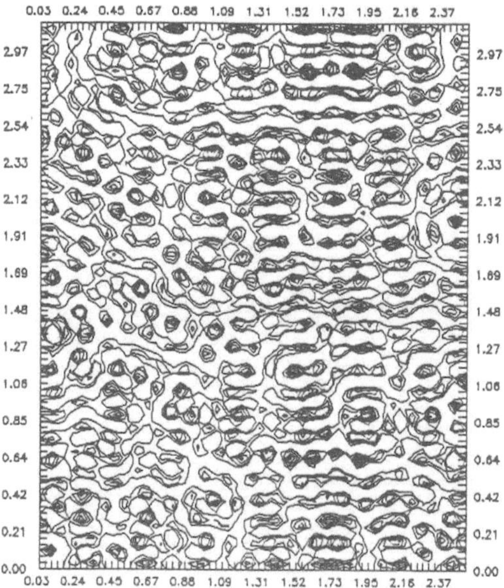

Fig.10

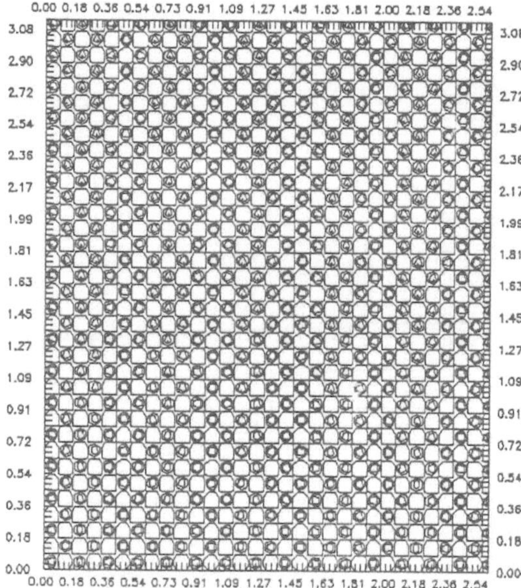

Fig.11

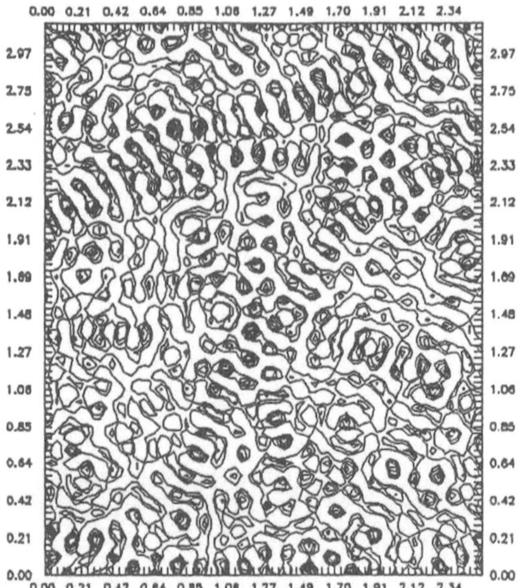

Fig.12

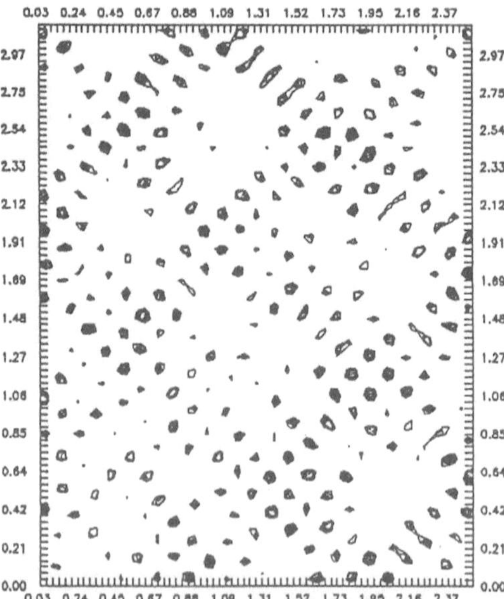

Fig.13

Operator Theory:
Advances and Applications, Vol. 46
© 1990 Birkhäuser Verlag Basel

RELEVANCE OF THE LOCALIZATION TO QUASIENERGY STATISTICS IN QUANTUM CHAOTIC SYSTEMS

F.M.Izrailev

Statistical properties of quasienergy spectra and structure
of eigenfunctions are investigated for some model which is
strongly chaotic in the semiclassical limit. The main attention
is paid to the influence of the localization on the spacing distribution of the nearest levels depending on the degree of localization. Definition of localization length for chaotic states
is discussed for the finite basis with the limiting case of ful-
ly extended random states. Scaling properties of spectra are
studied in the intermediate region of suppressed quantum chaos.

1. INTRODUCTION

At present, much attention is paid to the so-called "quantum
chaos" (see, e.g. reviews /1-2/). The latter term is commonly
used in literature to describe a situation when investigating
the properties of quantum systems which are chaotic in the classical limit. One should emphasize that these systems are assumed to be dynamical in the sense that their behaviour is rigorously defined by the motion equations which have no random parameters and chaotic properties appear only under some conditions.
It is well known (see, e.g. /3-4/) that chaos in classical systems is closely related to the local instability of the motion,
the property which appears to be strongly suppressed in quantum
systems even if all semiclassical parameters are quite large.
This peculiarity creates the problem in the description of quantum behaviour in semiclassical region for the case of classical
chaos. It turns out that the time of complete correspondence to
the classical motion is extremely short and quantum interferen-

ce effects play essential role in the behaviour of systems. In particular, the so-called quantum suppression of classical chaos has been discovered /5-6/ which may be very important in real situations.

One of the interesting problem is to study how classical chaos manifest itself in spectra and eigenfunctions of the correspondent quantum systems. There are many approaches and results in this field (see /1-2/), here we discuss some relatively simple model which, in the classical limit, is well investigated both analytically and numerically. The advantage of this model is that it allows to follow the full transition from one extreme case of classically integrable system to another one, corresponding to the strongly chaotic system.

2. DESCRIPTION OF THE MODEL

The model under consideration is related to the so-called "kick rotator" /7,1-2/:

$$\hat{H} = -\frac{\hbar^2}{2I} \frac{\partial^2}{\partial\theta^2} + \varepsilon_0 \cos\theta\, \delta_T(t) \tag{2.1}$$

where perturbation is given in the form of δ-kicks with the period T. Here ε_0 is perturbation strength, $\hbar$ is Plank constant and I is moment of inertia. The behaviour of this model is totally discribed by Shroedinger equation:

$$i\hbar\frac{\partial\Psi}{\partial t} = -\frac{\hbar^2}{2I}\frac{\partial^2\Psi}{\partial\theta^2} + V(\theta)\cdot\delta_T(t); \quad V(\theta) = \varepsilon_0\cdot\cos\theta \tag{2.2}$$

which can be reduced to the mapping for Ψ function by making use of integration over one period of perturbation:

$$\Psi(\theta, t+T) = \hat{U}\,\Psi(\theta, t) \tag{2.3}$$

$$\hat{U} = \exp\left\{i\frac{T\hbar}{4I}\frac{\partial^2}{\partial\theta^2}\right\}\exp\left\{-i\frac{\varepsilon_0}{\hbar}\cos\theta\right\}\exp\left\{i\frac{T\hbar}{4I}\frac{\partial^2}{\partial\theta^2}\right\}$$

Here the value of Ψ function is determined in the middle of
the rotations, between two successive kicks. It is seen that
all dynamics is described by the evolution operator $\hat{U}$ which
depends only on two parameters

$$k \equiv \frac{\varepsilon_0}{\hbar} \; ; \qquad \tau \equiv \frac{\hbar T}{I} \tag{2.4}$$

It is useful to take as independent parameters k and $K = \tilde{\tau}k =$
$= \varepsilon_0 T/I$. Then the first one, k, can be treated as semiclassical parameter, while the second, K, appears to be of pure
classical nature. The classical counterpart of the model (2.1)
is the well known "standard mapping" /5-6/ which for relatively
large values of K ($K \gtrsim 5$) reveals strongly chaotic motion. In
particular, the energy of this model turns out to increase, in
average, linearly in time, a consequence of the diffusive grows
for the momentum of rotator.

Unlike unbounded (in momentum space) diffusion in the classical model (for $K \gg 1$), the quantum model (2.1) in a deep
semiclassical region $(k \gg 1 ; K = 5)$ shows typical behaviour numerically discovered in /7/: after some time $t_D \sim k^2$ the diffusive increase of the rotator energy $E(t)$ start to deviate more from the classical rate being suppressed by quantum interference effects. Numerious data indicate that for $t \gg t_D$ diffusion practically stops and only finite number of unperturbed states are involved in dynamics providing some sort of stationary
oscillations of energy. This effect has been extensively studied in /5-6/ with the conclusion that the mechanism of such suppression is closely related to the structure of quasienergy eigenfunctions in momentum space. To discuss this mechanism in more
detail we write (2.3) in the momentum representation

$$A_n(t+T) = \sum_{m=-\infty}^{\infty} U_{nm} A_m(t)$$

$$U_{nm} = e^{i\frac{\tilde{\tau}}{4}n^2} \, i^{n-m} \, J_{n-m}(k) \, e^{i\frac{\tilde{\tau}}{4}m^2} \tag{2.5}$$

Here Fourier coefficients $A_n(t)$ of the time-dependent wave function $\Psi(\theta,t)$ are essentially amplitudes of unperturbed eigenstates which interact to each other for $\varepsilon_0 \neq 0$. By iteration of the mapping (2.5) one can numerically find the dynamics of these amplitudes in time which is measured in units of perturbation periods (the rotator energy is then $E(t) = \sum_{-\infty}^{\infty} \frac{n^2}{2} |A_n(t)|^2$.

The quasienergy eigenfunctions $\Phi_n(\varepsilon,t)$ for our model (2.1) can be written in the form

$$\Phi_n(\varepsilon,t) = e^{i\frac{\varepsilon}{T}t} \, \varphi_n(\varepsilon,t) \tag{2.6}$$

where $\varphi_n(\varepsilon,t)$ are periodic in time, $\varphi_n(\varepsilon,t+T) = \varphi_n(\varepsilon,t)$ and ε stand for the quasienergies. Since we are interested in the values of Ψ function in instant times ($t=MT$ with M integer) t can be omitted in the notation of $\varphi_n(\varepsilon)$. It is seen now that $\varphi_n(\varepsilon)$ and ε are eigenfunctions (EF) and eigenvalues of the unitary matrix $\hat{U}$:

$$e^{i\varepsilon} \varphi_n(\varepsilon) = \sum_{m} U_{nm} \varphi_m(\varepsilon) \tag{2.7}$$

As a result, the time behaviour of the rotator can be explained by the properties of eigenfunctions and eigenvalues of U_{nm}. The main point of the approach /5-6/ is that for non-resonant values of the rescaled period $\tilde{\tau}$ (for $(\tilde{\tau}/4\pi) \neq r/q$ with r,q integers) all eigenfunctions $\varphi_n(\varepsilon)$ are exponentially localized in momentum space:

$$\varphi_n \sim \exp\left(-\frac{|n-n_0|}{\ell_\infty}\right); \quad n \to \pm\infty \tag{2.8}$$

with ℓ_∞ standing for the localization length of EF. The data show that the values $\ell_\infty(\varepsilon)$ are strongly fluctuated but the average localization length is proportional to the classical diffusion coefficient $D_{c\ell}$ (see, details in /5-6/):

$$\ell_\infty \approx \frac{D_{c\ell}}{2\tau^2} \approx \frac{k^2}{4} \; ; \; D_{c\ell} \approx \frac{K^2}{2} \quad (k \gg 1, \; K \gg 1) \qquad (2.9)$$

This remarkable expression is very important for the establishing relation between classical and quantum properties in the region of strong classical chaos.

The concept of localization allows to understand the mechanism of the suppression of classical diffusion. Indeed, for a generic initial distribution $A_m(0)$ which has to be localized in momentum space, only finite number of perturbed eigenstates are exited according to (2.9). On the other hand, each specific eigenfunction $\varphi_n(\varepsilon)$ is also localized, due to the same relation (2.9). It means that after long time only finite number $\Delta n \sim k^2$ of unperturbed states will be involved in the dynamics. Therefore, localization of EF leads to the localization of excitation in unperturbed states. This effect can be also explained in terms of quasienergy spectrum /5-6/. It turns out that the exponential localization is related to the discretness of the quasienergy spectrum, the fact is not rigorously proved but seems to be true according extensive numerical simulations. The opposite case of extended states and, correspondingly, the continuous spectrum of quasienergies appears for specific values of τ , when $\tau/4\pi = r/q$ (r , q are integers). This case is known as "the quantum resonance" /7/ which has been investigated both analytically and numerically in /8/.

Typical example of exponentially localized eigenstate (for $K = 5$; $k \approx 9.2$, $\tau \approx 0.54$) is presented in Fig. 1, where the components $w_n = |\varphi_n(\varepsilon)|^2$ of some EF is plotted in the unperturbed (momentum) states in logariphmic scale. The symmetry $|\varphi_n|^2 = |\varphi_{-n}|^2$ reflects the underlying symmetry of the Hamiltonian (2.1), $H(\theta) = H(-\theta)$, which results in the symmetric $(\varphi_n = \varphi_{-n})$

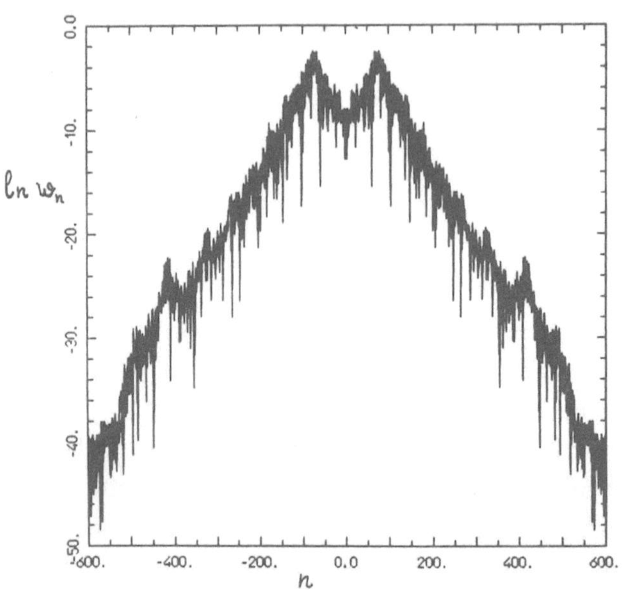

Fig. 1.

or assymmetric $\left(\varphi_n = -\varphi_{-n} \right)$ form of all EF.

It is clear, that in the given model (2.1) with infinite momentum space the localization of EF for any finite k always takes place, therefore, we can not expect the appearence of the maximal quantum chaos, which is associated with the delocalized chaotic states /6,9/. For this reason, it is convenient to pass to another model which, in some sense, may be treated as generalization of the model (2.1). Derivation of this model is based on the specific property of the quantum resonance (see details in /6,10/). Namely, for $\tau/4\pi = r/q$ all EF appear to be of Bloch states in momentum space, $\varphi_{n+q} = \exp(i\theta_0 q)\cdot\varphi_n$ with $0 \leq \theta_0 < 2\pi/q$. Therefore, by specifying the value of Bloch parameter θ_0 equal to zero, one can obtain the model for periodic states (with period $N \equiv q$). Physically, this model corresponds to the standard mapping with the phase space being a torus. The unitary matrix U_{nm} which describes this model, is now of finite size $N \times N$, the result rigorously proved in /8/. Essentially, this procedure is some sort of quantization of the classical model on the torus. As a result, we have (see /6,9/):

$$U_{nm} = \frac{1}{N} \exp\left(i\frac{\tilde{\iota}n^2}{4}\right) \sum_{p=-N_1}^{N_1} \exp\left(-ik\cos\frac{2\pi}{N}p\right)\exp\left(i\frac{2\pi}{N}p(n-m)\right)\exp\left(i\frac{\tilde{\iota}m^2}{4}\right) \quad (2.10)$$

where $n,m = -N_1,\ldots,N_1$ and $N = 2N_1+1$. It is important to note that this model can be also considered as some conservative system with a finite number of energy levels on the closed energy surface. Our main interest is now to study how the structure of EF and the spectrum of quasienergies are dependent on the semiclassical parameter $k \sim 1/\hbar$ when the classical parameter is large enough to have strong chaos in the corresponding classical system (in what follows, we keep $K = 5$).

3. MAXIMAL QUANTUM CHAOS

Due to the symmetricity ($\varphi_n = \pm\varphi_{-n}$) one can consider only the components of EF for $n > 0$. Then, it is naturally to expect that physical parameter determining the degree of chaos in the model (2.10) is the ratio of localization length ℓ_∞ to the size of momentum space, $\Lambda = \ell_\infty/N_1$. When the parameter Λ is small, the boundness of the basis is not important for almost all EF, due to exponential localization (see Fig. 1). On the other hand, when Λ is much larger than the total size of momentum space, all eigenstates are expected to be strongly extended. This is, indeed, the case, as it is seen from Fig. 2. Here, the components w_n with $n > 0$ are plotted versus n for $K = 5$, $k \approx 240$, $N_1 = 600$ (with $\Lambda \approx 24$). The structure of the state looks very chaotic. To check the random character of EF, we search the distribution of components φ_n by considering all EF of the matrix U_{nm}. It is well known that random matrix theory (RMT) /11-13/ predicts the "microcanonical" distribution of φ_n - components which can be obtained by the integration over all but one components of φ_n, uniformly distributed in N_1 -dimensional Hilbert space:

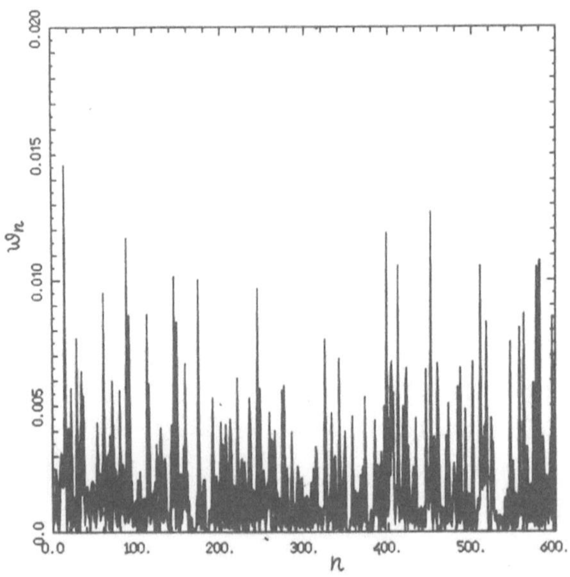

Fig. 2.

$$W_{N_1}(\varphi_n) = \frac{\Gamma\left(\frac{N_1}{2}\right)}{\sqrt{\pi}\;\Gamma\left(\frac{N_1-1}{2}\right)}\left(1-\varphi_n^2\right)^{\frac{N_1-3}{2}} \tag{3.11}$$

For $N_1 \to \infty$ distribution W_{N_1} takes the **gaussian** form reflect-
ing the random nature of eigenstates.

To improve statistics, in /9/ the averaging over number NG
of matrices $U_{n,m}$ has been performed with a slightly different
values of k (with the step $\Delta k \ll k$). The data are shown in
Fig. 3 where $N = 51$, $NG = 20$, $\bar{k}=5$, $k=20$; $\tau = 16\pi/N$, $\Delta k = 0.1$,
$N_1 = 25$ (correspondingly, $\Lambda \approx 5.2$). The distribution $W(\varphi_n)$
in normalized variables, $\tilde{\varphi}_n^2 = N_1 \varphi_n^2$ ($\langle \tilde{\varphi}_n \rangle = 0$, $\sigma^2 = \langle \tilde{\varphi}_n^2 \rangle = 1$ looks
very close both to (3.1) (curve I) and to the gaussian distribu-
tion (curve II). Nevertheless, the statistical χ^2-probe clear
indicates that correspondence to the gaussian distribution is
very bad (χ^2 for 38 intervals is equal $\chi^2_{38} \approx 98$ with very
low confidence level, $P_w < 10^{-6}$) while the correspondence with
the prediction (3.1) of RMT is quite good ($\chi^2_{38} \approx 56$, with

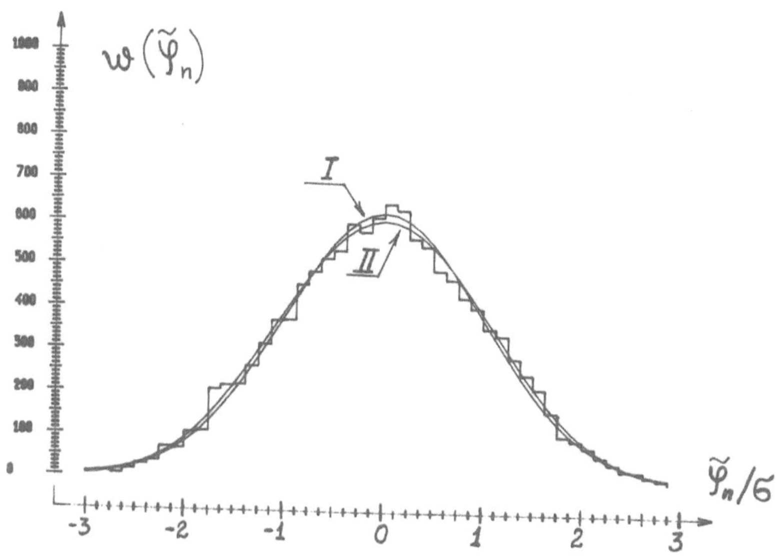

Fig. 3.

high confidence level, $P_w \approx 5\%$). This result shows that under
two conditions, strong classical chaos ($K \gg 1$) and extended states ($\Lambda \gg 1$), all EF can be treated as random ones.

Another manifestation of chaos in quantum systems is known
to be the special type of fluctuations in the spectrum. One of
the commonly used quantity to measure these fluctuations is the
so-called spacing distribution $P(s)$ of nearest levels of energy (quasienergy). The RMT predicts that for ensemble of random
matrices this distribution has the form, which is very close to
the Wigner-Dyson surmise (see details in /11-13/):

$$P(s) = A\, s^{\beta} \exp\left(-\beta s^2\right) \tag{3.2}$$

Here A and B are normalized parameters given by the conditions

$$\int_0^{\infty} P(s)\,ds = 1 \; ; \qquad \int_0^{\infty} s\, P(s)\,ds = 1 \tag{3.3}$$

The parameter β in (3.2) characterizes the degree of the repulsion of nearest levels and equals to 1; 2 or 4, in dependence on the symmetry of the matrices. In our case of symmetric ($U_{nm} = U_{mn}$) unitary matrix one can expect the distribution $P(s)$ in the form (3.2) with $\beta = 1$ in the limit case, when $\Lambda \gg 1$. Detailed analysis of the distribution $P(s)$ for very large K and k has been done in /14/ both for the model (2.10) and for other models with different symmetries of U_{nm}. In particular, numerical data for (2.10) have shown very good agreement with Wigner-Dyson distribution (3.2) for $K \gg 1$ and $k \gg N$. It turns out /9/ that even if Λ is not very large, the distribution $P(s)$ is quite close to (3.2), see Fig. 4. Here, $N = 199$, $N_1 = 99$, $k \approx 60$, $\tau \approx 12\pi/N$, $K \approx 11.4$, $\Delta k = 1.0$ and $NG = 5$ matrices have been used with different k, to get more confident results (all $P(s)$ are summed, assuming they are independent). The spacings s are measured in Fig. 4 in the units of average spacing, $\Delta = 2\pi/N_1$. The χ^2 - probe shows good agreement with the analytical dependence (3.2), giving $\chi^2_{24} \approx 20.1$, with the confidence level $P_w \approx 30\%$ (the total number of quasienergy levels in Fig. 4 equals $N \cdot NG = 995$).

Thus, the fluctuations of quasienergy spectra are correspondent to the RMT when $K \gg 1$ and $\Lambda \gg 1$. It is interesting to note that our unitary matrix U_{nm}, by the construction is not random and depends only on two parameters, τ and k, which are dynamical parameters of our model. Nevertheless, under above two conditions this matrix appears to be very close to random one. In this respect, it is important to study, to what extent such dynamical matrices can be described by Random Matrix Theory.

4. CHAOTIC LOCALIZED STATES

As we could see from the above discussion, maximal quantum chaos has properties which are well described by the RMT. Now let us discuss another situation. In our model of the kicked rotator on the torus (2.10) for the fixed degree of classical

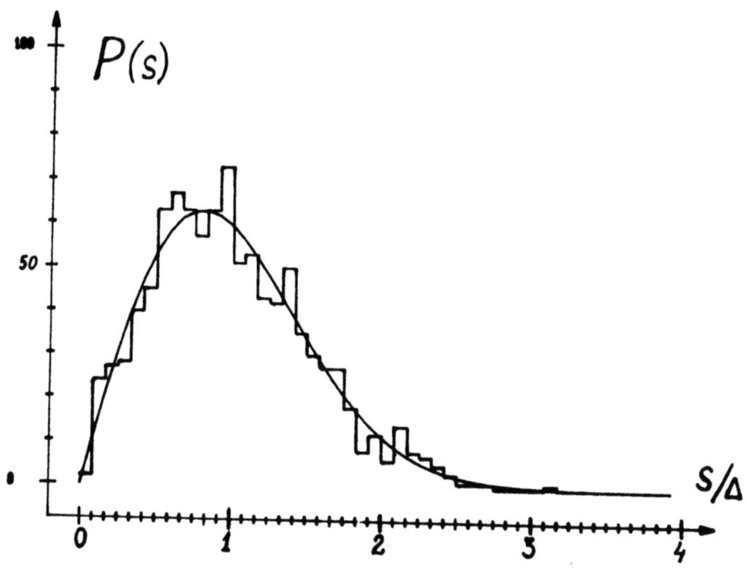

Fig. 4.

chaos (K = 5), the only parameter which affects the statistical properties is the degree of localization of EF. When the parameter $\Lambda \sim$ 1 we may expect that EF are no more extended.
This can be seen in Fig. 5a where, as a typical example, 10
eigenstates are presented from the total set (N_1 = 399) for
the parameters $k \approx$ 28.9 and $K \approx$ 5 ($\Lambda \approx$ 0.52). It is seen
that some eigenstates are localized on the scale less than the
size of the basis, while the others still can be treated as ful-
ly extended. One should note quite strong fluctuations in the
sizes of EF. This peculiarity is typical for the intermediate
situation between completely localized ($\Lambda \to$ 0) and delocalized
($\Lambda \to \infty$) states. With the decrease of Λ the size of EF (effective number of unperturbed states occupied by exact EF)
also decreases (see Fig. 5b with $k \approx$ 8.6, $\Lambda \approx$ 0.05). Nevertheless, on some scale in the unperturbed basis all EF seems to
be quite "chaotic". For this reason, one can use the term "localized chaotic states" to stress the chaotic structure of EF on
some scale.

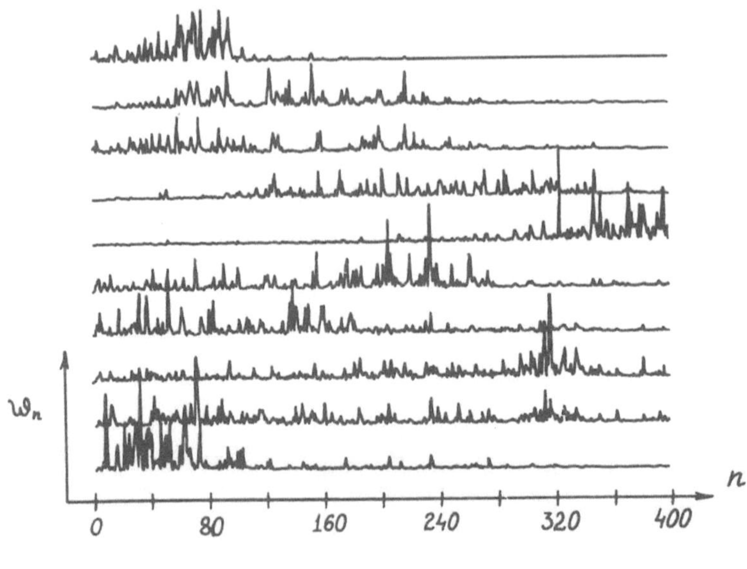

Fig. 5a.

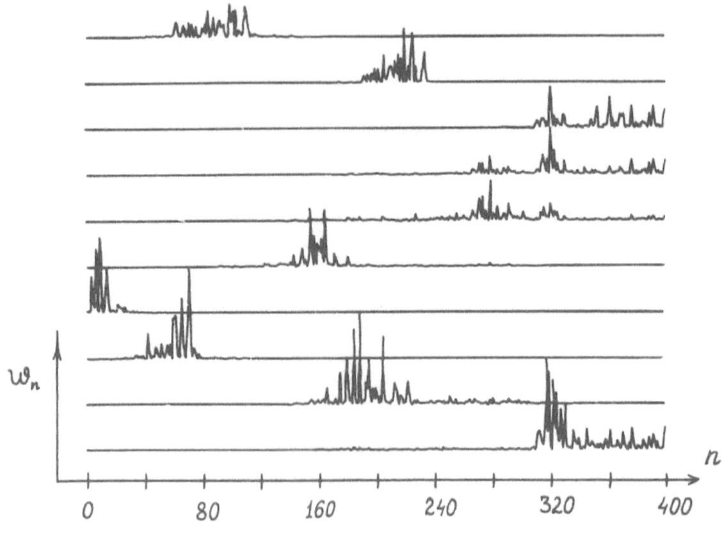

Fig. 5b.

It is natural now to introduce some quantity to measure effective "size" of EF in the unperturbed basis. As we could see above, for localized states the common definition of ℓ_∞ (by the rate of decay of amplitudes φ_n for $|n| \to \infty$) is a quite good measure. However, for the case when ℓ_∞ is compared to the total size of the basis, N_1, the above meaning of localization length is no more valid. For this reason, in /10/ another definition of localization length is proposed which is based on the computing the information entropy

$$\mathcal{H} = -\sum_{n=1}^{N_1} w_n \ln w_n \; ; \qquad w_n \equiv |\varphi_n|^2 \qquad (4.1)$$

for an eigenstate. It is seen that $\mathcal{H}$ is essentially the logarithm of the number of sites significantly populated by the given EF. In the case when all sites are equally populated, $\mathcal{H}$ equals to $\ln N_1$. If, instead, φ_n are exponentially localized around some site n_0 (see (2.8)), then $\mathcal{H} = 1 + \ln \ell_\infty + O(1/\ell_\infty)$. If eigenstate is completely delocalized and chaotic, then φ_n are distributed according to (3.1) and for $N_1 \gg 1$ one can obtain $\mathcal{H}_R = \ln(\alpha N_1/2) + 1/N_1 + O(1/N_1^2)$ which is $\approx \ln(2)$ less than $\ln N_1$ (here $\alpha = 4 \cdot \exp(\gamma - 2)$, $\gamma \approx 0.577$, see /10/). It means that if we define localization length $\ell_{\mathcal{H}}$ as

$$\ell_{\mathcal{H}} = N_1 \exp(\mathcal{H} - \mathcal{H}_R) \qquad (4.2)$$

with $\mathcal{H}_R$ standing for the entropy $\mathcal{H}$ of EF in the extreme case of random matrices, then the maximal value of $\ell_{\mathcal{H}}$ is exactly N_1. Therefore, the normalizing parameter $\exp(-\mathcal{H}_R)$ takes into account random character of the states. With this definition, in the opposite case of pure exponential localization we find the relation $\ell_{\mathcal{H}} = 2\ell_\infty \cdot e/\alpha + O(1)$. Finally, to reduce very large fluctuations, we define the average localization length d according to the expression

$$d = N_1 \exp(\langle \mathcal{H} \rangle - \mathcal{H}_R) \qquad (4.3)$$

where averaging is performed over all EF of one matrix U_{nm} .
Numerical data /10/ show that d has smooth dependence on k^2,
inspite of very strong fluctuations in $l_{\mathcal{H}}$ for different eigenstates. This new quantity, d , seems to be very convenient in
describing of the global properties of chaotic states.

5. INFLUENCE OF LOCALIZATION ON THE SPECTRUM STATISTICS

Numerical data show that when decreasing Λ , the spacing
distribution $P(s)$ changes from Wigner-Dyson (3.1) to the distribution which turns out to be very close (but different, see
details in /15/) to the Poissonian one ($P(s) \sim exp(-s)$).
To relate the level spacing distribution $P(s)$ with the degree of localization, we need some analytical description of $P(s)$.
In the literature one can find two analytical formulae in describing the transition from Poissonian to Wigner-Dyson statistics
(see reviews /1-2/), one of which is the well known Berry-Robnik dependence /16/. This dependence is commonly used to describe the intermediate statistics of spectrum for the situation
when the corresponding classical system has significant regions
of the stable motion in the phase space. The only parameter in
Berry-Robnik dependence is exactly the ratio of the area with
the stable motion to that of chaotic. It is clear, that for our
case of strong classical chaos this formula is not valid. In some sense, the Berry-Robnik expression deals with the situation
which is opposite to our case. Namely, the deviation of Berry-
-Robnik distribution from the limiting Wigner-Dyson distribution
is entirely caused by the classical effect (the existence of
stable regions) while in our case the intermediate statistics
is of pure quantum nature (localization).
Another known expression for $P(s)$ is the so-called Brody
distribution /17/ which is nothing but some approximate dependence with the only fitting parameter. Since the latter dependence is
wrong when repulsion is larger than one (for $\beta > $ 1), in /10/
another expression has been proposed for $P(s)$:

$$P(s) = A s^{\beta} \exp\left\{ -\frac{\beta \pi^2}{16} s^2 - \left(C - \frac{\beta}{2}\right)\frac{\pi}{2} s \right\}$$

(5.1)

Here A and C are normalizing parameters determined by the same conditions (3.3). This dependence, unlike Wigner-Dyson one, has the right limit for large spacings, $s \to \infty$. On the other hand, it appears quite close to (3.2) for $\beta = 1;2;4$ if s not too large. In addition, for $\beta = 0$ the dependence (5.1) is Poissonian with the correct values of A and C . The difference of (5.1) from the exact dependence, obtained in RMT numerically, (see /12/) can be seen only when the total number of states exceeds $\sim 10^4$. Therefore, the proposed dependence (5.1) may be used for the description of the intermediate statistics, in our case, for $0 \leq \beta \leq 1$ (another example is discussed in /10/ where the maximal repulsion is $\beta = 2$).

The main idea is to relate the repulsion parameter in the spacing distribution $P(s)$ with the degree of localization of the chaotic states. For this, in /10/ extensive numerical simulation has been performed with the model (2.10). Namely, the average localization length d of matrix U_{nm} and spacing distribution $P(s)$ for eigenvalues ε have been computed independently for a wide range of semiclassical parameter $k \gg 1$ with the fixed classical parameter $K = 5$. To improve the statistics, the summing of $P(s)$ for a number of matrices U_{nm} has been also used, as for the extreme case of maximal quantum chaos. As a result, good linear relation was found between the fitting parameter β in (5.1) and the ratio d/N_1 . This result seems to be important in attempts to relate the properties of the spectra and eigenfunctions in the region of classical chaos. It is interesting to study to what extent this relation for β is general.

6. UNIVERSAL SPECTRUM FLUCTUATIONS

As it is known from RMT, the spectral fluctuations in the

case of completely random matrices are universal in the sense
that they are independent on the size of matrices for $N \gg 1$.
According to above results, it is natural to assume that in the intermediate region of partly suppressed quantum chaos some sort
of universality may also exist. Indeed, if the spacing distribution essentially depends on the repulsion parameter β only,
then $P(s)$ has the same scaling properties as the ratio of the
average localization length to the size of the basis. This situation is similar to that one known in solid state physics,
where some scaling theory is developed for the models with random potentials (see, e.g. /18-19/). Namely, there is conjecture
of the existence of universal relation between rescaled localization length ℓ_∞/N (determined for infinite sample) and rescaled localization length ℓ_N/N (determined for finite sample with
the same potential), $\ell_N/N = f(\ell_\infty/N)$. Since for the strong classical
chaos in our model (2.10) the interaction between unperturbed
states in the range $\Delta n \approx 2k$ may be treated as random /5-6,
10/, it means that we may expect some scaling dependence of d/N_1
on the parameters of our model. This conjecture has been recently studied numerically /20/. The data have shown quite good scaling behaviour for d . Namely, the ratio d/N_1 with d determined according to (4.3) turns out to be the same if the ratio
k^2/N is fixed. One should note that k^2 is proportional to
the localization length ℓ_∞ , therefore, this result may be
regarded as the manifestation of universal scaling

$$\beta \equiv \frac{d}{N_1} = f\left(\frac{\ell_\infty}{N_1}\right)$$

$$(6.1)$$

which has the same meaning as in solid state models. These pre-
liminary data /20/ are still waiting theoretical analysis.

From the above discussion it is naturally to conclude that
the spacing distribution $P(s)$ is dependent only on the ratio
ℓ_∞/N_1 (or k^2/N_1 , if the classical parameter is fixed). Extensive numerical data with the model (2.10) has been performed
for different N_1 = 200, 398 and 600 (for $k \gg 1$ and K = 5).

The value of β was determined by the χ^2 -fitting of (5.1).
In Figs. 6a-6c some examples are given where the values k^2/N are approximately the same ($k^2/N \approx$
≈ 0.5) but k and N_1 are different (here (a) presents $P(s)$
for $N_1 = 200$, $NG = 10$, $k \approx 9.95$; (b) - $N_1 = 398$, $NG = 5$,
$k \approx 14.4$; (c) - $N_1 = 600$, $NG = 3$, $k \approx 17$). The full lines

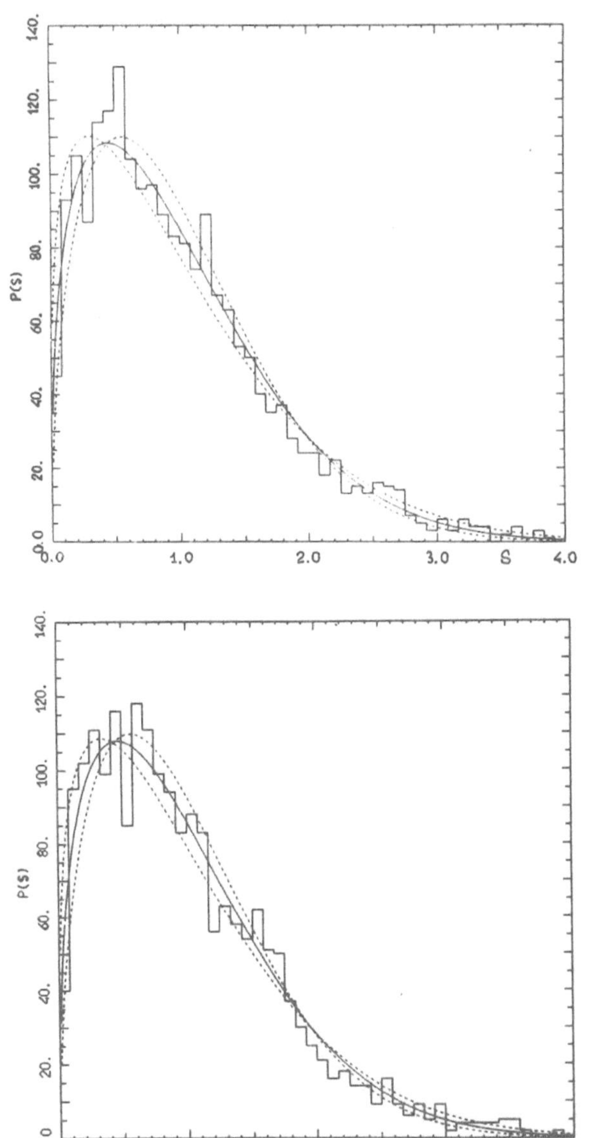

Fig. 6a.

Fig. 6b.

correspond to the best fitting and the dashed lines to the deviations with 1% confidence level. The data for β are: (a) – 0.25, 0.39, 0.52; (b) – 0.29, 0.41, 0.55; (c) – 0.22, 0.37, 0.53, for β_{min} , β_{fit} , β_{max} , correspondingly. It is seen that there is good scaling for $P(s)$. Quite large fluctuations in the form of $P(s)$ are probably provided by the existence of correlations between matrices U_{nm} with different k , due to not too large Δk ($\approx$ 0.05; 0.5; 1.0 for (a), (b), (c), corresondingly). Similar scaling has been also obtained for k^2/N 0.125; 0.25; 1.0. All these results indicate that scaling of $P(s)$, with the scaling parameter ℓ_∞/N_1 , seems to exist. Recently, similar scaling has been found for Band Random Matrices /21/.

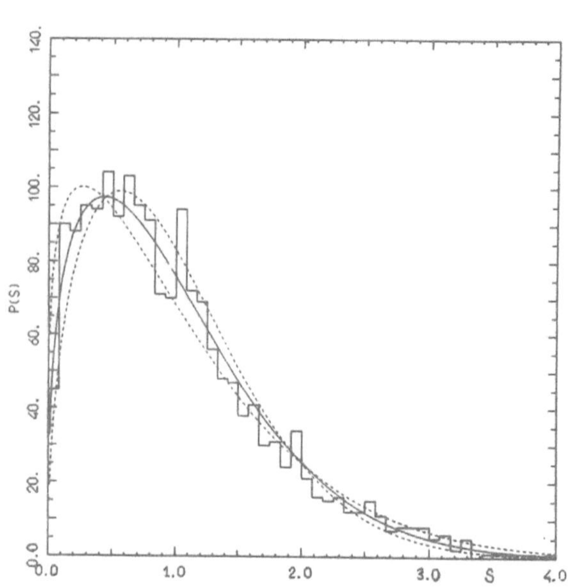

Fig. 6c.

REFERENCES

1. B.Eckhardt, Phys. Rep., 163 (1988) 205.
2. P.V.Elyutin, Usp. Fis. Nauk, 155 (1988) 397.
3. B.V.Chirikov, Phys. Rep. 52 (1979) 263.
4. A.J.Lichtenberg and M.A.Lieberman, Regular and Stochastic Motion (Springer, Berlin, 1983).
5. B.V.Chirikov, F.M.Izrailev and D.L.Shepelyansky, Sov. Sci. Rev. C2 (1981) 209.

6. B.V.Chirikov, F.M.Izrailev and D.L.Shepelyansky, Physica D33 (1988) 77.
7. G.Casati, B.V.Chirikov, J.Ford and F.M.Izrailev, Lecture Notes in Physics 93 (1979) 334.
8. F.M.Izrailev and D.L.Shepelyansky, Dokl. Akad. Nauk SSSR, 249 (1979) 1103 (Sov. Phys. Dokl. 24 (1979) 996); Teor. Mat. Fiz. 43 (1980) 417 (Theor. Math. Phys. 43, 553).
9. F.M.Izrailev, Phys. Lett. A125 (1987) 250.
10. F.M.Izrailev, Phys. Lett. A134 (1988) 13; J. Phys., A22 (1989).
11. F.J.Duson, J. Math. Phys., 3 (1962) 140, 157, 166.
12. M.L.Mehta, Random Matrices, Academic, 1967.
13. T.A.Brody, J.Flores, J.B.French, P.A.Mello, A.Pandey, S.S. M.Wong, Rev. Mod. Phys., 53 (1981) 385.
14. F.M.Izrailev, Phys. Rev. Lett. 56 (1986) 541.
15. G.Casati, I.Guarneri, F.M.Izrailev, Phys. Lett. A 124 (1987) 263.
16. M.V.Berry and M.Robnik, J. Phys. A19 (1986) 649.
17. T.A.Brody, Lett. Nuov. Cim., 7 (1973) 482.
18. P.A.Lee, T.V.Ramakrishnan, Rev. Mod. Phys. 57 (1985) 287.
19. J.L.Pichard and G.J.Sarma, J. Phys. C14 (1981) L127.
20. G.Casati, I.Guarneri, F.Izrailev and R.Scharf, "Towards a scaling theory of localization in quantum chaos", appear in Phys. Rev. Lett., 1989.
21. G.Casati, L.Molinari, and F.Izrailev, "Scaling properties of Band Random Matrices", to be published, 1989.

Institute of Nuclear Physics, 630090, Novosibirsk, USSR

Operator Theory:
Advances and Applications, Vol. 46
© 1990 Birkhäuser Verlag Basel

SINGULAR CONTINUOUS QUASI-ENERGY SPECTRUM IN THE KICKED ROTATOR WITH SEPARABLE PERTURBATION : ONSET OF QUANTUM CHAOS?

B.Milek and P.Šeba

We prove that the quasi-energy spectrum of the kicked quantum rotator model with a separable potential which has been recently introduced by M.Combescure is singularly continuous under certain conditions. The time evolution of this system is numerically investigated in detail.

1. INTRODUCTION

Recently a big amount of work has been devoted to time-dependent quantum systems beyond the usual perturbation theory [1]. The motivation of these papers was to investigate to what extent is the chaotic behavior, which is displayed in the classical time-dependent systems, present in their quantum counterparts. In this line of research the numerically most thoroughly investigated model is the so-called kicked quantum rotator which corresponds to the classical map of Chirikov and Taylor [2]. The result obtained for it was, however, disappointing : it appeared that for small coupling constant (being on the other hand big enough

for the classical system to be completely chaotic) this system
displays recurrent behavior associated with the pure pointness of
its quasi-energy spectrum [3]. The non-recurrent pattern appeared
only if the driving pulses were rationally connected with the
internal frequency of the free rotator [3]. This "quantum
resonances" which are connected with the absolutely continuous
part of the quasi-energy spectrum represent, however, a
completely different behavior then the expected chaotic one. They
simply describe the resonant energy accumulation inside the
system. The chaotic behavior (if any) is expected to take place
for non-resonant frequencies only.

These desultory numerical results (known also as the
quantum suppression of chaos) obtained a clear heuristic
explanation in the work of the Maryland group (Prange, Fishman
and Grempel [4]) which showed that there is a close connection
between the behavior of the quantum rotator and the Anderson
localization in disordered solids. Using this connection it can
be easily shown that the recurrent and resonance patterns in the
quantum rotator correspond to localized and extended states in
the Anderson model, respectively. What concerns the chaotic
behavior of the rotator it is expected that <u>it must be
connected with the singularly continuous component of the Floquet
operator</u>, and hence with the appearance of exotic states in the
corresponding Anderson model. The singularly continuous spectrum
is ,however, a subject which is mathematically very subtle.
Trying to prove its existence for the quantum rotator one is led
beyond the range of applicability of the Maryland model. The
point is that the similarity between the quantum rotator and the
Anderson model (no matter how illustrative it may be) does not
endure a rigorous mathematical analysis and cannot be used if
such fragile objects like singularly continuous spectrum are
concerned.

On the other hand there are two arguments which make
the search for the singularly continuous states in the quantum
rotator quite promosing. The first of them is an abstract
mathematical result obtained by Casati and Guarneri [5] who
showed that for a "generic" potential there is some continuous

spectrum left even in the non-resonant case. They were, however, not able to prove that this spectrum is in fact singularly continuous although they conjectured that it was the case. The biggest disadvantage of their result is, however, that it holds only for a "generic" potential and one is in fact not able to check whether a given potential belong to this generic class or not.

The second argument in favor of the singularly continuous states comes from the numerical results [3,6]. It has been shown that the recurrent non-resonance pattern which occurs for small couplings is replaced by a non-recurrent behavior as soon as the coupling becomes strong. This results has been interpreted by some authors as a manifestation of the singularly continuous spectral component of the Floquet operator [3,6]. Their arguments are, however, not completely convincing. First of all it is not quite clear that the behavior is non-recurrent. It is possible that we have to do with a recurrent pattern with a very long period and hence with localized states. The computer results do not offer a possibility to distinguish which spectral type is in fact present. The second difficulty is connected with the fact that it is not clear how the singularly continuous states would manifest themselves in the course of computer simulation. In the ideal case one expects, of course, a non-recurrent pattern. But the real result represents an interplay between the presence of exotic states on one side and the unavoidable computer errors on the other side. Taking now into account the extreme complexity of the singularly continuous eigenfunctions (note that this spectrum lives on something like a Cantor set [7]) it is possible that the numerical errors dominate after a few iterations and that the expected non-recurrent pattern will be replaced by an oscillatory one.

In this situation it appeared to be of interest to investigate a more simple model in which the mathematical as well as the numerical analysis go hand in hand up to the end. Our aim here is to demonstrate how one can "see" the presence of the singularly continuous states in a standard numerical simulation of the time evolution of the system. For this purpose we choose a simplification of the quantum rotator model introduced recently by M.Combescure [8]. In this model the local potential of the standard rotator is replaced by a separable potential of rank one. The advantage of this procedure is that the spectral properties of the corresponding Floquet operator can be thoroughly analyzed and in particular the presence of the singularly continuous component can be <u>rigorously</u> proven. We describe the corresponding results in Section 2. Section 3 contains numerical results. We conclude the introduction by quoting from a paper by Prof. G. Casati [1] : "Mathematical rigour (in quantum chaos) though desirable, is a rare occurrence : only few are the landmarks in this "terra incognita". "

2.MATHEMATICAL PRELIMINARIES

Before proceeding further we recall the basic mathematical results concerning the instability of a time-dependent quantum system. Let $H(t)$ denote a time-dependent quantum Hamiltonian defined on a Hilbert space $\mathcal{H}$. We will suppose that $H(t)$ is of the form

$$H(t) = H_0 + V(t) \tag{2.1}$$

with H_0 being a below-bounded self-adjoined operator with the discrete spectrum. The potential $V(t)$ is assumed to be time-periodic,

$$V(t) = V(t+T) \ , \ T > 0 \ , \ t \in \mathbb{R} \tag{2.2}$$

and such that the resulting Hamiltonian $H(t)$ is reasonably defined. The dynamics of the corresponding quantum system is described by the evolution operator $U(t)$ which solves the time-dependent Schroedinger equation

$$i\partial_t U(t) = H(t)U(t)$$
$$U(0) = 1 \qquad (2.3)$$

For investigating the stability/instability of the system it is enough to investigate the stroboscopic picture of the time evolution at the times t_ν which are equal to multiples of the period T,

$$t_\nu = \nu T \qquad (2.4)$$

This "stroboscopic" evolution is governed by the one-cycle Floquet operator U

$$U = U(T) \qquad (2.5)$$

$$\psi(\nu T) = U^\nu \psi(0) \qquad (2.6)$$

($\psi(0)$ and $\psi(\nu T)$ denote the wave function at the times 0 and νT, respectively) [1].

Of central importance for the quantum evolution of the system is the spectral nature of U. In order to illustrate this statement let us compute the probability $p_{n,m}(\nu)$ to excite the n-th state of H_0 after ν cycles to the m-th state

$$p_{n,m}(\nu) = |P_{n,m}(\nu)|^2 \qquad (2.7)$$

with $P_{n,m}$ being the probability amplitude

$$P_{n,m}(\nu) = \langle n|U^\nu|m\rangle \qquad (2.8)$$

Decomposing the Floquet operator U into its eigenvectors

$$U|\omega> = e^{i\omega} |\omega> \tag{2.9}$$

we get

$$p_{n,m}(\nu) = \sum_{\omega} e^{i\nu\omega}<n|\omega><\omega|m>. \tag{2.10}$$

Hence the transition $n \rightarrow m$ is possible only if at least one quasi-energy state $|\omega>$ connects these two states, i.e., if $<n|\omega><\omega|m> \neq 0$ for at least one ω. Now the importance of the spectral nature of U becomes apparent : if the spectrum of U is pure point all the quasi-energy states $|\omega>$ are localized. Consequently, they can connect only a few states of the original Hamiltonian H_0 and a strong recurrence is expected (this is the essence of the well known theorem by Hogg and Huberman [9]). In the case of a continuous spectrum of U the quasi-energy states $|\omega>$ are not normalizable. They must therefore connect an infinite number of the original states $|n>$ leading in such a way to non-recurrent (unstable) evolution. Summarizing this heuristic arguments we can say that the pure pointness of the Floquet operator means stability while the occurrence of the continuous spectrum implies instability of the system.

Therefore the only promising systems (from the point of view of quantum chaos) are those with continuous quasi-energy spectrum and we are going to discuss them in some more details.

From the abstract point of view we can divide the continuous spectrum into two parts: the absolutely continuous (a.c) and the singularly continuous (s.c.) one. Let us assume that the spectrum of U is purely a.c. We get then for the probability amplitude

$$P_{n,m}(\nu) = \int e^{i\omega\nu} d\mu_{n,m}(\omega) = \int e^{i\omega\nu} f_{n,m}(\omega)d\omega \tag{2.11}$$

with $f_{n,m}(\omega) \in L^1(0,2\pi)$ (the spectral measure $d\mu_{n,m}$ is continuous with respect to the Lebesque measure). Using now the Lebesque lemma [10] we find

$$\lim_{\nu \to \infty} P_{n,m}(\nu) = 0 \tag{2.12}$$

for all n,m. Consequently, the probability to find the system after ν oscillations at a state m tends to zero as $\nu \to \infty$. In other words, this means that the system is continuously accelerated and excites to higher and higher states in the course of time. The mean energy is supposed to grow very quickly with time. This type of behavior is usually associated with some kind of resonance phenomena. For example for the kicked quantum rotator with resonance frequency the energy growth is quadratic in time.

In the singularly continuous case the probability $P_{n,m}(\nu)$ decreases very slowly toward zero as $\nu \to \infty$. We find in this case

$$P_{n,m}(\nu) = \int e^{i\omega\nu} \, d\mu_{n,m}(\omega) \tag{2.13}$$

with the a measure $d\mu_{n,m}$ being singular with respect to the Lebesque measure. This implies that (see the RAGE theorem [11])

$$\lim_{M \to \infty} \sum_{\nu=0}^{M} P_{n,m}(\nu) = \infty \tag{2.14}$$

together with

$$\lim_{M \to \infty} \frac{1}{M} \sum_{\nu=0}^{M} P_{n,m}(\nu) = 0 \tag{2.15}$$

The system spends now an infinite amount of time in the "lower" states and the sum (2.14) diverges. In average, however, the

system escapes any finite state which leads to (2.15). The
evolution along the states $|n>$ is now "recurrently pulsing" with
larger and larger "amplitude". The energy growth is slow and
mimics the diffusive acceleration known from classical chaotic
systems. This type of quasi-energy spectrum is assumed to be
responsible for the "true" quantum chaotic evolution.

Let us now return to the model which has been recently
introduced by M.Combescure. It is described by the time-dependent
Hamiltonian

$$H(t) = H_0 + \Delta(t)P \qquad (2.16)$$

where H_0 is the kinetic energy operator of the free rotator

$$H_0 = -\frac{1}{2}\frac{\partial^2}{\partial\theta^2} \qquad (2.17)$$

defined on Hilbert space

$$\mathcal{H} = L^2(0,2\pi) \qquad (2.18)$$

with periodic boundary conditions. Here P denotes a separable
potential of rank one,

$$P = |f><f| \; ; \; f \in \mathcal{H} \qquad (2.19)$$

and $\Delta(t)$ is a periodic sequence of kicks,

$$\Delta(t) = \sum_{\nu\in\mathbb{Z}} \delta(t-\nu T). \qquad (2.20)$$

In this case the Floquet operator has a particularly simple form of a quantum map

$$U = e^{-iH_0T} e^{-iPT} \qquad (2.21)$$

the spectral properties of which can be simply analyzed.

THEOREM 1 (M. Combescure)
(i) *Suppose that $f \in L^2(0,2\pi) \cap L(0,2\pi)$. Then the operator U has a pure point spectrum for almost every T.*
(ii) *If $f \in L^2(0,2\pi)$ but $f \notin L(0,2\pi)$ then*
  (a) *the spectrum of U is pure point if T/π is a rational number,*
  (b) *the spectrum of U is purely continuous if T/π is a Diophantine number (i.e., a irrational one which is badly approximable by the rationals).*

Combescure based the proof of this theorem on her generalization of results of Simon and Wolff which concern the stability of the dense pure point spectra of self-adjoined operators under rank-one perturbations [12]. The part (i) of the theorem holds for all T which are rational multiples of π and for almost every T which are irrational multiples of π (see [8] for more details).
For our purpose we need to know more about the continuous spectrum ; we are going to prove prove that this spectrum is in fact <u>purely singularly continuous.</u>

THEOREM 2 *Assume that the conditions of the part (ii) of Theorem 1 are fulfilled with T/π being a Diophantine number. Then the spectrum of U is purely singularly continuous.*

PROOF: We know from Theorem 1 that the spectrum of U is continuous. It is therefore enough to prove that its absolutely continuous part is empty. To this aim we use the fact that P is a rank-one operator which enables us to express exp(-iPT) as

$$\exp(-iPT) = 1 + \frac{1}{\|f\|^2}\left(\exp(-i\|f\|^2 T) - 1\right) P \qquad (2.22)$$

Inserting this formula into (2.21) we get

$$U = U_0 + R \qquad (2.23)$$

where U_0 is the free evolution

$$U_0 = \exp(-iH_0 T) \qquad (2.24)$$

and R is an operator of rank one. We use now the scattering theory. Let us assume that the spectrum of U is absolutely continuous. In this case the wave operators

$$\Omega_\pm = \operatorname*{s-lim}_{\nu \to \pm\infty} U^\nu U_0^{-\nu} \qquad (2.25)$$

exist (because of the rank-one perturbation, see for instance [13,14] for the proof) and hence the absolutely continuous spectrum of U_0 contains the spectrum of U

$$\sigma_{a.c.}(U) \subset \sigma_{a.c.}(U_0) \qquad (2.26)$$

On the other hand we know that $\sigma_{a.c.}(U_0) = \emptyset$ (the operator U_0 can be trivially diagonalized), and therefore $\sigma_{a.c.}(U) = \emptyset$.

Let us now proceed to numerical simulations in order to illustrate the above theorems.

3. NUMERICAL INVESTIGATIONS

Here we use a simple representation of the state $|f>$ in terms of the unperturbed basis $|n>$ of H_0 ,

$$|f> = \sum_{n=-N}^{N} a_n |n> \qquad (3.1)$$

with

$$a_n = |n|^{-\gamma} . \qquad (3.2)$$

Studying the sums

$$S_1 = \sum_{n=-N}^{N} |a_n|^2 \qquad (3.3)$$

and

$$S_2 = \sum_{n=-N}^{N} |a_n| , \qquad (3.4)$$

one can easily see that in the limit of an infinite number of basis states, $N \to \infty$, the state f belongs to $L^2(0,2\pi)$ for $\gamma > 0.5$. But for $0.5 < \gamma < 1.0$ the sum S_2 tends in the considered limit to infinity. Hence $|f>$ tends to a vector which does not belong to $L(0,2\pi)$ and one should expect the singularly continuous spectrum to start manifest itself in the evolution of the system. Due to the Theorem 1 we will throughout the paper refer to the cases with $\gamma > 1.0$ and $0.5 < \gamma < 1.0$ as to the cases (i) and (ii), respectively.

The solution of the Schroedinger equation for the considered model can be found expanding as usually the wave function in the unperturbed basis

$$|\psi(t)> \; = \; \sum_{n=-N}^{N} \; c_n(t) \; |n> \; . \qquad (3.5)$$

The unknown coefficients are governed by a recursion formula based on the quantum map (2.21) (for projection operators one has $P = P^k$, $k = 0,1,2,\ldots$)

$$U \; = \; e^{-iH_0T} \; (1 \; + \; (e^{-iTS_1} - 1) \; P/S_1) \qquad (3.6)$$

according to

$$c_n(\nu+1) \; = \; e^{-iE_nT} \left(c_n(\nu) \; + \; A \; a_n \sum_{m=-N}^{N} a_m \; c_m(\nu) \right) \qquad (3.7)$$

with the eigenvalues of the rotator states

$$E_n \; = \; n^2/ \; 2 \qquad (3.8)$$

and

$$A \; = \; (e^{-iTS_1} - 1) \; / \; S_1 \; . \qquad (3.9)$$

Here $c(\nu)$ denotes the coefficients just before the ν-th kick. Now the quantum mapping (3.7) can be iterated numerically for any given initial condition $c(0)$. In the current investigations we have initially localized the particle in the center of the unperturbed basis, i. e. $c_0(0) = 1$ and $c_n(0) = 0$ for all other n.
        An interesting quantity from the viewpoint of quantum chaos is the time autocorrelation function [1]

$$\left| \; \lim_{\tau \to \infty} (1/2\tau) \int_{-\tau}^{\tau} <\psi(s)|U(t+s,0)|\psi(0)> \; ds \; \right| \qquad (3.10)$$

which can be easily expressed for the considered system with the chosen initial conditions using the expression (2.7) as $(p_{0,0}(t))^{-1/2}$ with

$$p_{0,0}(\nu) = |<0|U^\nu|0>|^2 \ . \tag{3.11}$$

It can be expected from eqs. (2.11-15) that the different qualitative nature of the p.p. or s.c. quasi-energy spectra can be seen also in the numerical results, calculated from (3.11). From eq. (2.13) one has to state that the autocorrelator is Fourier transform of a singular continuous measure! However, due to the very weak time decrease (2.14-15), the detailed time behavior of the autocorrelation function is rather unpredictable.

Before proceeding to the numerical results let us say a few words about the time evolution of the averaged energy. In the perturbation under consideration the excitation of all basis states takes place already after the first kick,

$$c_n(1) \cong A \ a_n \ a_0 \ , \ n \neq 0 \ , \tag{3.12}$$

and is essentially given by the chosen distribution of the state f over the basis. From this we have to show a quite uncommon property of the model. The averaged energy

$$<E> = \sum_n n^2/2 \ |c_n|^2 \tag{3.13}$$

is after the first kick proportional to $\sum |n|^{2(1-\gamma)}$, which means that the energy diverges for $\gamma < 1.5$ in the limit of an infinite number of basis states. In this sense, one should remember for the interesting limit $N \rightarrow \infty$ that a finite expectation value of H_0 can be realized in case (ii) only by a reduction of the states to a finite basis.This is done by the physical equipment during the measuring process. Roughly speaking in a real experiment the measurement is associated with an observable which is different from the operator H_0. We measure in fact the energy from some chosen interval, which depends on the apparatus used in the experiment. The measured energy is therefore given by

$$<E> = <\psi|\hat{E}|\psi> \tag{3.14}$$

with the observable $\hat{E}$ given by

$$\hat{E} = \sum_{n=n_1}^{n_2} \frac{n^2}{2} |n><n| \tag{3.15}$$

and n_1, n_2 dependent on the energy interval in which the measurement is performed (see, for instance, [15]). In this sense one need worry about the infinite matrix element of H_0 which appear for $N \longrightarrow \infty$.

In order to see in the numerical calculations some non-recurrent behavior one is forced to choose a large number N. This number must be large enough (depending on the maximum kick number), in order to avoid a domination of the rescattered flux which comes from the reflection on the borders in the basis at $\pm N$. One expects, however, that for rational ratios T/π the wave function must be localized (the spectrum is pure point) and hence rather independent on the size of N.

In accordance with the standard investigations of the kicked rotator (see for example [9]) the crucial parameter for the dynamics has been chosen as

$$x = T/4\pi \quad . \tag{3.14}$$

For $\gamma > 1$ this parameter should not play any role in the qualitative features of the numerical solutions because according to part (i) of Theorem 1 the spectrum is always pure point. But for $0.5 < \gamma < 1$ one expects localization phenomena for rational values of x and delocalization for irrational (Diophantine) values which are governed by the singularly continuous spectrum of the system. In practical calculations the parameter x has been in the irrational case chosen as the golden mean $x=(\sqrt{5}-1)/2$ $=0.6180339...$, (the "most irrational" number since it is most difficult to approximate it by rationals). The example of a rational x has been realized by $x=6/10$ which is close to the golden mean.

The Schroedinger equation has been solved up to 500 kicks with a total number of basis states up to $2N+1=5001$. The calculations have been performed at IBM-AT compatible personal computers (CPU 80286/386) at JINR, Dubna and at the IBM main frame at GSI, Darmstadt within an accuracy of 16 valid digits. This guarantees a conservation of the norm after the maximum kick number within an uncertainty of 10^{-14}. As an additional check we considered the time reverse after some small time, using the operator U^{-1}, and found the same degree of accuracy.

On Fig.1 the averaged energy $<E>$ for the case (ii) $(\gamma=0.7)$ has been drawn versus the kick number for $x = 0.6$ (dashed curve) and $x = 0.618...$(solid curve). In an excellent agreement with the theorems from chapter 2 one can observe a qualitative difference : for the rational x quantum interference ensures a periodic behavior of the averaged energy, not exceeding an upper limit, but for the irrational case the energy grows non-recurrently in the considered time interval. The increase in energy is not monotonic. One can observe a plateau-like structures which have been discussed also for the usually perturbed rotator with strong coupling [3,6] and have been interpreted there as a manifestation of the singularly continuous component of quasi-energies. The plateaus are connected with the "pulsating" spreading of the initial state during the evolution. From the global point of view one can see a more or less linear increase which is a direct manifestation of the "energy diffusion" due to the "random walk-like" spreading of the wavefunction which is typical for the singularly continuous case,

see for instance [16]. However, due to the fact that in the
presented calculations the difference in the x value is only 3
percent, both curves nearly coincide during some short time.

Repeating the same calculations, but now with γ = 3.0
(case (i), not shown in the figures), one can state that the
averaged energy is a periodic function of time for both values of
x as a consequence of the pure-pointness of the quasi-energy
spectrum connected with a strong localization effect in the wave
function around the initially occupied basis state.

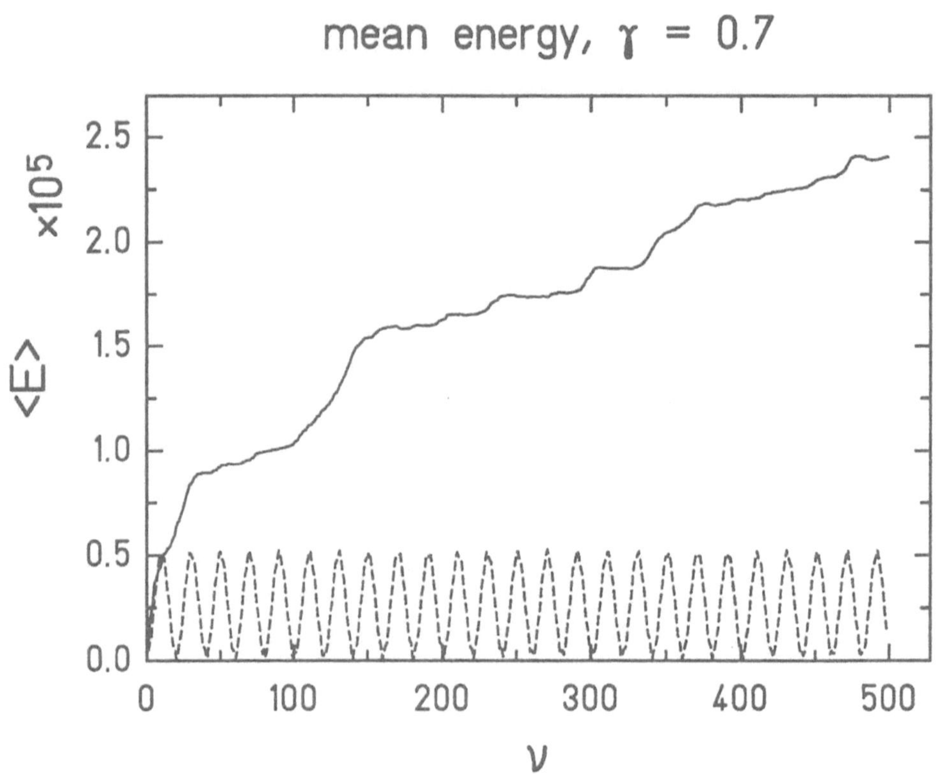

Fig.1 *Averaged energy of the kicked rotator versus kick
number for the case (ii) of Theorem 1 (γ=0.7). Solid curve: for an
irrational frequency ratio (x is the golden mean) with a singular
continuous quasi-energy spectrum, dashed line: for a rational x
value close to the golden mean (x=0.6) with a pure point
quasi-energy spectrum. The rotator basis covers 5001 states
(N=2500).*

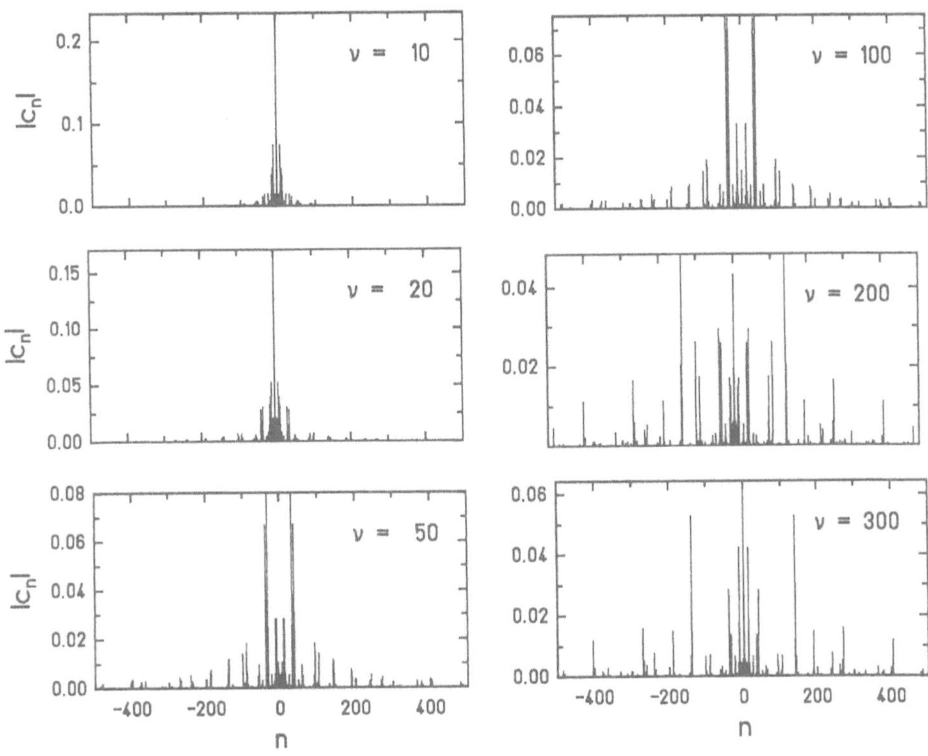

Fig.2 *Time evolution of the wave packet in the rotator basis for the case ii) with x equal to the golden mean (solid curve of Fig.1). The growing kick number is indicated inside the figure. Please note that the scale of the diagrams changes with time and that only a part of the basis states are incorporated.*

In Fig.2 for the case (ii) with x equal to the golden mean the magnitudes of the expansion coefficients $|c|$ of the wave function are shown on a sequence of diagrams for growing time (the kick number is indicated inside the figures). Instead of a monotonic spread of the wave function over the unperturbed states during the the course of kicking, expected for an absolutely

continuous spectrum, one observes once more a "pulsing" spread
which is directly related to the observed plateau-like structure
of the averaged energy (Fig.1). For comparison we demonstrate in
Fig.3 the results for the case (ii) but now for x = 0.6. The
manifestation of the localization phenomenon is obviously a
consequence of the pure point quasi-energy spectrum. The
analogous calculations for case (i) (not shown in the figures)
yield qualitatively the same results but with an even stronger
localization.

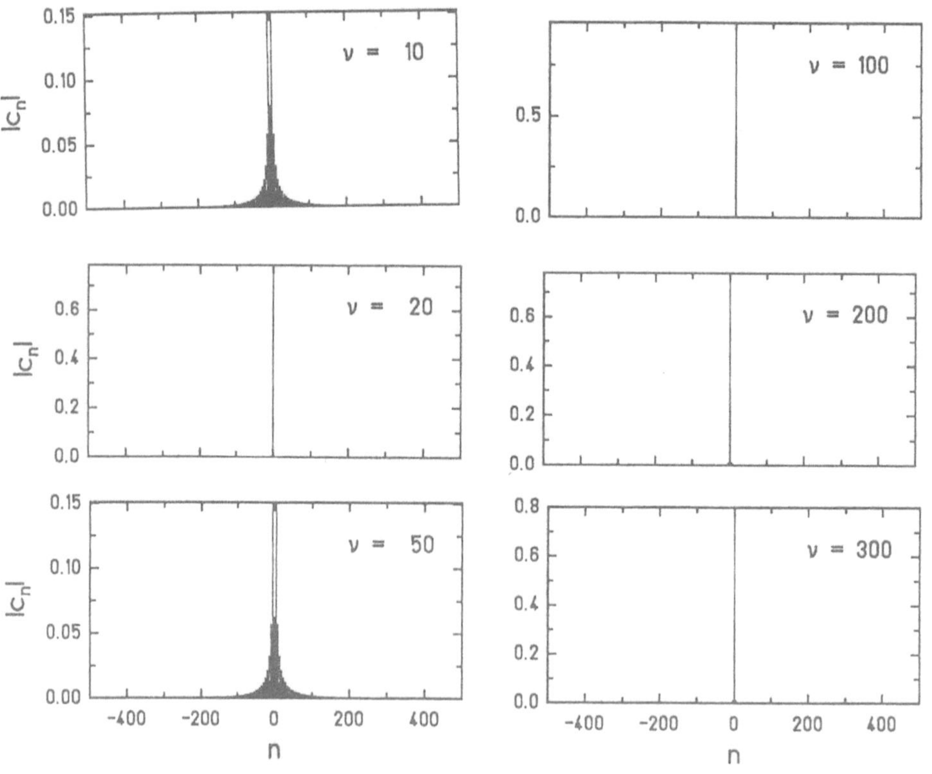

Fig.3 *The same as in Fig.3 but for the rational case (x=0.6,
dashed line of Fig.1).*

Fig.4 shows the autocorrelation function $(p_{0,0}(\nu))^{-1/2}$ = $|c_0|$ (3.10) for the case (ii) with γ = 0.7. The slow and irregular decrease which is characteristic for a singularly continuous quasi-energy spectrum is demonstrated (x equal to the golden mean) and compared with the strong recurrence which appears for the pure point case (x = 0.6). Because of the delicate mathematical nature of the autocorrelation function (see eqs. (2.13-15) and (3.11)) it is very difficult to comment in detail on its calculated long-time behavior. But in any way one can see a series of "bumps" which mean a series of comparatively large "comeback" probabilities during the time evolution of the system with a decreasing heigth. The time average over some time interval of the heigts h of the bumps will finally decrease to zero. This type of behavior should be expected from the complementary results on the averaged energy and the wave function.

4. CONCLUDING REMARKS

We have demonstrated the occurrence of a singularly continuous spectrum of the Floquet operator for the periodically kicked quantum rotator with a separable perturbation. In distinction to previous papers on the similar subject we have been able to base our statements on rigorous mathematical proofs. The exotic quasi-energy eigenstates accompanying the singularly continuous spectrum have been manifested in standard numerical calculations applying a special version of the separable interaction. If T is an irrational multiple of π (strictly speaking a Diophantine number) the numerical quantities exhibit qualitatively all the expected features like the plateau-like structures in the growing time-dependent averaged energy of the rotator, non-recurrent behavior of the wave function , and a decreasing, however very weakly and irregularly, autocorrelation function with respect to time.

Fig.4 *Autocorrelation function in dependence of the kick number for the case (ii) of Theorem 1. Lower part: for the irrational frequency ratio x with a singularly continuous quasi-energy spectrum, Upper part: for the rational x value close to the golden mean (x=0.6) with a pure point quasi-energy spectrum. The other parameters are the same as in Fig.1.*

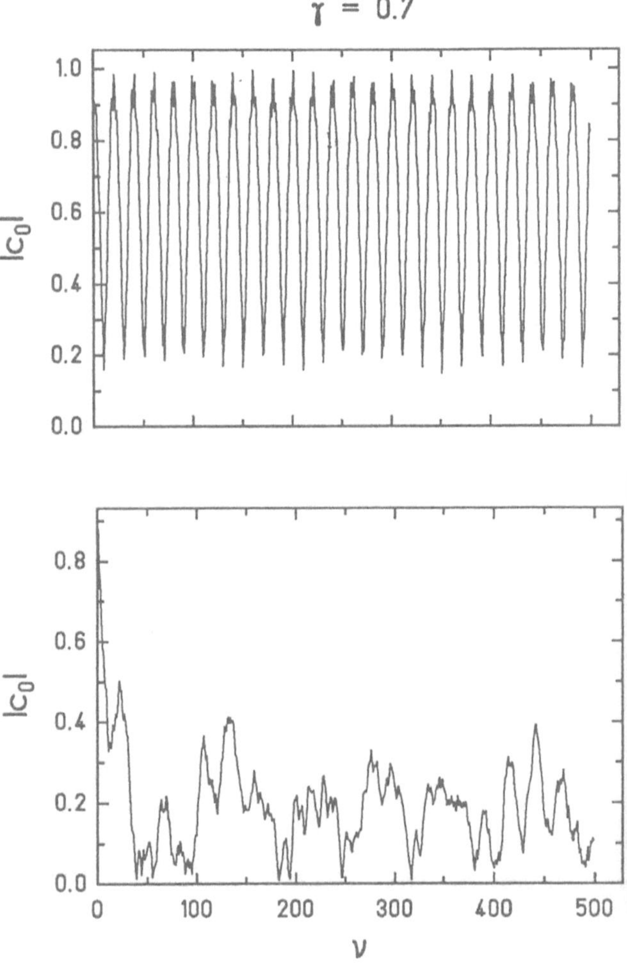

ACKNOWLEDGMENT

We are very grateful for stimulating discussions with Professors M.Combescure, B.V.Chirikov, P.Exner, D.L.Shepelyansky and J.A.Smorodinsky.

REFERENCES

1.   G.Casati, L.Molinari : Quantum chaos with time-periodic Hamiltonians ; in *New Trends in Chaotic Dynamics of Hamiltonian Systems,* to appear in Suppl.to Prog.Theor.Phys.
2.   G.Casati, B.V.Chirikov, J.Ford, F.M.Izrailev : in *Stochastic Behaviour in Classical and Quantum Hamiltonian Systems* ; Como, June 1977, Lecture Notes in Physics 334, Springer Verlag, Berlin 1979

3.   B.Dorizzi, B.Grammaticos, Y.Pomeau: J.Stat.Phys.37 (1984) 93
4.   S.Fishman, D.R.Grempel and R.E.Prange : Phys.Rev.Lett.49 (1982) 509 ; Phys.Rev.A 29 (1984) 1639
5.   G.Casati and I.Guarneri : Comm.Math.Phys.95 (1984) 121
6.   G.Casati: A review of progress in the kicked rotator problem in *Proceedings of the Conference "Chaos, Noise, Fractals"*, Como, Italy 1986
7.   J.Bellissard, D.Bessis, P.Mousa: Phys.Rev.Lett.49 (1984) 701 H.Kunz, R.Livi and A.Sütö : Commun.Math.Phys.122 (1989) 643
8.   M.Combescure : Spectral properties of a periodically kicked quantum Hamiltonian ; Preprint LPTHE Orsay 89/17, May 1989
9.   T. Hogg, B.A.Hubermann : Phys.Rev.Lett.48 (1982) 711
10.   M. Reed, B.Simon : Methods of Modern Mathematical Physics, vol.II, Academic Press, New York 1975
11.   M. Reed, B.Simon : Methods of Modern Mathematical Physics, vol.III, Academic Press, New York 1979
12.   B.Simon, T.Wolff : Comm. Pure Appl.Math.39 (1986) 75
13.   J.S.Howland : Indiana Univ.Math.J. 28 (1979) 471
14.   M.S.Birman, M.G.Krein : DAN SSSR 144 (1962) 475
15.   P.Exner : Rep.Math.Phys.17 (1980) 275
16.   D.B.Pearson : Comm.Math.Phys.40 (1975) 125

Laboratory of Theoretical Physics
Joint Institute for Nuclear Research
141980 Dubna, USSR

Operator Theory:
Advances and Applications, Vol. 46
© 1990 Birkhäuser Verlag Basel

RELATION BETWEEN CORRELATION FUNCTIONS AND SPECTRUM STATISTICS
IN THE REGION OF QUANTUM CHAOS

G.P.Berman[1] and F.M.Izrailev[2]

One of the most popular model for investigation of the quantum dynamical chaos is the so-called kicked rotator (see, e.g. [1])

$$\hat{H} = -\frac{\hbar^2}{2}\frac{\partial^2}{\partial\theta^2} + \mathcal{E}_0 \cos\theta \cdot \delta_T(t); \quad \delta_T(t) = \sum_{n=-\infty}^{\infty} \delta(t-nT) \tag{1}$$

Here $\partial/\partial\theta$ is the angular momentum, θ is angular displacement, $\mathcal{E}_0$ is the kick strength and $\delta_T(t)$ is the periodic delta-function with T being the period of the kicks. This model is very convenient to study because its dynamics can be described in terms of the map for the Ψ -function written after one period of perturbation:

$$\Psi(\theta,t+T)=\hat{U}\Psi(\theta,t); \quad \hat{U}=\hat{G}\hat{B}; \quad \hat{G}=\exp(i\frac{\tilde{\tau}}{2}\frac{\partial^2}{\partial\theta^2}); \quad \hat{B}=\exp(-ik\cos\theta) \tag{2}$$

Here $\hat{G}$ is the operator of the free rotation between successive kicks and $\hat{B}$ describes the kick. The behaviour of the model depends on two parameters, $K=\mathcal{E}_0 T$, which is a classical parameter and $k=\mathcal{E}_0/\hbar$, which is a pure quantum parameter. In the classical limit ($k\to\infty$, $\tilde{\tau}=\hbar T\to 0$, K =const) this model is known as standard mapping which depends, in distinction to the quantum model (1), on the parameter K only. The motion of the classical model is known to exhibit strong chaotic properties for $K \gtrsim 5$ [2]. Therefore, one of the most important problem is to understand the statistical properties (properties of "quantum chaos") that the correspondent quantum model (1) has in dependence on the value of the parameter k (we assume K large and fixed).

One of the distinctive property of chaotic motion in the classical model is the linear increase of energy $E(t)$, due to the diffusion of momentum, $(\Delta p)^2 = Dt$, with the diffusion rate $D \approx k^2/2$. Unlike the classical model, in quantum system this diffusion was found to be suppressed after some time $t \gtrsim t_D \sim k^2$ by quantum interference effects. The origin of this suppression is similar, in some sense, to the well-known Anderson localization. More specifically, it turns out that for $\tau/4\pi \neq r/q$ all eigenfunctions of the evolution operator $\hat{U}$ (quasienergy functions) are localized in the unbounded momentum space (or, in other words in the unperturbed basis, $\varepsilon_0 = 0$) with the mean localization length $\ell \approx \frac{D}{2} \approx k^2/4$ [1]. The approach developed in [1] on the basis of quantum localization in the region of classical chaos appears to be effective in describing the dynamical and statistical properties of some real physical models (see, e.g. [3]), however, many questions are still waiting for solutions.

In this note we present some results dealing with attempts to relate dynamics of the model (1) with its spectral properties. To be more precise, we give some analytical description of the energy growth $E(t)$ by making use of the relation between the time decay of correlation functions and the repulsion of the neighboring quasienergy levels (see also [4]).

Let us write the change of the rotator energy after one kick and rotation:

$$\tilde{E}_{t+1} = \tilde{E}_t + \frac{k^2}{4} - \frac{k^2}{4} F_1(t) + \frac{k}{2} F_2(t) \tag{3}$$

where $\tilde{E} = E/\hbar^2$. Here

$$F_1 \equiv \langle \cos 2\theta \rangle_t ; \quad F_2 \equiv i \langle \sin \theta \cdot \frac{\partial}{\partial \theta} + \frac{\partial}{\partial \theta} \sin \theta \rangle_t \tag{4}$$

are correlation functions which determine entirely the dynamics.
It can be shown [4] that the generic correlation functions depend essentially on the spacing distribution of the nearest
neighbour quasienergy levels, $S \equiv \lambda_{i+1} - \lambda_i$. It is known that for
the eigenstates which are partly overlapped, there is some re-
pulsion between the nearest levels. Assuming that such a repulsion is described by the dependence $P(s) = A s^\beta \exp(-B s^2 - C s)$, with
the repulsion parameter β (see details in [5,6]) we may obtain that correlation functions decrease in time as $F \sim \frac{1}{t^{1+\beta}}$ on
some time scale, $t \gtrsim t^*$. As a result, it is reasonable to assume that

$$\Delta \tilde{E}_t \equiv \tilde{E}_{t+1} - \tilde{E}_t = \frac{A_o}{(t + t^*)^{1+\beta}} \tag{5}$$

with the parameters A_o and t^* which can be defined from
the known expression for $\tilde{E}(t)$ in the limits $t \ll t_D$ and $t \gg t_D$
(see [4]). This allows us to obtain the analytical dependence
of $\tilde{E}(t)$ which can be finally written in the form

$$\tilde{E}(t) = \frac{D_{c\ell}^2(K)}{2\tau^4} \left\{ 1 - \frac{1}{\left(1 + \frac{\tau^2 t}{D_{c\ell}(K)\beta} \right)^\beta} \right\} \tag{6}$$

where $D_{c\ell} = \tau^2 D$ is the classical diffusion coefficient and t
is given in number of kickes. Numerious simulations [4] in the
wide range $10 \leqslant k \leqslant 40$ and $K = 5$ show a good agreement of numerical data with the proposed dependence (6). The data show
that the only undefined parameter β in (6) fluctuates depending on different initial conditions giving for the repulsion parameter $0.1 \leqslant \beta \leqslant 0.3$. The typical example is shown in Fig. 1
where the smooth curve corresponds to (6) in dimensionaless va-
riables $Y \equiv \tilde{E} / \left(\frac{D_{c\ell}(K)}{4\tau^3} \right)$ and $X \equiv (2\tau^3 t)/D_{c\ell}(K)$ for $K = 5$; $k = 40$
(I is classical diffusion; II is the proposed dependence
(6) with $\beta = 0.3$).

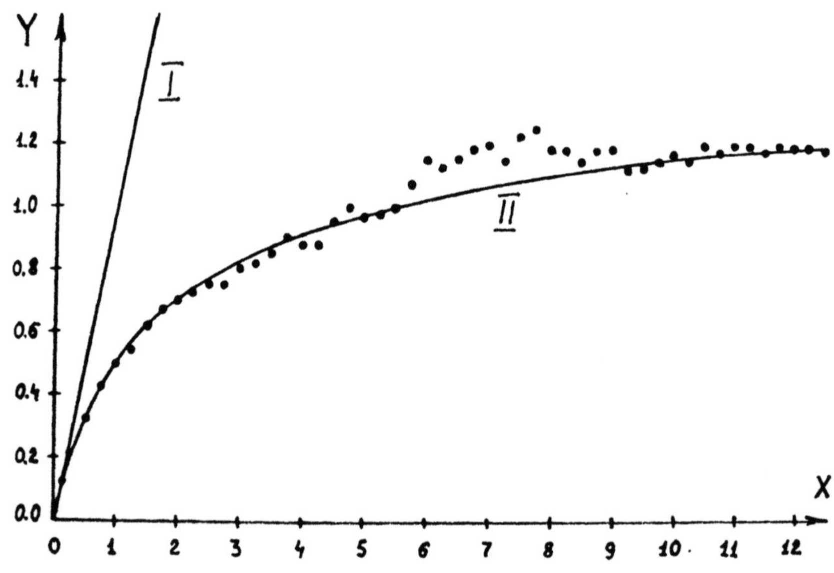

References

1. B.V.Chirikov, F.M.Izrailev and D.L.Shepelyansky, Physica D33 (1988) 77.
2. B.V.Chirikov, Phys. Rep. 52 (1979) 263.
3. G.Casati, B.V.Chirikov, I.Guarneri and D.L.Shepelyansky, Phys. Rep. 154 (1987) 78.
4. G.P.Berman and F.M.Izrailev, Preprint 497F, Krasnoyarsk, USSR, 1988.
5. F.M.Izrailev, Phys. Lett. A134 (1988) 13.
6. F.M.Izrailev, J. Phys. A22 (1989) 865.

1) Institute of Physics, 660036 Krasnoyarsk, USSR
2) Institute of Nuclear Physics, 630090 Novosibirsk, USSR

Part 4

OTHER TOPICS

Operator Theory:
Advances and Applications, Vol. 46
© 1990 Birkhäuser Verlag Basel

LOCALIZATION EFFECTS IN

NON-HOMOGENEOUS DIELECTRICS

P.B.Kurasov, B.S.Pavlov

It it well known that dielectrics with the reconstruction of zone structure ($Pb_{1-x}Sn_xTe$, for example) has two-dimensional layers with a specific electron properties. A corresponding relativistic model has been discussed, in particular, by B.A.Volkov and O.A.Pankratov [1]. The general nature of the effect under consideration can be demonstrated by in a simple model setting, and our aim here is to construct such an exactly solvable model. We regard a two-dimensional solid as an infinite number of one-dimensional layers.

First of all, let us consider the simplest discrete one-electron model. The self-adjoint operator

$$H = \begin{pmatrix} \lambda_1 & 0 \\ 0^1 & \lambda_2 \end{pmatrix} + \Gamma_+ \, T_+ + \Gamma_- \, T_-$$

acts in the Hilbert space $\mathcal{H} = \ell_2(\mathbb{C}^2)$. Here $T_\pm$ are shifts in ℓ_2 and $\Gamma_\pm$ are 2×2 matrices. The self-adjointness and hysotrophy of the operator means

$$\begin{cases} \Gamma_+ = \Gamma_-^* \\ \Gamma_+ = \Gamma_- \end{cases} \rightarrow \Gamma_\pm = \Gamma, \ \Gamma = \Gamma^*, \ \Gamma = \begin{pmatrix} a & \gamma \\ \bar{\gamma} & b \end{pmatrix}$$

The spectrum of H is purely continuous and consists of two zones. All eigenfunctions are of Bloch type, $U_{n+1} = U_n \, e^{it}$ with $U \in \mathbb{C}^2$. The corresponding dispersion equation is the following

$$(\lambda_1 + 2a \cos t - \lambda)(\lambda_2 + 2b \cos t - \lambda) = 4 \, |\gamma|^2 \cos^2 t$$

1. NON-INTERACTING MODES, $\gamma=0$

In this case we have two dispersion equations, corresponding to two different modes

$$\lambda_1 + 2a \cos t = \lambda$$
$$\lambda_2 + 2b \cos t = \lambda$$

We have the usual picture of Fermi surfaces if ab<1 (see Fig.1).

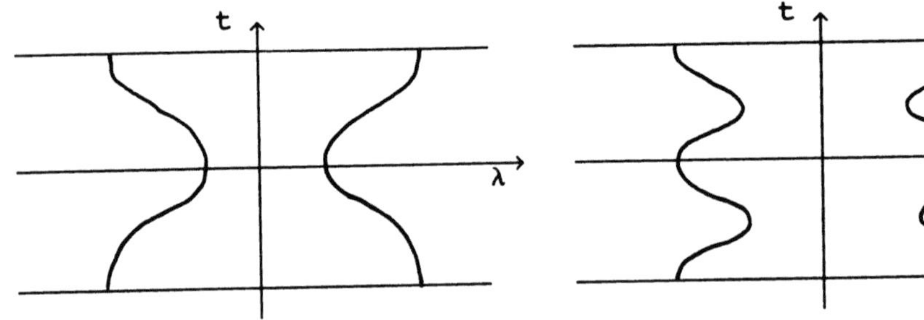

Fig.1   Fig.2

Fermi surfaces for   Fermi surfaces for
non-interacting modes   strongly interacting modes

2. STRONGLY INTERACTING MODES, a=b=0

The dispersion equation $(\lambda_1-\lambda)(\lambda_2-\lambda) = 4|\gamma|^2\cos^2 t$ defines the
Fermi surfaces sketched on Fig.2 for the case $\lambda_1 = -\lambda_2$ when

$$\lambda = \pm\sqrt{\lambda_1^2 + 4|\gamma|^2\cos^2 t} \cong \pm\lambda_1\left(1 + \frac{2|\gamma|^2\cos^2 t}{\lambda_1^2} + \dots\right)$$

3. THE GENERAL CASE

If we put $\lambda_1=-\lambda_2$ and a=-b, then

$$\lambda = \pm\sqrt{\lambda_1^2 + 2a\lambda_1\cos t + 4(a^2+|\gamma|^2)\cos^2 t} \underset{\lambda\gg|\gamma|}{\approx}$$

$$\approx \pm\left(\lambda_1 + 2a\cos t + \frac{2|\gamma|^2\cos^2 t}{\lambda_1} + \dots\right)$$

In the case $\lambda_1 = 0$ we have a non-gap model of a solid, $\lambda = \pm 2\sqrt{|\gamma|^2+a^2}\cos t$ (see Fig.3)

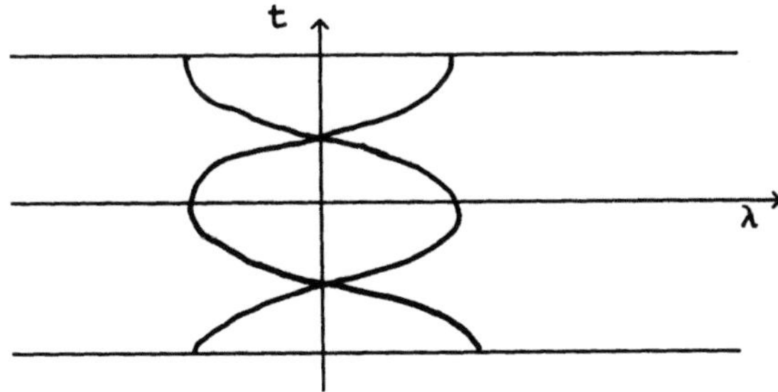

Fig.3

Fermi surfaces for
a non-gap dielectric

There is an approximately linear dependence between energy and quasi-momentum, while in the general case ($\lambda_1 \neq 0$) the dependence has the form of a quadratic function.

Let us now construct a model of non-homogeneous dielectric using the self-adjoint operator

$$L = -\frac{\partial^2}{\partial x^2} \otimes I + I \otimes H(x)$$

acting in the Hilbert space $L_2(\mathbb{R}) \otimes \ell_2(\mathbb{C}^2)$.

The main question to be discussed is whether there exist the so-called waveguide eigenfunctions, i.e., eigenfunctions of the continuous spectrum localized near the surface of the non-gap solid. We shall choose for the operator-valued function $H = H(x)$ the simplest linear form corresponding to the eigenvalues

$$\lambda_1 = \alpha x, \quad \lambda_2 = -\alpha x \ .$$

By means of Fourier transformation we get the following equation

$$-\frac{\partial^2}{\partial x^2} \hat{U} + \begin{pmatrix} \alpha x + 2a \cos t & 2\gamma \cos t \\ 2\bar{\gamma} \cos t & -\alpha x - 2a \cos t \end{pmatrix} \hat{U} = E\,\hat{U}$$

for $t \in [-\pi,\pi]$. The matrix potential has two eigenvalues (Fig.4)

$$\lambda_\pm = \pm \sqrt{(\alpha x + 2a \cos t)^2 + 4|\gamma|^2 \cos^2 t}$$

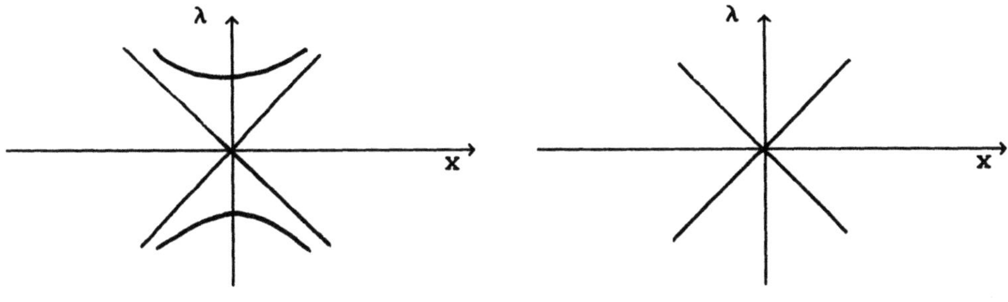

Fig.4

The adiabatic potential
for t ≠ π/2

Fig.5

The adiabatic potential
for t = π/2

The corresponding normalized eigenvectors are

$$X_1^{\pm} = \frac{2\,\gamma\,\cos t}{\sqrt{2}\sqrt{(\alpha x+2a\cos t)^2+4|g|^2\cos^2 t\pm(\alpha x+2a\cos t)\sqrt{(\alpha x+2a\cos t)^2+4|\gamma|^2\cos^2 t}}}$$

$$X_2^{\pm} = \frac{\pm\sqrt{(\alpha x+2a\cos t)^2+4|\gamma|^2\cos^2 t}-(\alpha x+2a\cos t)}{\sqrt{2}\sqrt{(\alpha x+2a\cos t)^2+4|g|^2\cos^2 t\pm(\alpha x+2a\cos t)\sqrt{(\alpha x+2a\cos t)^2+4|\gamma|^2\cos^2 t}}}$$

In accordance with the standard adiabatic-approximation procedure
[2] let us rewrite the equation in the moving basis $\{X^{\pm}\}$:

$$\left(-\left(\frac{\partial}{\partial x}+\mathcal{A}\right)^2+\Lambda\right)\hat{U}=E\,\hat{U}\,,\quad \Lambda=\begin{pmatrix}\lambda_+ & 0\\ 0 & \lambda_-\end{pmatrix}$$

The connection $\mathcal{A}$ represents a 2×2 matrix and can be easily
calculated,

$$\mathcal{A}_{ij}=\langle\partial_x X^j,\,X^i\rangle$$

Its non-diagonal elements are

$$\mathcal{A}_{\pm\mp} = -\frac{\alpha}{2} \frac{4\,|\gamma\,\cos t|}{\sqrt{(\alpha x + 2a\cos t)^2 + 4|\gamma|^2\cos^2 t}} \times$$

$$\left((\alpha x + 2a\cos t)^2 + 4|\gamma|^2\cos^2 t \pm (\alpha x + 2a\cos t)\sqrt{(\alpha x + 2a\cos t)^2 + 4|\gamma|^2\cos^2 t} \right)^{-1/2}$$

In order to solve the equation let us suppose that the connexion is equal to zero. Than there are two separated stationary Schroedinger equations with the adiabatic potentials $\lambda_\pm(x)$. In a neighbourhood of the zero point the "upper" one can be approximated by a harmonic-oscillator equation:

$$\lambda_+ = 2\,|\gamma\,\cos t|\left(1 + \frac{\alpha^2 x^2}{8\,|\gamma\,\cos t|^2} + \dots \right)$$

(we put here a=0). For this equation the lowest eigenstate is the following :

$$\Psi^+ = \left(\frac{\alpha}{\pi\sqrt{4\,|\gamma\,\cos t|}} \right)^{1/4} \exp\left\{ -\frac{\alpha^2}{4\,\sqrt{|\gamma\,\cos t|}}\,x^2 \right\}$$

The localisation lenght must be very small, $|\gamma|/\alpha^2 \ll 1$. In the approximation of weak coupling between the modes we can evaluate the connection

$$\mathcal{A} = \begin{pmatrix} \mathcal{A}_{++} & \mathcal{A}_{+-} \\ \mathcal{A}_{-+} & \mathcal{A}_{--} \end{pmatrix}$$

$$|\mathcal{A}_{+-}| \approx \frac{\alpha}{8}\,\frac{|\gamma|}{a^2\,|\cos t|}$$

For $t \neq \pi/2$ we obtain the second condition necessary for existence of the waveguide functions :

$$\alpha \, |\gamma| \, / \, a^2 \ll 1$$

Both the conditions can be satisfied simultaneously, and therefore one can find surface eigenfunctions in a non-homogeneous dielectrics with a weak coupling between the modes. Such states can be observed also in strongly non-homogeneous solids, $\alpha \ll 1$. The simplest example of such a solid is the contact of two dielectrics with an inverse picture of terms.

The authors are gratefull to A.M. Ifiasov for many fruitfull discussions. This article can hardly be written without the friendly support we have got from Yu.A.Kuperin and Yu.B.Melnikov.

REFERENCES

1. Volkov B.A., Pankratov O.A., Massless two-dimensional electrons in an inverse contact, Sov.Phys.JETP Letters $\underline{42}$ (1985),145.
2. Kuperin Yu.A., Kurasov P.B., Melnikov Yu.B., Merkuriev S.P., Connections and effective S-matrix in triangle representation for quantum scattering, Preprint INFN-ISS 89/6, Roma 1989.

Department of Mathematical and Computational Physics, Institute for Physics, Leningrad University, 198904 Leningrad, USSR.

Operator Theory:
Advances and Applications, Vol. 46
© 1990 Birkhäuser Verlag Basel

A SELF-ADJOINT DILATION OF THE LINEAR BOLTZMANN OPERATOR

Yu.A.Kuperin

A self-adjoint dilation of the dissipative linear Boltzmann operator describing neutron transport or radiation transfer through a non-multiplicative medium is constructed.

The dilation theory [1-3] of non-selfadjoint operators reduces the investigation of their spectral properties to studying of the corresponding characteristic function (c.f.). In this way it is important to have some tool for the direct investigation of c.f. without taking into account the resolvent of the non-selfadjoint operator under consideration. Such a tool is given by a general Adamjan-Arov result [4] claiming that c.f. can be realized as a scattering matrix for a pair of operators: one of them is a self-adjoint dilation of the original non-selfadjoint dilation of the original non-selfadjoint operator and the other one is some special conjuction of the parts of this dilation in the incoming and outgoing subspaces. As a rule, this conjuction is trivial. Thus as a first step to a rigorous scattering theory for non-selfadjoint operators one has to construct their self-adjoint dilations.

In the present talk I am going to construct explicitly the self-adjoint dilation for a non-local dissipative operator important from the physical point of view, namely the one-speed

linear Boltzmann operator. From the mathematical side the new
feature of the suggested scheme is that it extends the well-known
dilation theory for non-local operators. Let us consider, in
particular, the one-speed Boltzmann operator [5]

$$L = i\mu\partial_x + i\sigma \left(\mathbb{1} - \frac{c}{2} \int_\Omega K(\mu,\mu')\ d\mu' \right) \tag{1}$$

acting in the Hilbert space $\mathcal{H} = L^2(\mathbb{R} \times \Omega)$. Here $\mu \in \Omega=[-1,1]$, σ
and c are some non-negative bounded functions on $\mathbb{R}$ and
$K(\mu,\mu')$ is a real-valued bounded function on $\Omega \times \Omega$ satisfying
the following conditions: $K(\mu,\mu')=K(\mu',\mu)$ (symmetry) and $\frac{1}{2}\int_\Omega Kd\mu =$
$\frac{1}{2}\int_\Omega Kd\mu' = 1$ (normalization). Then the evolution equation

$$- i\partial_t\psi = L\psi \qquad \& \qquad \psi\big|_{t=0} = \psi_0 \tag{2}$$

generates in the initial-data space $\mathcal{H}$ a strongly continuos
semigroup $T(t) = \exp(iLt)$, $t\geq0$, of bounded operators. Moreover,
if $K(\mu,\mu') > 0$ then the following lemma is valid [6].

LEMMA 1. *If* $c_0 \equiv \sup\limits_x c(x) < \sigma_1/\sigma_2$, *where* $\sigma_1 = \inf\limits_x$
$\sigma(x)$, $\sigma_2 = \sup\limits_x \sigma(x)$, *then the semigroup* $T(t)$, $t \geq 0$, *is*
contractive in the space $\mathcal{H}$ *with the "energy" metrics* $\|f\|_E^2 = \frac{1}{2} \int\limits_{T=\mathbb{R}\times\Omega} d\gamma$
f^2, $d\gamma = = dx \wedge d\mu$, *and*

$$T(t)_E \leq \exp\left\{ -\sigma_1 \left(1 - c_0 \frac{\sigma_2}{\sigma_1} \right) t \right\}, \ t \geq 0 .$$

It follows from the lemma that for a non-
multiplicative medium ($c_0 < \sigma_1/\sigma_2$) the generator L of the
semigroup $T(t)$ is a dissipative operator with the domain

$\mathcal{D}(L)=\mathcal{D}(L_0)$, where $L_0 = i\mu\partial_x$ and $\mathcal{D}(L_0)=\{f\colon f \ \& \ L_0 f \in \mathcal{H}\}$. Hence the task of dilating the operator L to a self-adjoint one is reduced to the dilation of $T(t)$ to a unitary group $U(t)$, $-\infty < t < \infty$. To this end let us consider functions $v_\pm(\xi,\gamma,t)$ and $u_\pm(\xi,\gamma,t)$ defined at any fixed t on $\mathbb{R}_\pm \times \Gamma_a$, $\Gamma_a \equiv \mathrm{supp}\ \sigma \times \Omega$, and satisfying the boundary conditions:

$$v_+(\sigma) - u_-(\sigma) = \alpha\, P_K\,\psi\,, \qquad (3)$$

$$v_-(0) - u_+(0) = \tfrac{1}{2} \int d\mu' K\, P_K\psi\,, \qquad (4)$$

$$v_+(0) - v_-(0) = u_+(0) - u_-(0)\,, \qquad (5)$$

$$\partial_t v_\pm(u_\pm) = -\,\partial_\xi v_\pm(u_\pm)\,.$$

Here the subspace $\mathfrak{K} = L^2(\Gamma_a)$ is naturally embedded into H, $\alpha^2 = 2c$ and P_K is an orthoprojection from $\mathcal{H}$ on $\mathfrak{K}$.

Let us introduce the vector-functions $\Psi = \mathrm{col}(\phi_-,\psi,\phi_+)$ where $\phi_\pm = \mathrm{col}(v_\pm,u_\pm)$. Then the dilation of the problem (2) is given by the equation

$$-i\partial_t\Psi = \left\{ \begin{array}{l} i\partial_\xi\phi_- \\[4pt] L_0\Psi + \dfrac{i\alpha}{2}\,(v_+(0)+u_-(0)) - \dfrac{i\alpha c}{4}\displaystyle\int_\Omega d\mu' K(v_+(0)+u_-(0)) \\[4pt] i\partial_\xi\phi_+ \end{array} \right\} = \mathcal{L}\Psi \quad (7)$$

with the boundary conditions (3)-(6).

Denote as $\psi_0,\ v_{\pm 0}$ the initial data for the time-dependent problem (3)-(7), and define the initial-data space $\mathfrak{H} = \mathfrak{D}_- \oplus \mathcal{H} \oplus \mathfrak{D}_+$, $\mathfrak{D}_\pm = \mathcal{D}_\pm \oplus \mathcal{D}_\pm$, $\mathcal{D}_\pm = L^2(\mathbb{R}_\pm,\mathfrak{K})$, with the new "energy" metrics

$$\|f\|^2_{E,N} = \frac{1}{2}\int_\Gamma d\gamma\left[\int_{-\infty}^{0} d\xi\,(v^2_{-0} + u^2_{-0}) + \psi^2_0 + \int_0^\infty d\xi\,(v^2_{+0} + u^2_{+0})\right], \qquad (8)$$

$$f = \mathrm{col}\,(\phi_{-0},\ \psi_0,\ \phi_{+0}) \in \mathfrak{H}\ .$$

The following theorem is valid

THEOREM 1. *Let $\mathcal{L}$ be the operator given by Eq.(7) which acts on the set of vector-functions $f = \mathrm{col}(\phi_{-0},\psi_0,\phi_{+0}) \in \mathfrak{H}$, where $v_{\pm 0}$, $u_{\pm 0} \in W_2^1(\mathbb{R}_+,\mathfrak{K})$ and $\psi_0 \in \mathcal{D}(L)$ whose components satisfy also the boundary conditions (3)-(6). Then :*

1. $\mathcal{L}$ is the generator of a one-parameter unitary evolution group $U(t)=\exp\{i\mathcal{L}t\}$ of the initial data.

2. $\mathcal{L}$ is a self-adjoint dilation of the operator L ; i.e., $P_{\mathcal{H}}(\mathcal{L}-\lambda)^{-1}P_{\mathcal{H}}f = (L-\lambda)^{-1}\psi_0$, $f\in\mathfrak{H}$, $\psi_0\in\mathcal{H}$, $\mathrm{Im}\lambda < 0$.

A complete proof of this theorem one can find in [6].

References.

1. P.D.Lax, R.S.Phillips. Scattering Theory. Academic Press, New York 1967.
2 B. Sz.-Nagy, C.Foias. Analyse Harmonique des Operateurs de L'espace de Hilbert. Masson Et Cie, Szeged-Bucharest 1967.
3. B.S.Pavlov. A Self-Adjoint Dilation of Dissipative Schroedinger Operator and Eigenfunction Expansion. Func. Anal. v.9, N 2, p.87 (in Russian).
4. V.M. Adamjan, D.Z. Arov. On Unitary Conjuctions of Semi-Unitary Operators. Math.Investig. v.1, N2, 1966 (in Russian).
5. K.M. Case, P.F. Zweifel. Linear Transport Theory. Addison-Wesley 1967.
6. Yu. A. Kuperin. A Self-Adjoint Dilation of One-Speed Transport Operator. Preprint INFN-ISS 89/10, Roma, 1989.

Department of Mathematical & Computational Physics,
Institute for Physics, Leningrad University,
Leningrad 198904 , USSR .

Operator Theory:
Advances and Applications, Vol. 46
© 1990 Birkhäuser Verlag Basel

KATO PROBLEM FOR FUNCTIONAL-DIFFERENTIAL EQUATIONS AND DIFFERENCE SCHRÖDINGER OPERATORS

G.A.Derfel

We study the problem of existence of bounded solutions of functional-differential equations proposed by T.Kato by means of the spectral theory for difference Schrödinger operators.

1. INTRODUCTION

Functional-differential equations with linearly transformed argument arise in various applications. For instance, in a well-known paper V.A.Ambartsumjan [1] showed that the equation

$$y'(t) = ay(\alpha t) + by(t) \ , \ \alpha > 1 \tag{1}$$

describes the absorption of light in the Milky way. Earlier the analogous equation but with $\alpha < 1$ was investigated by K.Mahler and N.de Bruign in connection with one problem of the number theory [2]. An outstanding analysis of the equation (1) is given in the paper by T.Kato and J.McLeod [3]. Among a lot of other applications let us recall the probability theory on algebraic structures and the theory of one-dimensional dynamical systems [4]. The problem of existens of bounded (particularly, almost periodic) solutions represents a special interest in physical applications.

1. COMMENSURABLE STRETCHING OF THE ARGUMENTS

The question about bounded solutions of the equation

$$y''(t) = y(qt) + y(t/q) + \lambda y(t), \quad q>1 \tag{2}$$

or the equation

$$\lambda y(t) = y(qt) + y(t/q) + \sigma[y(t+1)+y(t-1)] \tag{3}$$

may be reduced to the problem about purely point spectrum of the Schrödinger difference equations

$$c_{n+1} + c_{n-1} + [\lambda+q^{2n}\omega^2]c_n = 0 , \quad \omega \in [1,q] \tag{4}$$

and

$$c_{n+1} + c_{n-1} + 2\sigma \cos(2\pi q^n\omega)c_n = \lambda c_n , \tag{5}$$

respectively, the latter being of almost-Mathieu type. This problem was studied in [5,6]. The equation (2) generalizes to

$$y''(t) = \sum_{j=0}^{1} a_j\, y(\alpha_j t) + \lambda y(t) , \tag{6}$$

where α_j ($\neq 1$) are multiplicatively commensurable values, that is $\alpha_j=q^{r_j}$, where $q >1$, r_j ($\neq 0$) are rational numbers.

THEOREM 1. [7] *Let α_j be multiplicatively commensurable. Then there exists K>0 such that (i) if $\lambda<-K$, the equation (6) has a non-trivial almost-periodic solution, (ii) if $\lambda<-K$, any bounded solution is almost-periodic, (iii) if $\lambda>K$ the equation (6) has no solution bounded on the whole axis.*

2. UNCOMMENSURABLE STRETCHING

Let us consider the model equation

$$y''(t) = y(\alpha_1 t) + y(t/\alpha_1) + y(\alpha_2 t) + y(t/\alpha_2) + \lambda y(t) , \quad (7)$$

where α_1, α_2 ($\neq 1$) are multiplicatively uncommensurable. The Kato problem for the equation (7) may be reduced to the problem about purely point spectrum of the two-dimensional difference Schrödinger equation

$$H[C](n) = - \Delta C_{m,n} - e^{\beta m + \gamma n} C_{m,n} = \lambda C_{m,n}$$

with uncommensurable β, γ ($\beta = \ln \alpha_1$, $\gamma = \ln \alpha_2$).

THEOREM 2. [7] *For almost all β, γ with respect to the Lebesgue measure, there exists $K>0$ such that if $\lambda<-K$ the spectrum of the operator H is purely point and dense.*

References

1   V.A.Ambartsumian, Sov.Math.Doklady <u>44</u> (1944), 223-226 (in Russian).
2   K.Mahler, J.London Math.Soc.<u>15</u> (1940), 115-123.
3   T.Kato, J.B.McLeod, Bull.Amer.Math.Soc.77 (1971), 891-937.
4   E.Yu.Romanenko, A.N.Sharkovsky, in *Asymptotic Behaviour to Solutions of Differential Difference Equations,* Institute of Mathematics, Kiev 1978, pp.5-39 (in Russian).
5   G.A.Derfel, in *Dynamical Systems and Differential-Difference Equations,* Institute of Mathematics, Kiev 1986, pp.14-20 (in Russian).
6   G.A.Derfel, S.A.Molchanov, Sov.Math.Uspekhi <u>42</u> (1987), 126 (in Russian).
7   G.A.Derfel, S.A.Molchanov, Sov.Math.Uspekhi <u>44</u> (1989), 211-212 (in Russian).

Department of Mathematics
Karaganda State University
Kirova 48-110
Karaganda,USSR

Operator Theory:
Advances and Applications, Vol. 46
© 1990 Birkhäuser Verlag Basel

AN EXISTENCE THEOREM FOR SOME NONLINEAR NONLOCAL SCHROEDINGER OPERATORS AND THE SOLITON-LIKE SOLUTIONS FOR THE CORRESPONDING DYNAMIC SYSTEMS

S.I.Petrukhnovsky

An existence theorem for a certain class of nonlinear nonlocal Schroedinger operators is presented and the sketch of the proof is given. In addition, it is shown that there exist soliton-like solutions for nonlinear nonlocal systems describing generalized dynamics of an inhomogeneous electron gas in a crystal.

The idea of attempting to represent the ground state (and perhaps some of the excited states as well) of atomic, molecular and solid-state systems in terms of one-body reduced density matrix $\rho(x)$ is an old one. Its realisation gave birth to nonlinear nonlocal models such as Hartree, Hartree-Fock and Hartree-Fock-Slater equations [1], the so-called self-consistent-field models. This approach has been justified in the Hohenberg-Kohn conception of the density functional [2] and has been developed in different ways in many papers (see, for instance, [3]). There are many works devoted particularly to solid-state applications of the method, however, existence theorems are practically never discussed in them. Among the exceptions, let us remind that the existence of dynamics for Hartree-type time-dependent systems has been studied by Maslov [3].

Let us consider Schroediger operator $H = -\Delta+V(x)$ in $L_2(\mathbb{R}^3)$ with $V(x)$ being a periodic function from $L_{2loc}(\mathbb{R}^3)$ of period 2π in each variable. Spectral representation of this operator has the form

$$H = \int_{\Omega_B} \sum_j \lambda_j(k)\,(.\,,\psi_j(x,k))\,\psi_j(x,k)\,dk,$$

where $\Omega_B=[0,1]^3$; for every k the sequece $\lambda_j(k)$ is increasing with the only accumulation point at infinity, the generalized eigenfunctions have the form $\psi_j(x,k) = \hbar\upsilon_j(x,k)e^{ikx}$, $\upsilon_j(x,k)$ being 2π-periodic in the variables x and belonging to Sobolev space $W_2^2(\Omega)$, $\Omega=[0,2\pi]^3$, while $V(x)$ is supposed to belong to $L_2(\Omega)$ - cf.[5].

Let $q(x),G(x)$ be certain real functions belonging to $L_2(\Omega)$, and suppose further that a,b are real and s,α are positive constants, $\alpha\in[0,1/2]$. We consider the following nonlinear eigenvalue problem:

$$H\psi \equiv (-\Delta+V(x))\psi = \lambda\psi \tag{1}$$

$$V(x) = q(x) + a\int_\Omega G(|x-y|)\rho(y)dy + b\rho(x)^\alpha \tag{2}$$

$$\rho(x) = \int_{\Omega_B} \sum_{\lambda_j(k)\le\lambda_F} |\psi_j(x,k)|^2 dk,$$

$$\lambda_F = \sup\left\{\lambda \,\Big|\, \sum_j \int_{\lambda_j(k)\le\lambda} dk < s\right\} \tag{3}$$

$$\int_\Omega \rho(x)\,dx = s \tag{4}$$

The system (1)-(4) is called the Hartree-Fock-Slater model. It is easy to see that there holds the following inequality

$$\|V\|_{L_2(\Omega)} \le \|q\|_{L_2(\Omega)} + |a|s\|G\|_{L_2(\Omega)} + |b|s^\alpha (2\pi)^{1/6} \tag{5}$$

The inequality (5) implies that solutions of the problem (1)-(4) should be looked for among the solutions of the problems of the following family

$$H_v\psi \equiv (-\Delta + v(x))\psi = \lambda\psi \ , \quad \|v\| \leq V_0 \tag{6}$$

where V_0 is the value of the *rhs* in (5). The spectral representations of the operators H_v have the form mentioned above, but the eigenvalues and eigenfunctions may be uniformly estimated under the condition (6). This fact together with the spectral absolute continuity of the investigated operators [5] imply a uniform estimate for Fermi levels λ_F calculated for the operators H_v according to (3) and this estimate proves finiteness of the number of bands $\lambda_j(k)$ which have to be analyzed for the problem (1)-(4). As a consequence, uniform estimates for $|\lambda_j(k,v)-\lambda_j(k,v')|$, $|\lambda_F(v)-\lambda_F(v')|$ and $\|\rho_v(x)-\rho_{v'}(x)\|_{L(\Omega)}$ are obtained: all these expressions converge uniformly to zero as $\|v-v'\|_{L(\Omega)} \to 0$, $j=1,..,N$ [5].Also, one can establish that for all H_v under the condition (6) the densities $\rho_v(x)$ are uniformly bounded in Sobolev space $W_2^1(\Omega)$ with the norm $(\|\rho\|^2_{L_2(\Omega)}+\|\rho\|^2_{L_2(\Omega)})^{1/2}$.

Let us now consider the set $E_M = \{\rho(x)| \ \rho(x)\geq 0 \ , \ \int_\Omega \rho(x)dx = s, \ \|\rho\|_{W_2^1(\Omega)} \leq M \}$. The convexity of this set and its compactness are obvious. Let us define the mapping F which assigns to each $\rho\in E_M$ the potential $V(x)$ defined by (2). Then the eigenvalues and the eigenfunctions of the problem (6) with v = V are defined and they can be used to define the function $\eta(x)= \int_\Omega \sum_{\lambda_j(k,v)\leq\lambda_F(k,v)} |\psi_j(x,k;v)|^2 dk$; it is nothing else than the image of the element ρ under the mapping F. It can be proved that there exists a number M_0 such that for any $M \geq M_0$ the mapping F transforms continuously the convex compact E_M in $L_2(\Omega)$ into itself. Thus according to Schauder priciple [7] we have

THEOREM. *For any periodic real* L_{2loc} *q(x), G(x), real numbers a,b, positive s and* $\alpha \in [0,1/2)$ *there exists a solution of the integral-differential system (1)-(4).*

Finally, let us discuss the unitarily nonlinear [4] dynamical system

$$id/dt\ \psi(x,t) = (-\Delta+q(x)+Q[\rho])\psi(x,t),\tag{7}$$

where $Q[\rho] = a \int_{\Omega} G(|x-y|)\rho(y)dy + b\rho^{\alpha}(x,t),$

$$\rho(x,t) = \int_{\Omega_B}\Sigma_{\lambda_j(k)\leq\ \lambda_F}|\psi_j(x,k,t)|^2dk,\tag{8}$$

where the wave functions $\psi_j(x,k,t)$ are defined by (7) with the following initial condition : $\psi_j(x,k,t)$ are eigenfunctions of the nonlinear problem (1)-(4). Then the theorem proved above enables to establish existence of an integral of motion, $d\rho/dt=0$ Thus we may say that the system (7)-(8) admits soliton-like solutions.

References

1   S.Lundquist, N.H.March, eds.*Theory of Inhomogeneous Electron Gas*, Plenum Press, New York-London 1983.
2   P.Hohenberg, W.Kohn, Phys.Rev.B 136, 864-871 (1964).
3   E.H.Lieb, Intern.J.Quant.Chem.24, 243-277 (1983).
4   V.P.Maslov, Contemporary Problems of Mathematics 11, 153-234 (1978) (in Russian).
5   M.Reed, B.Simon, *Methods of Modern Mathematical Physics*, vol.4, Academic Press, New York 1978.
6   S.I.Petrukhnovsky ,Preprint No 84/3, Institute of Metal Physics, Ural Scientific Centre of the Soviet Academy of Sciences, Sverdlovsk 1984 (in Russian).
7   J.Schauder, Math.Zetschr.26, 47-65 (1927).

Institute of Electrophysics
Soviet Academy of Sciences, Ural Division
620219 Sverdlovsk, USSR.

Operator Theory:
Advances and Applications, Vol. 46
© 1990 Birkhäuser Verlag Basel

THE STURM-LIOUVILLE PROBLEM WITH A POTENTIAL LINEAR IN SPECTRAL PARAMETER

V.N.Pivovarchik

Consider the boundary problem

$$-y'' + \left[\lambda^2 + U(x) + 2\lambda Q(x)\right] y = 0, \qquad (1)$$

$$y(\lambda, 0) = 0, \qquad (2)$$

where $y \in L_2(0, \infty)$ is an unknown function, λ is the spectral parameter. The following conditions are assumed to be satisfied throughout this paper:

1) $U(x)$ is real a function continuous on the semiaxis $(0, \infty)$ and $\int_0^\infty |U(x)| x dx < \infty$,

2) $Q(x)$ is real a function continuously differentiable on $[0, \infty)$, $\int_0^\infty |Q(x)| x dx < \infty$, $\int_0^\infty |Q'(x)| x dx < \infty$,

3) $Q(x) \geq 0$.

Various problems in mechanics may be reduced to (1), (2) by using the Liouville transformation [1]; in that case the condition 3) expresses dissipation of energy under the influence of damping. The essential spectrum of the problem is continuous and covers the imaginary axis of the λ-plane [2]. There are no eigenvalues embedded in the continuous spectrum. Normal (i.e., isolated Fredholm) eigenvalues are possible. Their geometric multiplicities are equal to one [2]. The following theorem is a consequence of more general results obtained in [3].

THEOREM 1. The right half-plane eigenvalues of (1), (2) are simple (i.e., of the algebraic multiplicity one), positive and of a finite number. This number coincides with the number of right half-plane eigenvalues of (1), (2) for $Q(x) = 0$.

THEOREM 2. If there exist numbers $C > 0$ and $s > 0$ such that

$$\left| U(x) - Q^2(x) - Q'(x) \right| < C\, e^{-sx}$$

for all $x \in (0, \infty)$, then the number of eigenvalues of (1), (2) is finite.

THEOREM 3. If

$$\int_0^\infty x \left| U(x) - Q^2(x) - Q'(x) \right| dx \, \exp \int_0^\infty \left[3Q(x) + x \left| U(x) \right| \right] dx < 1$$

then the problem (1), (2) possesses no eigenvalues.

Proofs of Theorems 2 and 3 see in $\left[4 \right]$.

Denote by

$$q = \int_0^\infty Q(x)dx, \quad p = \int_0^\infty x \left| U(x) \right| dx, \quad r = \int_0^\infty x \left[\left| U(x) \right| + \left| Q'(x) \right| \right] dx.$$

THEOREM 4. If

$$2r \left[e^q - 1 \right] e^{p+2q} + \frac{1}{8} r^2 \left[e^{4q} - 1 \right] e^{2(p+q)} < 1$$

then all the eigenvalues of (1), (2) are simple real and the number of left half-plane eigenvalues coincides with the number of right half-plane eigenvalues.

The proof of the Theorem 4 see in $\left[5 \right]$.

References

1. M.Jaulent, J.Math.Phys. $\underline{17}$(1976),1351-1360.
2. M.Jaulent, C.Jean, Comm.Math.Phys. $\underline{28}$(1972),117-220.
3. V.N.Pivovarchik, Funk.Anal.Appl. $\underline{23}$(1989),80-81 (in Russian).
4. V.N.Pivovarchik, to appear in Siberian Math.J.
5. V.N.Pivovarchik, Diff.Eq. $\underline{24}$(1988),705-708 (in Russian).

Department of Higher Mathematics
Civil Ingeneering Institute of Odessa
Didrihson Street, 4.
270029 Odessa, USSR

Operator Theory:
Advances and Applications, Vol. 46
© 1990 Birkhäuser Verlag Basel

STOCHASTIC MODEL OF TREE GROWTH

M.A.Antonets, I.A.Shereshevsky

Let T^N be a Caley tree, i.e. a partially ordered set
with a unique minimal element $\bar{0}$ such that for any element $a \neq$
$= \bar{0}$ there exists only one element $\hat{a}$ immediately preceding a
and for any a the set $C(a)$ of immediately following elements
contains exactly N of them. We shall call the subsets of T^N
configurations as in the statistical mechanics. To every element $a \in T^N$ and any instant of discrete time $t \in \{0,1,\dots\}$ we
shall ascribe the logical variable $X(a,t)$, which is "true" if
a belongs to the configuration at the time t and "false" in
the opposite case. Let us consider the stochastic model of tree
growth, defined by the following Boolean equations

$$
\begin{cases}
X(a,t+1) = \bigvee_{b \in C(a)} X(b,t) \vee (\bar{\eta}(a,t) \wedge X(a,t)) \vee (\xi(a,t) \wedge \bar{X}(a,t) \wedge X(\hat{a},t)) \\
\qquad\qquad \bar{0} \neq a \in T^N, \ t \geq 0, \\
X(\bar{0},t) = \text{"true"}, \ t \geq 0
\end{cases}
\tag{1}
$$

with some initial field of logical variables $X(a,0)$, $a \in T^N \setminus \{\bar{0}\}$.
In these equations η and ξ are "external" random fields of
logical variables such that $\xi(a,t)$, as well as $\eta(a,t)$ are
jointly independent for all a and t and the dependence between ξ and η can be given by the condition

$$\xi(a,t) \Rightarrow \overline{\eta}(\widehat{a},t), \quad a \in T^N \setminus \{\overline{0}\}. \tag{2}$$

The tree growth model described by the relations (1), (2) is the Markov sequence of random-logical-variable fields on T^N ; it can be shown that it satisfies the continuity-of-the-configuration principle expressed by the following relations

$$(\forall a,b \in T^N, \ a \le b \Rightarrow (X(b,t) \Rightarrow X(a,t))) \Rightarrow$$
$$\Rightarrow (\forall a,b \in T^N, \ a \le b \Rightarrow (X(b,t+1) \Rightarrow X(a,t+1))). \tag{3}$$

The continuity of configurations means that if an element a belongs to a configuration then any preceding one belongs to it too. The continuous configurations are just the ideals of poset T^N , or equivalently, the subtrees with the root $\overline{0}$. Thus the equation (1) describes the growth process such that the configuration at any time is a subtree if it holds for the initial one. It follows from Eq.(1) that the configuration changes only near its boundary at every time step.

The random continuous logical field $X(a,t)$ determines a measure μ_t on the lattice of ideals $\mathcal{L}(T^N)$ in T^N with the σ-algebra Z generated by subsets of the following kind:

$$\mathcal{F}_y = \{ K \in \mathcal{L}(T^N) : K \supseteq y \}, \quad y \in \mathcal{L}(T^N).$$

On the other hand, such a measure is defined uniquely by its correlation functions

$$G(y,t) = \mu_t(\mathcal{F}_y).$$

THEOREM. Let ξ and η be homogeneous and stationary random logical fields, satisfying the relations (2), and let p be the probability of the fact that $\eta(a,t)$ is true and r is the same for ξ . Then
(i) For any N , $N \ge 2$, and $\beta = r/p \le \beta_N^* = (N-1)^{N-1}/N^N$ the process described by Eq.(1) has three invariant states with correlation functions given by the relations

$$G_j(\mathfrak{I}) = \nu_j^{|\mathfrak{I}|-1}, \qquad j = 1,2,3,$$

where $|\mathfrak{I}|$ is the number of elements in configuration $\mathfrak{I}$ and ν_j are the roots of the algebraic equation

$$(1-\nu)\left(\beta - \gamma(1-\nu)^{N-1}\right) = 0 \tag{4}$$

such that $\nu_1 = 1$, $0 < \nu_2 \le 1/N \le \nu_3 < 1$.

Furthermore, if the initial configuration is finite than the limiting state of the process at $t \to \infty$ corresponds to the correlation function G_2.

(ii) If $\beta > \beta_N^*$, then there exists a unique positive root $\nu_1 = 1$ of Eq.(4) and the limiting state corresponds to the correlation function G_1.

In this way, for $\beta < \beta_N^*$ the configuration of the limiting state is finite with probability 1 if it holds for the initial state, while for $\beta > \beta_N^*$ the limiting configuration is T^N.

For $N = 1$ our model is a kind of random walk. In this case $\beta_1^* = 1$ and the resulting limiting configuration T^1 for $\beta > 1$ means that a particle leaves any bounded region with probability 1 at $t \to \infty$.

The growth model considered is related to the so-called Stavskaya models of active media [1]. The Eqs.(1) were proposed in [2] to describe the blood flow in a small vessel network (see also [3]). The proofs of the assertions formulated above can be found in [4] where a more general case of inhomogeneous trees was also considered.

REFERENCES

1. Stavskaya O.N., Pjatetsky-Shapiro I.I. On the homogeneous networks of the spontaneous-active elements. Problems of Cybernetics, N20, 1963, p.91-106 (Russian).
2. Antonets V.A., Antonets M.A., Shereshevsky I.A. The statistical dynamics of blood flow in a small vessel network. In: Medical Biomechanics, v.4, Riga, 1986, p.37-43 (Russian).
3. Antonets V.A., Antonets M.A., Shereshevsky I.A. Stochastic dynamics of pattern formation in discrete systems. In: Nonlinear Waves. Physics and Astrophysics, ed. M.I.Rabinovich, A.V.Gaponov, Y.Engelbrecht, Springer Verlag, 1989 (to appear).

4. Antonets M.A., Shereshevsky I.A. Analysis of Stochastical
   Tree Growth Model. Preprint N231, Institute of Applied Physics, Acad. of Sci., Gorky 1989, p.48 (Russian).

Institute of Applied Physics, Academy of Sciences of the USSR,
Uljanov Street 46, 603600 Gorky, USSR

Operator Theory:
Advances and Applications, Vol. 46
© 1990 Birkhäuser Verlag Basel

CRITICAL PHENOMENON IN THE STATIONARY MODEL
OF RANDOM GROWTH

M.A.Antonets, I.A.Shereshevsky

Let Γ be a countable partially ordered set (poset) with a unique minimal element $\bar{0}$. A subset $\mathcal{I}$ containing together with any element a all the elements of Γ preceding a is called ideal (the necessary information on the poset theory see, for example, in [1]). By a stationary random growth model of Γ we shall understand a probability measure μ on the set $\mathcal{L}(\Gamma)$ of ideals in Γ defined on the 6-algebra Σ of subsets in $\mathcal{L}(\Gamma)$ generated by the sets of the type $\mathcal{F}_y = \{K \in \mathcal{L}(\Gamma): K \supseteq \mathcal{I}\}, \mathcal{I} \in \mathcal{L}(\Gamma)$. Notice that unlike the usual percolation models, not every configuration of Γ lattice elements is admissible in the situation considered, but only such of them which match the ordered structure of Γ. In a large class of systems exhibiting the growth phenomenon order arises in a natural way. For instance, in the process of growth of dendroid systems with a unique root (minimal element) neither of branches can be present in configuration if all branches connecting it with the root are not grown. There is another example: in the process of square-parquetry laying in a quadrant which starts from the angle only such configurations are possible where each square borders on other squares or walls at least by two sides. Notice that these configurations are exactly Young diagramms.

We shall suppose below that the following condition of local finiteness is valid for Γ : for any $a \in \Gamma$ the set $J_a = \{ b \in \Gamma , \; b \leq a \}$ and the set $C(a)$ of elements immediately following a in Γ are finite.

Let us consider a one-parameter family of random-growth models, i.e., measures μ_ν defined by the relation

$$\mu_\nu(\mathcal{F}_J) = \nu^{|J|-1} \tag{1}$$

where $0 < \nu \leq 1$ and $|J|$ is the cardinality of the set J . It is easy to show that the above relation determines a unique measure μ_ν on $\{ \mathcal{L}(\Gamma), \Sigma \}$. For $\Gamma = \mathbb{Z}_+$, where $\mathbb{Z}_+$ is the set of nonnegative integers with the natural order, these measures are nothing else than geometrical distributions of probabilities.

Let $p_J = \mu_\nu(\{J\})$ be a probability of the configuration J . It follows from the relation (1) that

$$p_J = \nu^{|J|-1} (1-\nu)^{|\partial \bar{J}|}$$

where $\partial \bar{J}$ consists of the minimal elements of the set $\bar{J} = \Gamma \setminus J$. Let

$$m_\Gamma(\nu) = \sum_{J \in \mathcal{L}_0(\Gamma)} p_J$$

where $\mathcal{L}_0(\Gamma)$ is the subset of finite ideals in $\mathcal{L}(\Gamma)$. The quantity $m_\Gamma(\nu)$ is the probability that the appeared configuration is finite.

THEOREM 1. There exists a unique number ν_Γ^* belonging to interval $[0,1]$ such that $m_\Gamma(\nu) = 1$ for $0 < \nu < \nu_\Gamma^*$ and $m_\Gamma(\nu)$ is less than 1 when $\nu_\Gamma^* < \nu \leq 1$.

In this way the number ν_Γ^* is the critical value of the following phenomenon: when $\nu > \nu_\Gamma^*$ the probability of the infinite-configuration set $\mathcal{L}(\Gamma) \setminus \mathcal{L}_0(\Gamma)$ becomes nonzero. It turns out that the value of ν_Γ^* depends on the power of branching of the Γ lattice.

THEOREM 2. (i) Let
$$\mathscr{X}_\Gamma = \inf_{\mathfrak{J} \in \mathcal{L}_0(\Gamma)} |\partial \mathfrak{J}| / |\mathfrak{J}|.$$

Then $\nu_\Gamma^* \leq (1 + \mathscr{X}_\Gamma)^{-1}$

(ii) If there exists a sequence $\{\mathfrak{J}_n\}$, $n = 0, 1, \ldots$ of finite ideals in Γ , which is monotonically increasing, exhausting Γ and such   that

$$\sup_n |\partial \overline{\mathfrak{J}}_n| < \infty,$$

then ν_Γ^* is equal to 1 (notice that in such a situation $\mathscr{X}_\Gamma$ is equal to zero).

The following examples illustrate the assertion of the Theorems.

Let $\Gamma = \mathbb{Z}_+^n$ with the order defined by the condition: $(m_1, \ldots, m_n) \leqslant (m_1', \ldots, m_n')$ if and only if $m_i \leqslant m_i'$ for every $i = 1, \ldots, n$. When $n = 1$ we return to the set of nonnegative integers with the natural order. When $n = 2$ the ideals of $\mathbb{Z}_+^2$ are just the Young diagramms. Let $\mathfrak{J}_m = \{a \in \Gamma : a \leqslant (m, \ldots, m)\}$. It is easy to see that the sequence $\{\mathfrak{J}_m\}$, $m = 1, 2, \ldots$ satisfies the conditions of part (ii) of Theorem 2, because of $\partial \overline{\mathfrak{J}}_m = \{(m+1, 0, \ldots, 0), \ldots, (0, \ldots, 0, m+1)\}$ and so $|\partial \overline{\mathfrak{J}}_m|$ is equal to n for any m. Consequently, $\nu_\Gamma^* = 1$, i.e., the support of the measure μ_ν is $\mathcal{L}_0(\Gamma)$ for any ν from the interval $[0, 1)$.

Furthermore, let $\Gamma = T^N$ be the Caley tree with branching multiplicity N . We say that a preceds b if it lies on the (unique!) way connecting the branch b with the root $\overline{0}$. It can be shown that for any finite ideal $\mathfrak{J}$ the equality $|\partial \mathfrak{J}| = (N-1)|\mathfrak{J}| + 1$ holds and it follows from Theorem 2 (i) that $\mathscr{X}_\Gamma = (N-1)$ and $\nu_\Gamma^* \leq 1/N$. In fact it turns out that $\nu_\Gamma^* = 1/N$ for Caley trees. The parameter ν of the considered growth model is the conditional probability of growing an element to a configuration. It follows from Theorem 2 that the possibility of infinite-configuration appearance depends on the structure of Γ . We see that dendroid systems are growing more willingly than the lattice-like ones. The proofs of the above formulated assertions can be found in [2].

REFERENCES
1. Birkhoff G. Lattice Theory, Providence, Rhode Island, 1967,
   p.565.
2. Antonets M.A., Shereshevsky I.A. On a critical phenomenon in
   a random growth model. Preprint of IAP N229, Gorky, 1989,
   p.21.

Institute of Applied Physics, Academy of Sciences of the USSR,
Uljanov Street 46, 603600 Gorky, USSR

Operator Theory:
Advances and Applications, Vol. 46
© 1990 Birkhäuser Verlag Basel

ON A QUANTUM-CLASSICAL CONNECTION, HIDDEN SYMMETRIES,

AND A MODEL OF JOSEPHSON JUNCTION

Pavel B o n a

It is shown how can a given kinematical symmetry group G of an infinite quantum system determine a set of classical-mechanical macroscopic variables with a natural Poisson structure on it. For the long-range interactions of the type introduced by Hepp and Lieb [Helvetica Phys.Acta **46**, 573 (1973)] the dynamics of the large quantal system [P.Bona, J.Math.Phys. **29**, 2223(1988)] leads to the classical Hamiltonian motion of the macroscopic quantities determined by a Hamilton function specified uniquely by the local quantal Hamiltonians. A model of Josephson junction is a nontrivial integrable example of this class of systems.

Some of main ideas presented at this Conference were based on a specific (and traditional) connection between quantum theory (QT) and classical mechanics (CM): QT of a specific physical system is obtained as a "quantization" of a "corresponding" CM-model which can be, in turn, recovered by taking a limit of the theory for Planck constant ℏ tending to zero. If a specific physical system described by QT contains, however, also variables behaving classically (e.g., the center-of-mass motion of a large system, variables describing macroscopic quantal phenomena), classical and quantum behaviors occur simultaneously at the nonzero value of ℏ. It is possible to describe such systems in a way, where instead of taking limit ℏ → 0 the infinite limit of the size |Λ| of the system is taken. Such a possibility was used in [2] as an effective tool for formulation and solution of dynamics of a wide class of quantum mean-field models, [1]. This way of understanding of macroscopic classical features of large quantal systems offers possibility of simultaneous description of the "underlying" quantal as well as the "corresponding" classical kinematics and dynamics in the framework of a unique theory.

Any large quantal system offers many different possibilities for a choice of a (reasonably restricted) set of macroscopic variables. A convenient choice can be given by a connected Lie group G of kinematical symmetry transformations $\sigma(g)$ ($g \in G$) of the large system. This group can represent motions of (macroscopic) measuring devices determining the relevant observable quantities. Corresponding classical kinematics is described by $\mathrm{Ad}^*(G)$-orbits [3] and their canonical symplectic structure (resp. Poisson structure on $\mathfrak{g}^*$, [4]). The transformations $\sigma(g)$ represent a kind of "hidden symmetries", which need not be symmetries of (unspecified, up to now) dynamics. If the dynamics is specified by a net of local mean - field quantum Hamiltonians Q^Λ, [1], [2], determining the time evolution τ^Λ of local algebras $\mathcal{A}_\Lambda$ (Λ is a finite subset of an infinite set Π of "elementary constituents" of the infinite system). The time evolution τ^Q of the total infinite system obtained from τ^Λ as a kind of "thermodynamic limit", [2], is an automorphism group of a C^*-algebra $\mathfrak{C}$ of continuous functions on an $\mathrm{Ad}^*(G)$-invariant closed convex subset E of $\mathfrak{g}^*$ with values in $\mathcal{A}$, $\mathcal{A} \equiv C^*$-inductive-$\lim_{\Lambda \nearrow \Pi} \mathcal{A}_\Lambda$, cf. [5]. The local Hamiltonians Q^Λ are determined by a unitary representation U(G), as well as by a given classical Hamilton function $Q \in C^\infty(\mathfrak{g}^*)$. The corresponding classical dynamics is the restriction of τ^Q to a central subalgebra $\mathcal{N}$ of $\mathfrak{C}$ consisting of scalar-valued functions on E. This restriction of τ^Q coincides with classical Hamiltonian evolution given by the Hamiltonian Q and by the canonical Poisson structure [4] of $\mathfrak{g}^*$. By such a description of mean-field theories (for compact G and finite-dimensional U(G)) we can determine the dynamics τ^Q of the infinite quantum system by solving two finite-dimensional ordinary differential equations one of which is linear, [2].

One of the simplest nontrivial examples of the considered type of models is the strong-coupling quasi spin formulation of a model of Josephson junction based on the BCS theory of superconductivity. The group G is here $SU(2)_a \times SU(2)_b$ (the index $c \equiv a,b$ distingushes between the two interacting superconductors), the (generalized) classical phase space g^* is six dimensional with coordinate functions F_{cj} ($c = a,b$; $j = 1,2,3$), and the Poisson brackets corresponding to the canonical Poisson structure on g^* are

$$\{F_{cj}, F_{dk}\} = -\delta_{cd}\varepsilon_{jkl}F_{cl} \text{ (no summation over c).}$$

Let $F_{c\pm} \equiv F_{c1} \pm iF_{c2}$. The Hamilton function Q is chosen, [1],[6]-[8]:

$$Q(F) \equiv - 2\varepsilon_a F_{a3} - \lambda_a F_{a+}F_{a-} - 2\varepsilon_b F_{b3} - \lambda_b F_{b+}F_{b-} + \kappa(F_{a+}F_{b-}+F_{a-}F_{b+}).$$

Here ε_c, λ_c, and κ are some real constants. The corresponding equations of motion have the form:

$$\dot{F}_{a+} = 2i[(\lambda_a F_{a3} - \varepsilon_a)F_{a+} - \kappa F_{a3}F_{b+}],$$

$$\dot{F}_{b+} = 2i[(\lambda_b F_{b3} - \varepsilon_b)F_{b+} - \kappa F_{b3}F_{a+}],$$

$$\dot{F}_{a3} = -i\kappa (F_{a+}F_{b-} - F_{a-}F_{b+}),$$

$$\dot{F}_{b3} = i\kappa (F_{a+}F_{b-} - F_{a-}F_{b+}).$$

These equations are completely integrable [3]. They have the following independent (global) integrals of motion:

$$Q, \quad F_3 \equiv F_{a3} + F_{b3}, \quad r_a^2 \equiv F_{a1}^2+F_{a2}^2+F_{a3}^2, \quad r_b^2 \equiv F_{b1}^2+F_{b2}^2+F_{b3}^2.$$

The integrals $r_{a,b}$ are of "kinematical character": they determine the Ad^*-orbit on which the motion is realized, i.e. the Cartesian product of two spheres S^2 with radii $r_{a,b}$ forming the symplectic manifold for the actual motion of the system. These integrals are present for an arbitrary Poisson flow on our g^*, hence they are independent of the Hamiltonian $Q \in C^\infty(g^*)$. The integrals Q and F_3 are the "dynamical" ones: their existence leads to the integrability of the system on the 4-dimensional symplectic manifold $S^2_{r_a} \times S^2_{r_b}$. Such a set of integrals corresponds also to the (integrable) modification of the model proposed in [8]. One can, however, prescribe an arbitrary classical motion of macroscopic quantities corresponding to the chosen group G by an arbitrary choice of the Hamiltonian Q, and subsequently define the quantum mean-field dynamics τ^Q leading (by its restriction to the subalgebra N of $\mathcal{C}$) exactly to the given classical motion. Hence, one can obtain also chaotic classical behavior by an arbitrary small perturbation of the considered Hamiltonian Q.

References

1. K.Hepp, and E.H.Lieb, Helvetica Phys.Acta **46**, 573 (1973).
2. P.Bona, J.Math.Phys. **29**, N̲o.10 (1988).
3. R.Abraham, and J.E.Marsden, *Foundations of Mechanics*, Second Edition (Bejamin/Cummings, Reading, Mass., 1978).
4. C.M.Marle, in *Bifurcation Theory, Mechanics and Physics* (Ed. C.P.Bruter, A.Aragnol, and A.Lichnerowicz) (D.Reidel, Dordrecht - Boston - Lancaster, 1983).
5. O.Bratteli, and D.W.Robinson, *Operator Algebras and Quantum Statistical Mechanics,* Vol.I and II (Springer, New York - Heidelberg - Berlin, 1979 and 1981).
6. E.Duffner, Z.Phys. **B** - condensed matter 63, 37 (1986).
7. P.Bona, J.Math.Phys. **30**, No.11 (1989).
8. T.Unnerstall and A.Rieckers: Quasispin-operator description of the Josephson tunnel junction and the Josephson Plasma Frequency, preprint of University Tübingen, 1988.

P.Bona, Department of Theoretical Physics,
Faculty of Mathematics and Physics, Comenius University,
842 15 Bratislava, Czechoslovakia.

Operator Theory:
Advances and Applications, Vol. 46
© 1990 Birkhäuser Verlag Basel

OPEN MULTIQUANTUM SYSTEMS. METHOD OF A
GENERATING FUNCTIONAL

V.R.Struleckaja

The algebraic approach to the description and representation of multiquantum (multicomponent) systems discussed in the present paper continues the approach developed in [5,9-12]. That is based on the generating multicomponent -state functional and the nonlinear completely positive map [3] functional method which extends in a natural way the variety of the processes regarded as quantum stochastic processes [2], including nonlinear ones with interaction in the system. The dynamics of such systems is considered in the Markovian approximation; the theory and classification of the corresponding "master equations" was constructed in [13]. In the nonlinear case we obtain in a natural way a nonlinear extension of the basic constructions and results of the representation theory and an extension to nonlinear structures (state and mapping) [3,8] which are good tools to describe systems with interaction.

The generating functional method for states and maps reduces the dynamics of a system with irreversible evolution to a pair of dual one-particle equations for an observable and a state of the system equivalent to a functional equation for the operator-valued functional of generating map [4] considered as a quantum stochastic process [2,4].

Let $\mathcal{H}_0, \mathcal{H}, \mathcal{Y}$ be separable Hilbert spaces, and $\Gamma(\mathcal{H}_0 \otimes \mathcal{H})$ be a Fock space of a multicomponent dissipative quantum system (the case when $\mathcal{H}$ is a space of stochastic integrals see in [12]), x, $z \in \mathcal{H}$. Let $\mathcal{A} \ni z$ be a W^*-algebra of observables of the one-particle system and $\mathcal{A}_*$ be a preconjugate space of trace-class operators x. Let $\langle x, z \rangle = \mathrm{Tr}\, zx$ be the pairing that induces on $\bigoplus_{n=0}^{\infty} \mathcal{A}^{(n)}$ the ultra weak topology with the dual topology on $\prod_{n=0}^{\infty} \mathcal{A}_*^{(n)}$. We consider a system with interaction having the intensity parameter $\mathcal{E}$. Let $a = \delta\alpha$, $a^+ = \delta\alpha^+$, $c_\# = \delta_\# k$ be variational derivatives of analytical functionals

$$\alpha_\mathcal{E}(x) = 1/2\, k_\mathcal{E}(x, I) + i\gamma_\mathcal{E}(x), \quad \alpha_\mathcal{E}^+(x) = 1/2\, k_\mathcal{E}(x, I) - i\gamma_\mathcal{E}(x),$$

$$\mathrm{Im}\, \alpha_\mathcal{E}(x) = \gamma_\mathcal{E}(x), \quad \alpha_\mathcal{E}(x) + \alpha_\mathcal{E}^+(x) = k_\mathcal{E}(x, I),$$

$$N_\mathcal{E}(x, z) = k_\mathcal{E}(x, z) - \alpha_\mathcal{E}(xz) - \alpha_\mathcal{E}^+(zx),$$

which is the generating functional of the generator of a dynamical semigroup of Lindblad type [6],

$$\gamma_\mathcal{E}(x) = \sum_{n=0}^{\infty} \mathcal{E}^n/n! \langle x^{\otimes n}, V^{(n)} \rangle,$$

$$k_\mathcal{E}(x, z) = \sum_{n=0}^{\infty} \mathcal{E}^n/n! \langle x^{\otimes n}, J^{(n)}(z) \rangle,$$

$x^{\otimes n} \in \mathcal{A}^{(n)}$, $V^{(n)}$ is the potential of n-particle interaction and $J^{(n)}(z)$ is the "collision integral" [8,10,13].

The limiting dynamics (as $\mathcal{E} \to 0$) of such a process is described by the canonical pair of kinetic equations [7,8,10]

$$\dot{x} + a(xz)\, x + x\, a^+(zx) = c_z(x, z) \tag{1}$$

$$\dot{z} + c_x(x, z) = z\, a(xz) + a^+(zx)\, z \tag{2}$$

(the series $\omega(x) = \sum_{n=0}^{\infty} \langle x_n, z^{\otimes n} \rangle$ is called a generating functional of the state $x = (x_n)_{n \geq 0}$).

Let us consider a crude model of a dissipative process, say, in nuclear physics or quantum optics based on the mean field approximation; it yields a description by nonlinear frictional Schrödinger equation [1]. The latter can be associated with the following pair of canonical equations for the observable z and the state x of the system

$$\dot{x} = \text{Re}\left\{b^+(zx)\left[\delta b(xz),x\right] + (b(xz)-b^+(zx))\delta b(xz)x\right\} +$$
$$+ i\left[x,\delta\gamma(xz)\right], \qquad x(0) = x_o, \qquad (3)$$

$$-\dot{z} = \text{Re}\left\{b^+(zx)\left[z,\delta b(xz)\right] + z\delta b(xz)(b(xz)-b^+(zx))\right\} +$$
$$+ i\left[\delta\gamma(xz),z\right], \qquad z(0) = z_o, \qquad (4)$$

where $\alpha(x)+\alpha^+(x)=b(x)b^+(x)$, and $k_\varepsilon(x,z)=b(xz)\exp\{\varepsilon\langle\delta_z,\delta_x\rangle\}b^+(zx)$, $\delta\gamma(x)$ is the functional of the Hamiltonian $H(x)$ of the system (in the Hamiltonian dynamics sense), δb is the variational derivative of an analytical functional, $x\in\mathcal{A}$.

Suppose the observable z is preserved, i.e. $\psi = I$, $\varphi\in\mathcal{Y}$, the eq. (4) is solved by $z_t(\psi\psi^+)=I$, and the canonical pair (3), (4) reduces to one nonlinear kinetic equation

$$\dot{x} = i\left[x,\delta\gamma(x)+ \text{Im}\delta b(x)\, b^+(x)\right], \qquad x(0) = x \qquad (5)$$

which generalizes the nonlinear frictional Schrödinger equation obtained in a phenomenological way [1].

If Re $k(x,I) = 0$, the equation (5) is the equation of nonlinear Hamiltonian dynamics [7].

References.

1. Alicki,R.,Messer,J., J.Stat.Phys., 32, N2, 299-312 (1983).
2. Accardi,L., Frigerio,A., Lewis,J.T. Publ. RIMS Kyoto Univ., 18, 97-133 (1982).
3. Evans,D.E., Lewis,J.T. Comm. of DIAS, N24, ser.A,119 (1977).
4. Hudson,R.L., Parathasarathy,K.R. Comm.Math.Phys., 93, 301-323 (1984).
5. Belavkin,V. DAN SSSR, 293, 18-21 (1987).
6. Lindblad,G. Comm.Math.Phys., 48, 119-130 (1976).
7. Maslov,V.P. TMPh, 33, N1, 17-37 (1977).
8. Struleckaja,V. In Proc. of Conf. on Select. Probl. Stat. Mech. and Field Theory, Kujbishev, 1987.
9. Struleckaja,V. Physica, 10, 89-92 (1988).
10. Struleckaja,V. Theses V Int. Conf. on Prob. Theory and Math. Stat., Vilnius, 1989.
11. Struleckaja,V. Theses Int. Conf. COSMEX'89, Wroclaw.
12. Struleckaja,V. Proc. II Int. Conf. Quant. Phys., Gdańsk, 1989 (to appear).
13. Gorini,V.,Frigerio,A.,Verri,M.,Kossakovski,A.,Sudarshan,E.C.G. Rep. Math. Phys., 13, 149-173 (1978).

Depart. Appl. Math., Moscow Institute
of Electronic Machineconstruction (MIEM),
B.Vuzovsky 3/12,
109028 Moscow, USSR

Operator Theory:
Advances and Applications, Vol. 46
© 1990 Birkhäuser Verlag Basel

GENERALIZED FUNCTIONS AND THEIR APPLICATIONS

Yu. V. Egorov

The distribution theory of L.Schwartz is practically non applicable to non-linear problems. Over the past decade, a new theory of generalized functions has been developed by J.Colombeau. It can work in non-linear theories but it is too complicated. Here another new theory is proposed which is more simple and more general than the Colombeau's one. Some applications to the theory of differential equations are indicated.

It was clear from the very beginning that the distribution theory was not applicable to solving of non-linear problems. P. Dirac introduced his famous delta-function in 1927 and tried to define its square but did not succeed. In 1954, L.Schwartz proved that it is impossible to define multiplication in the space of distributions as an associative operation. His proof is based on the simple observation that

$$((1/x) \cdot x) \cdot \delta(x) = \delta(x) \neq (1/x) \cdot (x \cdot \delta(x)) = 0.$$

One of the most powerful properties of distributions, namely their independence of the choice of approximations by smooth functions, turns out to be their Achilles' heel. It is impossible, for instance, to define the product $H(x) H'(x)$, where H is the Heaviside function (equal to 1 for $x > 0$ and to 0 for $x < 0$). Such a product appears frequently in the theories of gas dynamics, elasticity and other applications dealing with discontinuous solutions of partial differential equations. It could be interpreted as $(H^2(x))'/2$, being therefore equal to $\delta(x)/2$, if the function H' arising here was the derivative of the function H. This happens, for example, when we have a product of the type $u \cdot \partial u/\partial x$ and the function u has a jump at the point $x=0$. But when we have a product of the form $u \cdot \partial v/\partial x$, where u and v are mutually different functions both having a jump at this point and being equal to some constants outside, this product cannot be defined correctly unless we know the relations between u and v since it should be equal to $C\delta(x)$, with the constant C assuming any real value for different pairs of u and v. This observation motivated probably J.Colombeau in 1982 when he proposed a new theory of generalized functions; in it these functions are defined in dependence on the way in which they are approximated by smooth functions. The space of generalized functions in the Colombeau's theory is an algebra, and for any smooth function $F(z_1,\ldots,z_k)$ of a power growth and generalized functions $f_1,\ldots,f_k$ one can define the function $F(f_1,\ldots,f_k)$. As an application of his theory, Colombeau has obtained very interesting numerical results concerning problems of elasticity and gas dynamics. He has got also general theorems on the existence of solutions for linear and non-linear partial differential equations.

At the same time, however, Colombeau's theory seems to be too complicated and overloaded by unnecessary constructions that make it too clumsy and restricted. We are going to formulate here a more simple and general theory in which, for instance, it is possible to define functions of the form $F(f_1,\ldots,f_k)$ if F is an arbitrary function, even a generalized one. In such a theory, very general theorems on the existence of solution to the Cauchy problem can be proven. It is worth of mentioning also that all the constructions are very natural and reflect the actual process of computing.

2. Let Ω be a domain in the space R^n and f_k for $k=1,2,\ldots$ be C^∞-functions in Ω. Two sequences $\{f_k\}$ and $\{g_k\}$ are called equivalent if for any compact subset K in Ω there exists a

number N such that we have $f_k(x) = g_k(x)$ if $k > N$ and x belongs to K. The equivalence classes of sequences are called generalized functions and the space of generalized functions is denoted by $\mathcal{G}(\Omega)$. Considering the sequence $\{f_k\}$ of functions from $C^\infty(\bar{\Omega})$ and taking compacts K from $\bar{\Omega}$ we can define in the same way the space $\mathcal{G}(\bar{\Omega})$.

It is clear that if a function $f(x)$ belongs to $C^\infty(\Omega)$, then the sequence $f_k(x) = f(x)$ defines a generalized function and so the space $C^\infty(\Omega)$ is embedded in the space $\mathcal{G}(\Omega)$ of generalized functions.

We call a generalized function continuous if the sequence $\{f_1(x)\}$ converges uniformly on all the compact subsets in Ω. We say it belongs to the class $L_p(\Omega)$ if the corresponding sequence converges in this class. Similarly we define the Sobolev spaces $W_p^1(\Omega)$ and other functional spaces. Finally, a generalized function is called a distribution if the corresponding sequence converges weakly, i.e., for any smooth function $h(x)$ with a compact support in Ω the sequence of integrals

$$\int f_k(x)h(x)\ dx$$

has a finite limit when k goes to infinity.

Let $\omega(x)$ be a function from $C_0^\infty(R^n)$ assuming real non-negative values such that $\int \omega(x)dx = 1$. Furthermore, let $h_k(x)$ be a real non-negative function from $C_0^\infty(R^n)$, which is equal to 0 in the $1/k$-neighborhood of $\partial\Omega$ and to 1 outside the $2/k$-neighborhood of $\partial\Omega$. Let g be a distribution in Ω and

$$f_k(x) = \int g(y)h_k(y)\omega_k(x-y)\ dy\ ,$$

where $\omega_k(y) = k^n\omega(ky)$. It is easy to see that the sequence $\{f_k(x)\}$ converges in $\mathcal{D}'(\Omega)$ to $g(x)$ so the space $\mathcal{D}'(\Omega)$ is embedded in $\mathcal{G}(\Omega)$, however, the corresponding embedding operator is not unique: choosing another function ω we obtain another operator.

We can differentiate a generalized function $g(x)$ simply identifying its derivative of order α with the generalized function defined by the sequence $\{D^{\alpha}g_k(x)\}$, where the sequence $\{g_k(x)\}$ defines the generalized function $g(x)$.

The definition formulated above is applicable also to the case $n = 0$; then the "generalized functions" are called generalized complex numbers. Each sequence $\{c_k\}$ of usual complex numbers define a generalized complex number as a class of sequences equivalent to $\{c_k\}$. Two sequences are called equivalent if they coincide for large values of k. If $g \in \mathscr{G}(\Omega)$ then for any point x of Ω the value $g(x)$ is defined as a generalized complex number. In addition, for any compact subset K of Ω the integral $\int g(x)dx$ over K is defined as a generalized complex number.

If $\Omega' \subset \Omega$ the restriction of $g \in \mathscr{G}(\Omega)$ to Ω' can be defined as an element of $\mathscr{G}(\Omega')$. The equality $g = 0$ in Ω' means that $g_k(x) = 0$ in Ω' for all large enough values of k. So the support of g can be defined as a minimal closed set outside of which $g = 0$. Similarly the singular support of g is defined as a minimal closed set outside of which g is an infinitely differentiable function.

The space $\mathscr{G}(\Omega)$ is an algebra. The product fg of two generalized functions f and g is the generalized function defined by the sequence $\{f_k g_k\}$ where $\{f_k\}$ and $\{g_k\}$ are any representatives of the classes f and g, respectively. Similarly, if $F(z_1,\ldots,z_m)$ is a smooth function and $f_1,\ldots,f_m$ are generalized functions then $F(f_1,\ldots,f_m)$ is the generalized function defined by $\{F(f_{1k},\ldots,f_{mk})\}$ where $\{f_{jk}\}$ are representatives of the classes f_j.

3. Weak equality. The well known Hopf equation

$$\partial u/\partial t + u\, \partial u/\partial x = 0$$

has solutions of the form $u = a + bH(x-vt)$ if $b = 2v-2a$. This equation is equivalent to the equation

$$u\, \partial u/\partial t + u^2\, \partial u/\partial x = 0$$

but the last equation has solutions of the above form if and only if

$$3v(b+2a) = 2(b^2+ab+a^2) \ .$$

This means that these functions can be considered as solutions only in some weak sense as it is common in the distribution theory. Most of the differential equations in mathematical physics are fulfilled only in the weak sense; recall, for instance, the string equation which is obtained by dropping some nonlinear terms which are non-zero in the sense of generalized functions.

DEFINITION. The generalized functions f and g are called *weakly equal* if for any function φ from $\mathcal{D}(\Omega)$

$$\lim \int (f_k(x) - g_k(x))\varphi(x) \ dx = 0.$$

This definition is not "too weak" as it might seem; for example, the following result is valid.

THEOREM 1. *If* $y \in \mathcal{G}(I)$, $y' \approx 0$ *on* I, *where* I *is an interval of the real axis, and there exists a function* $h \in \mathcal{D}(I)$ *such that* $\int h(x)dx \neq 0$ *and the value of* $\int y(x)h(x)dx$ *is finite, then* $y \approx const$.

It follows from Theorem 1 that any system of ordinary differential equations $P(D)y \approx 0$ can have only classical solutions if the assumptions of the theorem are valid for all components of the vector y and their derivatives of some degree.

Consider a system of differential equations of Kovalevskaya type. It is well known that it is always possible to reduce the Cauchy problem for such a system to the Cauchy problem for the first-order system:

$$\partial u/\partial t \approx \sum a_j(t,x)\,\partial u/\partial x_j + b(t,x)u + f(t,x)\ ,$$

$$u(0,x) \approx \varphi(x)\ .$$

Let us assume that the elements of the matrices a_j and b are analytic functions in a neighborhood of the origin, f and φ are generalized functions.

THEOREM 2. *The Cauchy problem has a unique solution* u *of the class* $\mathscr{G}(R^{n+1})$ *in a neighborhood of the origin. In particular, if this problem has a classical solution or a weak solution of the class* $\mathscr{D}'(R^{n+1})$, *then the latter coincides with* u.

A similar theorem is valid for strongly hyperbolic systems and other evolutionary differential equations without the analyticity assumption on the coefficients.

4. The Cauchy problem. If we study some differential equation in the weak form it is useful to look whether the derivatives can be substituted by difference operators. Namely, let g be a generalized function defined by some sequence $\{g_k(x)\}$ of smooth functions, then the derivative $g_1 = \partial g/\partial x_1$ is the generalized function defined by the sequence $\{\partial g_k/\partial x_1\}$. Let e_1 be a unit vector which is parallel to the x_1-axis. Then the generalized function h_1 defined by the sequence $\{k(g_k(x+e_1/k)-g_k(x))\}$ is weakly equal to g_1. The same is true for the derivatives of any order. Thus for distributions the operators of differentiation and the finite-difference operators are equivalent.

Let us consider now the Cauchy problem

$$\partial u/\partial t \approx \sum_{|\alpha|\leq m} a_\alpha(t,x)D_x^\alpha\, u(t,x) + f(t,x),$$

$$u(0,x) \approx \varphi(x)\ .$$

It can have no solutions in the distribution class as it is true for the famous H.Lewy's equation. Assume that the coefficients a_α and their derivatives are uniformly bounded, and f and φ are generalized functions. Consider the following auxiliary Cauchy problem

$$\partial v/\partial t \approx \sum_{|\alpha| \leq m} a_\alpha(t,x)\, \Delta^\alpha_{x,k} v(t,x) + f_k(t,x) \ ,$$

$$v(0,x) \approx \varphi_k(x) \ .$$

Here f_k and φ_k are smooth functions defining the generalized functions f and φ, while $\Delta^\alpha_{x,k}$ is the difference operator of order α and step $1/k$ with respect to x. This Cauchy problem has always a solution and this solution is unique since the operator at the right hand side is bounded.

THEOREM 3. *If the solution v of the auxiliary Cauchy problem is a Schwartz distribution, then it is a usual weak solution u of the original Cauchy problem. If $v \in C^m$ with respect to the variables x , then it is a classical solution of the original Cauchy problem.*

If the assumptions of the uniform boundedness imposed on the coefficients and their derivatives are not fulfilled we can prove only the local solvability changing correspondingly the coefficients in a neighborhood of infinity. Of course, we can solve the problem also in the case when the coefficients are not smooth or even generalized functions.

The same method works also for nonlinear differential equations. Consider the Cauchy problem

$$\partial u/\partial t \approx F(t,x,u,\ldots,D^a u,\ldots) \ , \qquad |\alpha| \leq m,$$

$$u(0,x) \approx \varphi(x) \ .$$

We can once again replace the derivatives by difference operators

and prove the existence and uniqueness of the solution assuming
that

$$|F(t,x,\xi,\ldots,\xi_\alpha,\ldots)| \leq C(1 +|\xi|+\ldots+|\xi_\alpha|+\ldots).$$

Also in this case we can obtain a weak solution in the class of
generalized functions and to prove that if they are C^m with
respect to the variables x, then it will be a classical solution
to the Cauchy problem. If the growth condition for the function F
is not satisfied, we have to use other properties, for example,
positivity, monotonicity, convexity etc.

5. Finally, we state some simple theorems on the
generalized harmonic and analytic functions.

THEOREM 4. *Let Ω be a bounded domain in R^n with a
smooth boundary Γ, $u \in \mathcal{G}(\bar{\Omega})$ and $\Delta u = 0$ in Ω. If $u_\Gamma \in \mathcal{D}'(\Gamma)$, then
u is a usual harmonic function in Ω.*

THEOREM 5. *Let Ω be a bounded domain in R^2 with a
smooth boundary Γ, $u \in \mathcal{G}(\bar{\Omega})$ and $\partial u/\partial z = 0$ in Ω. If $u_\Gamma \in \mathcal{D}'(\Gamma)$,
then u is a usual analytic function in Ω.*

REFERENCES

1. J.F.Colombeau. *New Generalized Functions and Multiplication of
   Distributions*, North-Holland Math. Studies, 84, 1984.
2. J.F.Colombeau. *Elementary Introduction to New Generalized
   Functions*, North-Holland Math. Studies, 113, 1985.
3. J.F.Colombeau. *A New Theory of Generalized Functions*, in
   "*Advances of Holomorphy and Approximation Theory*", North-
   Holland Math. Studies, 125 (1988), pp. 57-66.
4. Yu.V.Egorov. On a new theory of generalized functions, *Moscow
   University Vestnik*, No.4 (1989), pp. 96-99.

Moscow State University
Faculty of Mechanics and Mathematics
119 899 Moscow, USSR

Operator Theory:
Advances and Applications, Vol. 46
© 1990 Birkhäuser Verlag Basel

LIST OF UNPUBLISHED CONTRIBUTIONS

Lectures

D.L.Shepelyansky (Novosibirsk): Quantum localization of dynamical
    chaos

Short contributions

M.B.Kadomtsev, B.L.Markovsky, A.A.Suzko, S.I.Vinitzky (Dubna) :
    Adiabatic representation of scattering amplitude for
    three-body Coulomb problem

A.M.Khorunzhy, L.A.Pastur (Kharkov) : Limiting state density of a
    block Schroedinger operator

S.M.Lakaev (Moscow) : Efimov effect for the discrete three-body
    Schrodinger operator

M.M.Malamud (Donetsk) : Boundary problems for Schroedinger
    operators with gaps

A.I.Mogilner (Sverdlovsk) : Quantum mechanics of systems with a
    non-conserved bounded number of particles on a lattice

Panel discussions

Deterministic and random Schroedinger operators (moderated by
    H.Neidhardt)

Nonstandard Schroedinger operators : contact interactions,
    graphs, waveguides (moderated by P.Exner)

Quantum chaos (moderated by P.Šeba)

Posters

J.I.Abdullaev (Samarkand) : Magnetic polaron in a ferromagnetic
        crystal

V.I.Belokon, S.V.Semkin (Vladivostok) : Peculiarities of the
        reflection coefficient for some classes of potentials

O.I.Gerasimov, N.N.Khudyntsev (Odessa) : Structural properties of
        disordered atomic links by the EELS method

V.A.Kondratyev (Moscow) : Schroedinger operators in weighted
        spaces

Yu.G.Kondratyev, Yu.V.Kozitsky (Kiev) : A phase transition in
        Curie-Weiss model with a transversal field

T.V.Tsikalenko (Kiev) : Stochastic quantization of lattice
        systems : existence and uniqueness of time evolution

LIST OF PARTICIPANTS

J.I.Abdullaev	Samarkand State University Samarkand, USSR
L.Andrej	Academic Computer Centre Prague, Czechoslovakia
M.A.Antonec	Institute of Applied Physics Gorky, USSR
M.S.Birman	Leningrad State University Leningrad, USSR
P.Bona	Comenius University Bratislava, Czechoslovakia
J.Brasche	University of Ruhr Bochum, FRG
V.S.Buslaev	Leningrad State University Leningrad, USSR
S.E.Cheremshantsev	Steklov Institute Leningrad, USSR
N.A.Chernyavskaya	Institute of Mathematics and Mechanics, Alma-Ata, USSR
G.A.Derfel	Karaganda State University Karaganda, USSR
J.Dittrich	Nuclear Physics Institute Rez, Czechoslovakia
L.A.Dmitrieva	Leningrad State University Leningrad, USSR
Yu.V.Egorov	Moscow State University Moscow, USSR
V.V.Evstratov	Leningrad State University Leningrad, USSR
P.Exner	Joint Institute for Nuclear Research, Dubna, USSR and Nuclear Physics Institute Řež, Czechoslovakia

358

V.Ya.Ivrii   Institute of Mining and
                                  Metallurgy
                                  Magnitogorsk, USSR

F.M.Izrailev   Nuclear Physics Institute
                                  Novosibirsk, USSR

A.N.Khorunzhy   Low Temperature Physics
                                  Institute, Kharkov, USSR

S.V.Khryashchev   Leningrad State University
                                  Leningrad, USSR

N.N.Khudyntsev   Odessa State University
                                  Odessa, USSR

A.N.Kochubei   Energosetproject
                                  Kiev, USSR

V.A.Kondratyev   Moscow State University
                                  Moscow, USSR

Yu.G.Kondratyev   Mathematical Institute
                                  Kiev, USSR

Yu.V.Kozitsky   Trade-Economy Institute
                                  Lvov, USSR

Yu.A.Kuperin   Leningrad State University
                                  Leningrad, USSR

P.B.Kurasov   Leningrad State University
                                  Leningrad, USSR

S.N.Lakaev   Moscow State University
                                  Moscow, USSR

K.A.Makarov   Leningrad State University
                                  Leningrad, USSR

M.Malamud   Polytechnical Institute
                                  Doneck, USSR

Yu.B.Melnikov   Leningrad State University
                                  Leningrad, USSR

B.Milek   Joint Institute for Nuclear
                                  Research, Dubna, USSR
                              and Technical University
                                  Dresden, GDR

A.I.Mogilner   Metal Physics Institute
                                  Sverdlovsk, USSR

H.Neidhardt — Joint Institute for Nuclear Research, Dubna, USSR and Mathematical Institute Berlin, GDR

F.Nitzschner — University of Ruhr Bochum, FRG

M.V.Novicky — Low Temperature Physics Institute, Kharkov, USSR

S.I.Petrukhnovsky — Metal Physics Institute Sverdlovsk, USSR

V.N.Pivovarchik — Civil Engineering Institute Odessa, USSR

I.Yu.Popov — Institute of Fine Mechanics and Optics, Leningrad, USSR

P.Šeba — Joint Institute for Nuclear Research, Dubna, USSR and Nuclear Physics Institute Řež, Czechoslovakia

S.V.Semkin — Far-East State University Vladivostok, USSR

D.L.Shepelyansky — Nuclear Physics Institute Novosibirsk, USSR

I.A.Sherestevsky — Institute of Applied Physics Gorky, USSR

L.A.Shuster — Institute of Mathematics and Mechanics, Alma-Ata, USSR

A.Sobolev — Steklov Institute Leningrad, USSR

P.Šťovíček — Joint Institute for Nuclear Research, Dubna, USSR and Czech Technical University Prague, Czechoslovakia

V.R.Struleckaja — Institute of Electric Engineering Moscow, USSR

T.V.Tsikalenko — Mathematical Institute Kiev, USSR

A.V.Vodichev — Metal Physics Institute Sverdlovsk, USSR

S.A.Vugalter   Institute of Applied Physics
                          Gorky, USSR

E.A.Yarevsky   Leningrad State University
                          Leningrad, USSR

B.N.Zakhariev   Joint Institute for Nuclear
                          Research, Dubna, USSR

G.M.Zhislin   Institute of Applied Physics
                          Gorky, USSR